Derivative Spectroscopy

Derivative Spectroscopy

By

Joseph Dubrovkin

Cambridge
Scholars
Publishing

Derivative Spectroscopy

By Joseph Dubrovkin

This book first published 2021

Cambridge Scholars Publishing

Lady Stephenson Library, Newcastle upon Tyne, NE6 2PA, UK

British Library Cataloguing in Publication Data
A catalogue record for this book is available from the British Library

ISBN (10): 1-5275-6348-0
ISBN (13): 978-1-5275-6348-3

TABLE OF CONTENTS

PART IV: QUANTITATIVE DERIVATIVE SPECTROMETRY

PART V: DIFFERENTIATION OF THE ANALYTICAL SIGNALS: INDUSTRIAL AND SCIENTIFIC APPLICATIONS

PREFACE

The second half of the 20th century witnessed impressive technological advances triggered by the arms and space races between the Soviet Union and the United States. This rivalry stimulated significant progress in physical and chemical sciences, which required the improvement of traditional analytical methods and the creation of high-speed, highly effective instrumental techniques. Analog electronic devices were replaced by digital computerized tools managing large quantities of information, which, in turn, gave impetus to the emergence of a new discipline - chemometrics.

Chemometrics uses statistical and mathematical methods for processing data obtained by analytical instruments to extract maximum useful information. The first stage of this process, named preprocessing, usually includes data denoising, decomposition of complex numerical objects (vectors and matrices) into independent components, and background elimination.

One of the first chemometrics applications was the differentiation of analytical signals in electrochemical analysis, spectroscopy, and chromatography. In parallel, physicists developed the theory and technology of modulation spectroscopy, which has numerous mutual features with the derivative method. Spectroscopy stimulated the implementation of the modulation technique in analytical instrumentation.

Derivative Spectroscopy (DS) originated in the 1950s with the development of the electronic differentiators and optical modulation devices. Later, digital differentiation has become a priority. The most attractive features of the DS are:

1. Artificial improvement of the resolution of the analytical instruments.
2. Appearance of peaks and zero points suitable for analysis in the presence of interferences
3. Suppression of background.

The main advantage of the modulation technique is a significant improvement of the analytical sensitivity due to the remarkable noise reduction.

Numerous researches in the field of the DS were summarized in our short monograph, published in 1988 in Russian [1]. This study included theory, techniques, and analytical applications of the derivative methods. It was intended for a broad audience of students and professionals in physics, chemistry, and signal processing. After a few years, G. Talsky published a new book that provided analysts with useful introductions to the subject [2]. For the last three decades, the DS has continued to develop, mainly, in the field of the digital differentiation and its combination with other processing procedures.

The derivative technique was briefly reviewed by the leaders in chemometrics [3-6].

Summing up the results of more than half a century of research in the field of DS, we decided to prepare a new monograph that includes all fundamental problems of this method.

The present study is an attempt to give a detailed explanation of the reached theoretical and numerical results. The goal (similar to our previous book [7]), was to avoid the blind faith of readers in the reliability of the conclusions and recommendations.

Theoretical discussions on this issue are illustrated by various examples supplied by a simple program code on MATLAB, which can be easily modified by non-professional users. The readers who may wish to study the problem further can validate numerical data, given in the book, using computer calculations. Thus, they will be able to understand the details of the algorithm and, if necessary, modify computer programs.

References

1. Dubrovkin, J., Belikov, V. *Derivative spectrometry: Theory, Technology, and Application*. Rostov: Rostov University. 1988. [Russian].
2. Talsky .G. *Derivative spectrophotometry: low and high order*. VCH. 1994.
3. Workman Jr., J. J., Mobley, P. R., Kowalski, B. R., Bro, R. (1996). Review of Chemometrics Applied to Spectroscopy: 1985-95, Part 1. Applied Spectroscopy Reviews, 31, 73-124.
4. Mobley, P. R., Kowalski, B. R., Workman Jr., J. J., Bro, R. (1996). Review of Chemometrics Applied to Spectroscopy: 1985-95, Part 2. Applied Spectroscopy Reviews, 31, 347-368.
5. Mark, H., Workman, Jr. *Chemometrics in Spectroscopy*, ELSEVIER. 2007.
6. Brereton, R. G. *Applied Chemometrics for Scientists*. Wiley. 2007.
7. Dubrovkin, J. *Mathematical processing of spectral data in analytical chemistry: A guide to error analysis*. Cambridge Scholars Publishing. 2018.

ABOUT THE STRUCTURE OF THE BOOK

The book is organized into five parts. In the first part, mathematical derivatives of analytical signals are discussed together with the resolution enhancement and information aspects of derivative spectroscopy. The second part is dedicated to analog and numerical differentiation (difference method, Savitzky-Golay filters, Fourier transform, spline, and wavelet processing). The subjects of the two chapters are modulation spectroscopy based on optical monochromators and lasers. In Parts III and IV, derivative qualitative and quantitative analyses, illustrated by numerous examples, are discussed. The final bibliography part briefly describes hundreds of applications of the derivative spectroscopic and non-spectroscopic methods in the industrial and research laboratories.

We sincerely apologize to all those researchers whose outstanding works are not cited because the book does not have enough free space to include a complete bibliography. The project "Derivative Spectroscopy" (https://www. researchgate.net/profile/Joseph_Dubrovkin) provides a bibliographical supplement, which will be updated as new information becomes available.

The author wants to give each chapter status of an article whose reading is independent of other sections. This material presentation allows readers to avoid cramming a previous text before moving on to another topic. MATLAB-based examples, which are focused on the subject matter, illustrate each chapter.

The book closes with appendices, which include supplementary materials necessary to facilitate the readers' ability to understand the theoretical problems, which are discussed in the main text more deeply. For example, a brief introduction to the mathematical method such as Fourier transform, spline approximation, and Tikhonov regularization is given. Technical features of the analog RC devices are discussed. The laser diodes working principles are reviewed. One appendix summarizes, in brief, the main principles and instrumentation of optical spectrometry. However, for clarity, some details of the issue are given in the main text.

Reading requires the knowledge of the secondary school courses on differential calculus, linear algebra, and statistics. To perform the exercises, readers must have programming skills for beginners in MATLAB.

Appendix SW includes some supplementary programs needed for the exercise. Also, the project "Derivative Spectroscopy" (https://www.researchgate.net/profile/Joseph_Dubrovkin) provides open-source MATLAB files.

A significant part of the book is based on the original research carried out by the author.

For simplicity, the captions of figures, tables, exercises, and expressions have the following structure: "part.Chapter-current number."

The author would be very grateful for the criticisms, comments, and proposals about this book, which he hopes to consider in his future work.

ABBREVIATIONS

AAS- Atomic Absorption Spectrometer (Spectrometry)
AC-Analytical Signal
AES- Auger Electron Spectra
AFS-Atomic Fluorescence Spectroscopy
AM-Amplitude Modulation
ATF-AAS Atom-Trapping Flame AAS
CARS- Coherent Anti-Stokes RS
CCD- Charge-Coupled Device
CFT- Continuous Fourier Transform
CVAAS- Cold Vapour AAS
CWT- Continuous Wavelet Transform
DBR- Distributed Bragg Reflector
DCP-Diclorophenol
DCVA-Derivative Cyclic Voltabsorptometry
DFB-Distributed Feedback Architecture
DFT-Discrete Fourier Transform
DHS-Double Heterostructure
DIR-Derivative IR
DLSVA-Derivative Linear Sweep Voltabsorptometry
DPP- Diphenylpropane
DSC-Differential Scanning Calorimetry
DSP-Digital Signal Processing
DWOD-Dual-Wavelength Optical Differentiation
EFM-Electrochemical FM
EOM-Electro-Optic Phase Modulator
EPR- Electron Paramagnetic Resonance
EXAFS- Extended X-Ray Absorption Fine Structure
FAAS-Flame AAS
FD- Fractional Derivative
FM-Frequency Modulation
FT-Fourier Transform
FTIR-Fourier Transform Infrared Spectroscopy
FWHM-Full Peak Width at Half-Maximum
GDH-Generalized Discrete Harmonics
GFAAS- Graphite Furnace AAS
GFT- Generalized Fourier Transform
HA-Harmonic Analysis
HGAAS- Hydride-generation AAS
IFT-Inverse FT
IM-Intensity Modulation
ICP-OES- Inductively Coupled Plasma Optical Emission Spectroscopy
IT- Information Theory
LC/MS- Liquid-Chromatography-Mass Spectrometry
IR- Infrared
LF-Laser Fluorimetry
LIBS-Laser-Induced Breakdown Spectroscopy
LMS-Laser Modulation Spectroscopy
LTCM- Linear Transform Coordinates Methods
MDSC- Modulation DSC
MPI -Multiphoton Ionization

MS- Modulation Spectroscopy
MW-Microwaves
NAS-Net Analyte Signal
NLDM-Natural Logarithm Derivative Method
oCPA-Orto-Chlorophenoxyacetic Acid
OLS-Ordinary Least Squares
OPO-Optical Parametric Oscillator
OTM-Orthogonal Transformation Method
QCL-Quantum-Cascade Lasers
PAS-Photoacoustic Spectroscopy
PCA-Principal Component Analysis
pCPA- para-Chlorophenoxyacetic Acids
PMG- Polynomial Modified Gaussian function
PT- Petrash Resolution Criterion
REMPI-Resonant Enhanced MPI
RDS-Raman Difference Spectroscopy
RF-Radio Frequency
RS-Raman Spectrum
RW-Radio Waves
SERDS-Shifted Excitation RDS
SFS-Synchronous Fluorescence Spectroscopy
SG-Savitzky-Golay
SGF-SG Filters
S/N-Signal-to-Noise Ratio
SP-Sparrow Resolution Criterion
SVD-Singular Value Decomposition
TR-Tikhonov Regularization
UV-Ultraviolet
VIS-Visible
WM-Wavelength Modulation
WT-Wavelet Transform
XAES-X-Ray Excited AES
XANES-X-ray Absorption Near-Edge Structure
XS-X-Ray Spectroscopy
XAS-X-Ray Absorption Spectroscopy
XDS-X-Ray Diffuse Scattering Spectroscopy
XES-X-Ray Emission Spectroscopy
XPS-X-Ray Photoelectron Spectroscopy (Spectrum)

PART I:

MATHEMATICAL DERIVATIVES OF ANALYTICAL SIGNALS

INTRODUCTION

According to Danzer [1], analytical signal (AS) in analytical chemistry is a response of the measurement system (analytical instrument) to the object under study [1]. Usually, the system's output is a linear or nonlinear mixture of the responses to different analytes, including noise and background.

In terms of signal processing theory, AS "refers to either a continuous or discrete measurement sequence which consists of a pure or undistorted signal corrupted by noise" [2]. In spectroscopy, this measurement sequence is usually a function of the frequency or wavelength. These arguments are, in turn, also time functions. In chromatography, the x-axis (abscissa) of the AS is often time. AS processing is carried out in the time scale of coordinate x (the time domain) or the transformed x-scale, e.g., using the Fourier transform (FT) (the frequency domain). In the last case, the y-coordinates of the AS are the intensity of the Fourier harmonics. Each harmonic is a linear combination of the AS ordinates.

The y-axis of the AS may be the derivatives of the AS relative to the x-argument. Therefore, signal processing in the time domain involves the extraction of useful information from the transformed y-coordinate system. This method, e.g., derivative spectroscopy or chromatography, is one of the Linear Transform Coordinates Methods (LTCM) [3].

Before studying the properties of the derivative methods, let us consider the shapes of the AS components and characteristics of the measurement noise. For simplicity, the AS structural elements, which have a bell-shaped form (symmetrical and asymmetrical), are named peaks.

CHAPTER ONE

SYMMETRICAL PEAK SHAPES IN SPECTROSCOPY AND CHROMATOGRAPHY

In spectroscopy [4] and chromatography [5], a conventional model of AS is a sum of elementary peaks and a baseline. The peaks have symmetrical (Gaussian, Lorentzian, and Voigt) and asymmetrical [6-7] bell-shaped forms. These shapes are due to the impact of physical and instrumental factors. The last effects are essential in chromatography. Some asymmetrical peak shapes and their derivatives are considered in the next chapter.

Spectral lines have "a natural width" due to the expansion of energy levels according to the Heisenberg uncertainty principle [8]. However, this width is negligible. The following effects cause the lines to become broader: the Zeeman Effect, thermal Doppler broadening, collisional, broadening, and velocity broadening [8].

A significant contribution to the spectral contour formation is made by the thermal motion and interactions between the particles of the object under study. The first factor causes Doppler broadening due to changes in the frequency of the radiation emitted or absorbed by the moving particles.

The instrumental distortions include:

- The limited maximum delay time in the interferogram, which is obtained using Fourier-transform infrared spectrometers, and the non-zero width of the instrumental function of monochromators.
- Inter- and/or intra-atomic and molecular interactions in the samples under study (e.g., Stark broadening in dense plasma [9]; spin-spin interactions in NMR spectroscopy [10]; and inter- and intra-molecular associations in IR spectroscopy of liquids [11]).

Doppler broadening forms the Gaussian contour [8] (Fig. 1.1-1):

$$F_G(\lambda) = F_0 \exp\{-4ln2[(\lambda - \lambda_0)/w]^2\}, \qquad (1.1-1)$$

where λ is the abscissa argument with a dimension (e.g., a wavenumber); F_0 is the value of the peak maximum whose position is λ_0; and w is the full peak width on the half-maximum (FWHM). The frequently used peak model in chromatography is the Gaussian type.

Interactions between particles form the Lorentzian (or Cauchy) contour [8] (Fig. 1.1-1):

$$F_L(\lambda) = F_0\,\{1/[1+4[(\lambda-\lambda_0)/w]^2\}. \quad (1.1-2)$$

In physics, the functions $F_G(\lambda)$ and $F_L(\lambda)$ are usually normalized to the unit area:

$$M_G \int_{-\infty}^{\infty} F_G(\lambda)d\,\lambda = 1.$$

Since $\int_{-\infty}^{\infty} F_G(\lambda)d\,\lambda = F_0\sqrt{\pi}\,w/2\sqrt{ln2}$ [12], $M_G = 2\sqrt{ln2/\pi}\,/(wF_0)$.

$$M_L \int_{-\infty}^{\infty} F_L(\lambda)d\,\lambda = 1.$$

Since $\int_{-\infty}^{\infty} F_L(\lambda)d\,\lambda = F_0\pi\,w/2$ [12], $M_L = (2/\pi)/(wF_0)$.

Each spectrum measured by a spectral instrument is disturbed by this device. From a mathematical point of view, the measured ($F_m(\lambda)$) spectrum is the convolution of the undistorted ("true") spectrum (F_T) with the instrumental function (I) [13]:

$$F_m(\lambda) = \int_{-\infty}^{\infty} I(\lambda+\lambda')F_T(\lambda')d\lambda' + \eta(\lambda), \quad (1.1-3)$$

where $\eta(\lambda)$ is the additive noise of the measurements.

Formally, the generalized function (I_g) (which includes instrumental factors) also considers the physical-chemical interactions caused by the distortions of the spectrum under study. According to [14], the undisturbed spectrum equals the measured one corrected by its weighted derivatives:

$$F_T(\lambda) = F_m(\lambda) + \sum_{n=1}^{\infty}(b_n/j^n)d^n F_m(\lambda)/d\lambda^n, \quad (1.1-4)$$

where constants b_n are defined by I_g, $j = \sqrt{-1}$.

Shapes of the spectral peaks, measured in practice, often differ from the "pure" functions (Eqs. (1.1-1, 1.1-2)) due to the combined impact of broadening factors and instrumental distortions. Therefore, the peaks are approximated by the Voigt profile, which is the convolution of Gaussian and Lorentzian shapes:

$$F_V(\lambda) = \int_{-\infty}^{\infty} F_G(\lambda')F_L(\lambda-\lambda')d\lambda'. \quad (1.1-5)$$

The precise approximation of the Voigt FWHM with an accuracy of 0.02% [15]:

$$w_V = 0.5346w_L + \sqrt{0.2166w_L^2 + w_G^2}. \qquad (1.1-6)$$

There were some attempts to obtain non-integral analytical expressions of the Voigt function. However, these expressions are cumbersome; for example, the pseudo-Voigt normalized profile [16], and its improved version [17]. Approximation of Voigt function combined two expressions, one of which was an asymptotic [18]. Software products used many rough but simplified combinations of Gaussian and Lorentzian profiles (e.g., [19]).

The Voigt function and its derivatives were represented by series in Hermite polynomials [20].

Fig. 1.1-1 shows the intermediate position of the Voigt peak between the Lorentzian and Gaussian profiles.

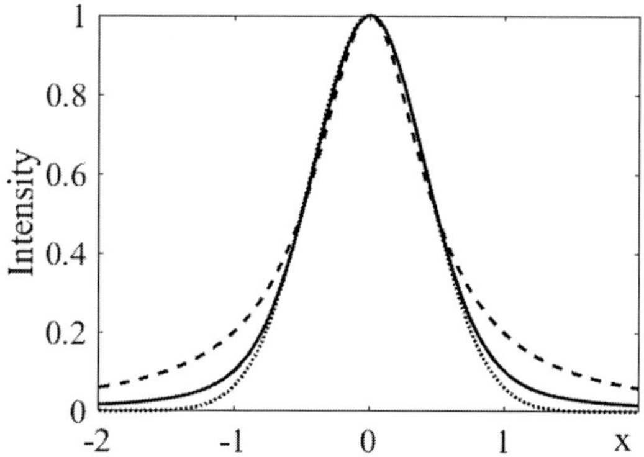

Figure 1.1-1. Lorentz, Gauss, and Voigt peaks (dashed, dotted, and solid curves, respectively). $F_0 = 1, x = (\lambda - \lambda_0)/w, w = 1$.

Exercise 1.1-1

The readers are invited to study the following:

1. The dependence of the Voigt peak width on the ratio w_L/w_G, using the program VoigtTest.m (Appendix Software SW1).
2. The rough approximation of Voigt peak: $F_L * F_G$. For this purpose, the program VoigtTest.m. must be slightly modified.

For the convenience of mathematical operations, Eqs. (1.1-1) and (1.1-2) are replaced by their Fourier transforms (FT) (Appendix A):

$$\tilde{F}_G(p) = F_0 w_G\left(\sqrt{\pi}/2\sqrt{ln2}\right) exp(-p^2/16ln2), \qquad (1.1-7)$$

$$\tilde{F}_L(p) = F_0 w_L(\pi /2)exp(-|p| / 2), \quad (1.1-8)$$

where $p = \omega w$, and ω is the angular (Fourier) frequency.

It is assumed that $\lambda_0 = 0$. If $\lambda_0 \neq 0$, then the FT is modified:

$$F(\lambda - \lambda_0) \rightarrow \tilde{F}(\omega) \exp(-j\omega\lambda_0). \quad (1.1-9)$$

The FT of the Voigt peak is the product of the FT-convolution components normalized to the maximum intensity or area:

$$\tilde{F}_V = \tilde{F}_G * \tilde{F}_L. \quad (1.1-10)$$

CHAPTER TWO

MATHEMATICAL DERIVATIVES OF TYPICAL PEAKS

Analysts usually perform modelling experiments using numerical data to study peculiarities of the qualitative and quantitative analysis based on the derivative methods. For example, the n-order derivative of a composition of the overlapping symmetrical Gaussian, Lorentzian, or Voigt peaks simulated a real analytical signal in spectroscopy. In chromatography, the asymmetrical shapes are usually used. Often the model is corrupted by random noise. These data obtained by the digital differentiation are disturbed by processing errors depending on the differentiation method, sampling interval, and noise. Therefore, only approximate information is extracted from the derivatives. Moreover, the resolution of the derivative peaks essentially depends on the accuracy of the differentiation.

To estimate the properties of the derivative methods accurately, first of all, we intend to carry out theoretical studies using analytical expressions of the mathematical derivatives of the typical peak shapes, both symmetrical and asymmetrical. Often, the roots and the extrema of the derivatives peaks are established analytically. However, generally, this accurate information is readily extracted numerically.

Despite the apparent simplicity, analytical differentiation of mathematical functions, describing model peaks, requires a smart algebraic technique.

Theoretical studies also use analytical expressions of the mathematical derivatives, e.g., for estimating the resolution limit.

Symmetrical peaks

Let us obtain accurate analytical expressions of the mathematical derivatives of the Gaussian (Eq. (1.1-1)) and Lorentzian (Eq. (1.1-2)) peaks. In the first case, we use the physicists' Hermite polynomials [1, 2]:

$$H_n(y) = (-1)^n exp(y^2)\frac{d^n}{dy^n}exp(-y^2). \quad (1.2-1)$$

$$H_n(y) = n!\sum_{m=0}^{[\frac{n}{2}]}\frac{(-1)^m}{m!(n-2m)!}(2y)^{n-2m}. \quad (1.2-2)$$

For example,

$$H_0(y) = 1,$$
$$H_1(y) = 2y,$$
$$H_2(y) = 4y^2 - 2,$$
$$H_3(y) = 8y^3 - 12y, \quad (1.2-3)$$
$$H_4(y) = 16y^4 - 48y^2 + 12,$$
$$H_5(y) = 32y^5 - 160y^3 + 120y,$$
$$H_6(y) = 64y^6 - 480y^4 + 720y^2 - 120,$$
$$H_7(y) = 128y^7 - 1344y^5 + 3360y^3 - 1680y.$$

Table 1.2-1. Zeros of Hermite polynomials [2]

Order					
1	**2**	**3**	**4**	**5**	**6**
0	±0.7071	0 ±1.22474	±0.52464 ±1.65068	0 ±0.95857 ±2.02018	±0.43607 ±1.33584 ±2.35060

From Eqs. (1.2-1) and (1.1-1) we have

$$F_0\frac{d^n}{dy^n}\exp(-y^2) = \frac{d^n}{dy^n}F_G(y) = (-1)^n H_n(y)F_G(y), \quad (1.2-4)$$

where $y = 2\sqrt{ln2}(\lambda - \lambda_0)/w$.

The derivative over λ:

$$\frac{d^n}{d\lambda^n}F_G(\lambda) = (2\sqrt{ln2}/w)^n\frac{d^n}{dy^n}F_G(y). \quad (1.2-5)$$

Instead of cumbersome direct sequential differentiation of the Lorentz shape, we first transform Eq. (1.1-2),

$$F_L(y) = 0.5F_0[1/(1+jy) + 1/(1-jy)], \quad (1.2-6)$$

where $j = \sqrt{-1}$.

The nth-order derivative of Eq. (1.2-6) is

$$F_L^{(n)}(y) = F_0(-1)^n(n!/2)[j^n/T_1 + (-j)^n/T_2] =$$
$$F_0(-j)^n(n!/2)\{[(-1)^n T_1 + T_2]/(1+y^2)^{n+1}\}, \quad (1.2-7)$$

where $T_1 = (1+jy)^{n+1}$, $T_2 = (1-jy)^{n+1}$.

Using the binomial formula [3], we transform the nominator of Eq. (1.2-7) (omitting F_0):

$$(n!/2)\sum_{k=0}^{n+1} C_{n+1}^{k} j^{k+n} y^{k}[(-1)^{k+n}+1], \qquad (1.2-8)$$

where $C_{n+1}^{k}=(n+1)!/[k!\,(n+1-k)!]$.

The term

$$(-1)^{k+n}+1=\begin{cases}0, k+n \text{ is odd},\\ 2, k+n \text{ is even}.\end{cases} \qquad (1.2-9)$$

Taking Eq. (1.2-9) into account, we have from Eq. (1.2-8):

n	**Eqs. (1.2-10)**
1	$C_2^1 j^2 y=-2y$
3	$3!\,(C_4^1 j^4 y+C_4^3 j^6 y^3)=24y-24y^3$
5	$5!\,(C_6^1 j^6 y+C_6^3 j^8 y^3+C_6^5 j^{10} y^5)=5!\,(-6y+20y^3-6y^5)$
7	$7!\,(C_8^1 j^8 y+C_8^3 j^{10} y^3+C_8^5 j^{12} y^5+C_8^7 j^{14} y^7)=$ $7!\,(8y-56y^3+56y^5-8y^7)=7!\,y(y^2-1)(-y^4+6y^2-1)$
2	$2(C_3^0 j^2+C_3^2 y^2)=-2+6y^2$
4	$4!\,(C_5^0 j^4+C_5^2 j^6 y^2+C_5^4 j^8 y^4)=24(1-10y^2+5y^4)$
6	$6!\,(C_7^0 j^6+C_7^2 j^8 y^2+C_7^4 j^{10} y^4+C_7^6 j^{12} y^6)=$ $6!\,(-1+21y^2-35y^4+7y^6)$

The roots of Eqs. (1.2-10) are readily obtained analytically except $n=6$.
The derivatives of Lorentzian peaks over λ:

$$\frac{d^n}{d\lambda^n}F_L(\lambda)=(2/w)^n\frac{d^n}{dy^n}F_L(y). \qquad (1.2-11)$$

Figure 1.2-1 represents the plots of the Gaussian and Lorentzian peaks and their derivatives up to the sixth order. Zero points of the derivatives $\Delta\lambda_r/w=(\lambda-\lambda_r)/w$ (Table 1.2-2) are the roots of Eqs. (1.2-3) and (1.2-10). They were obtained analytically and numerically.

The intensity of the nth-order derivative (Eqs. (1.2-5) and (1.2-11))

$$F^{(n)}\sim w^{-n}. \qquad (1.2-12)$$

So, the differentiation redistributes intensities in favor of narrow peaks.

It follows from Eqs. (1.2-3), (1.2-5), (1.2-10), and (1.2-11) that in the central point $y=0$ of the even-order derivatives,

$$F_G^{(n)}(0)=(-4ln2)^{n/2}n!/(n/2)!$$

$$F_L^{(n)}(0)=(-4)^{n/2}n! \qquad (1.2-13)$$

for the Gaussian and Lorentzian peaks, respectively.

The extrema $\Delta\lambda_e/w = (\lambda - \lambda_e)/w$ and $F_e^{(n)}w^n/F_0$ (Table 1.2-2) were evaluated numerically.

In further calculations, we use the following ratios estimated for the 3rd-, 5th-, and 7th-order derivatives in the extrema y_{e1} of the 1st-order derivatives of the Gaussian and Lorentzian peaks:

$$F_G^{(n)}(y_{e1})/F_G^{(1)}(y_{e1}) = 11.09,\ 184.56,\ 3413.4;$$
$$F_L^{(n)}(y_{e1})/F_L^{(1)}(y_{e1}) = 18.00,\ 0,\ 136.08 \text{ x } 10^3. \qquad (1.2-14)$$

Exercise 1.2-1

Readers are invited to do the following:

- Obtain the final formulas for the derivatives of the Gaussians and Lorentzians.
- Answer the following questions using Figure 1.2-1 and Table 1.2-2:
  a) What are the orders of the derivative curves which resemble the original peak?
  b) What information can you extract from zero and the extrema points?
  c) What can you tell about the properties of the central components of the even-order derivatives and additional side structures called satellites?
  d) Estimate the narrowing of the central peaks of the even-order derivatives.
  e) Estimate the form parameter [4]: the intensity ratios of the central peaks to the satellites of the even-order derivatives.

To estimate the signal-to-noise ratio of derivatives, we need the amplitude sweep, that is, the maximum distance between the extrema points $(F_e^{(n)})$:

$$\Delta\mathcal{D}^{(n)} = |\mathcal{D}_{Max}^{(n)}| + |\mathcal{D}_{Min}^{(n)}|, \qquad (1.2-15)$$

where $D_e^{(n)} = F_e^{(n)}w^n/F_0$.

From Table 1.2-2, we have

$$\Delta\mathcal{D}_G^{(1)} = 2 \text{ x } 1.4283 = 2.8566;$$
$$\Delta\mathcal{D}_G^{(2)} = 5.5452 + 2.4746 = 8.0198;$$
$$\Delta\mathcal{D}_G^{(4)} = 92.2470 + 57.0353 = 149.2823. \qquad (1.2-16)$$
$$\Delta\mathcal{D}_L^{(1)} = 2 \text{ x } 1.2990 \approx 2.6;$$
$$\Delta\mathcal{D}_L^{(2)} = 8 + 2 = 10;$$

$\Delta\mathcal{D}_L^{(4)} = 384 + 162 = 546.$

Let us use the Fourier transform (Appendix A) to estimate the integral intensity of the derivatives:

$$\tilde{F}^{(n)}(\omega) = \int_{-\infty}^{\infty} F^{(n)}(\lambda)\exp(-j\omega\lambda)\,d\lambda, \qquad (1.2-17)$$

where ω is the angular (Fourier) frequency, $j = \sqrt{-1}$.

Since $\tilde{F}^{(n)}(\omega) = (j\omega)^n\tilde{F}(\omega)$, $\tilde{F}^{(n)}(0) = 0$ if $n > 0$.

So, we have from Eq. (1.2-17):

$$\tilde{F}^{(n)}(0) = \int_{-\infty}^{\infty} F^{(n)}(\lambda)\,d\lambda = 0. \qquad (1.2-18)$$

That is, the integral intensity of the derivative curves is zero. Therefore, the areas under their positive and negative peaks are equal.

The value of the area under the unipolar derivative peaks was a measure of the analyte amount in the quantitative analysis (Chapter 4.1).

Asymmetrical peaks

There are a lot of mathematical models of asymmetrical peaks usually used in chromatography [5]. As examples, consider the polynomial modified Gaussian (PMG) and Lorentzian (Dobosz) functions.

The PMG function and its derivatives:

$$F_{PMG} = exp\left(-y^2/B_y^2\right). \qquad (1.2-19)$$

$$F_{PMG}^{(1)} = C_1 y F_{PMG}, \qquad (1.2-20)$$

$$F_{PMG}^{(2)} = C_1\left[(1-3\tau y)F_{PMG}/B_y + yF_{PMG}^{(1)}\right] = C_2 F_{PMG}\Phi, \qquad (1.2-21)$$

$$F_{PMG}^{(3)} = C_2\left[\left(F_{PMG}^{(1)} - 6\tau F_{PMG}/B_y\right)\Phi - F_{PMG}\left(4y + 6\tau^2 y B_y\right)\right], \qquad (1.2-22)$$

where $y = 2\sqrt{ln2}(\lambda-\lambda_0)/w$; $C_1 = -2/B_y^3; C_2 = -2/B_y^6$;

B_y is $= 1 + \tau y$; τ the asymmetry parameter; $\Phi = B_y^3 - 3\tau y B_y^2 - 2y^2$.

Exercise 1.2-2

Readers are invited to validate that Eqs. (1.2-19)-(1.2-22) describe the Gaussian peak and its derivatives if $\tau = 0$.

Analytical expressions of the derivatives, with an order higher than 3, are more cumbersome. Appendix B1 describes sequential high-order differentiation of the PMG peak.

Figure 1.2-2 shows that the right asymmetrical tail of the high-order derivatives of slightly skewed PMG peak ($\tau = 0.05$) is strongly distorted compared to the symmetrical wing of corresponding Gaussian shape derivatives. Table 1.2-3 summarizes the quantitative parameters of the PMG shape derivatives. Numerical errors caused small discrepancies between zeros and the extrema positions in the derivatives of $n + 1$- and n-order, respectively.

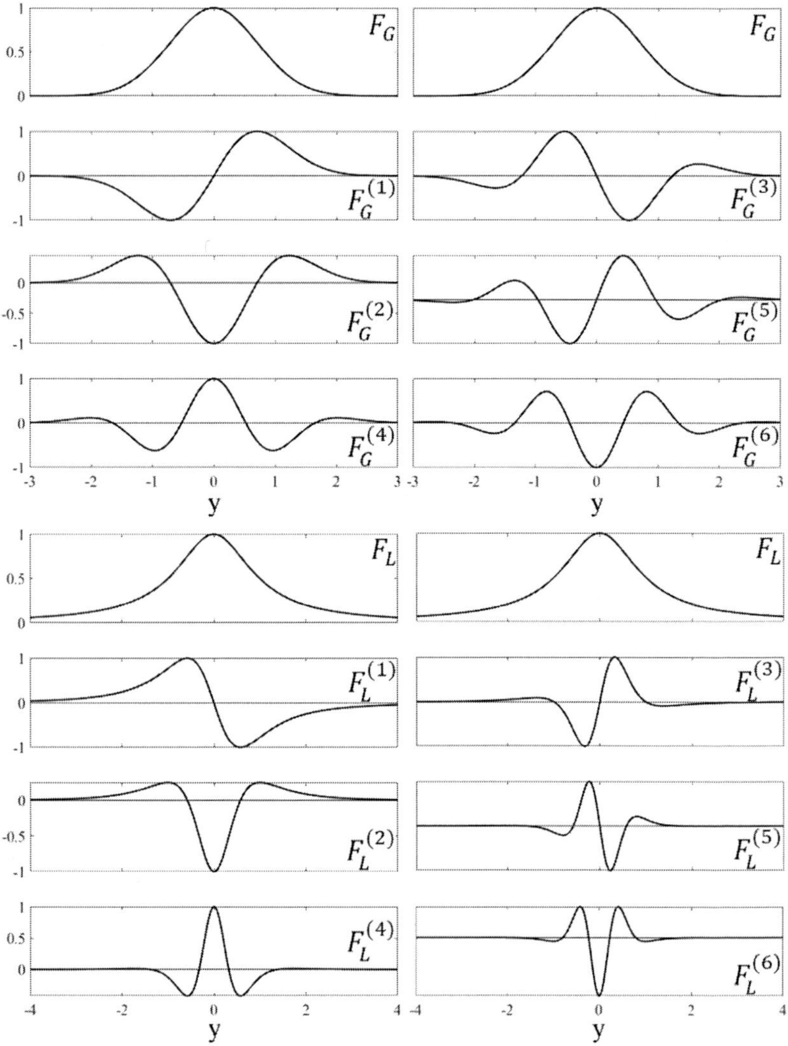

Figure 1.2-1. Plots of the normalized derivatives.

Table 1.2-2. Parameters of the derivatives

n	Peak							
	Gaussian				Lorentzian			
	$\frac{\Delta\lambda_r}{w}$	$\frac{\Delta\lambda_e}{w}$	$\frac{F_e^{(n)}w^n}{F_0}$	$\lvert\varphi^{(n)}\rvert^*$	$\frac{\Delta\lambda_r}{w}$	$\frac{\Delta\lambda_e}{w}$	$\frac{F_e^{(n)}w^n}{F_0}$	$\lvert\varphi^{(n)}\rvert^*$
1	0	±0.425	±1.428	-	0	±0.289	±1.299	
2	±0.425	0 ±0.736	-5.545 2.475	2.241	±0.289	0 ±0.500	-8 2	4
3	0 ±0.736	±0.315 ±0.992	±18.0 ±4.90	-	0 ±0.5009	±0.163 ±0.689	±37.349 ±3.368	
4	±0.315 ±0.991	0 ±0.576 ±1.213	92.3 -57.0 10.7	1.617 8.603	±0.163 ±0.688	0 ±0.289 ±0.867	384 -162 6	2.37 64
5	0 ±0.576 ±1.213	±0.262 ±0.802 ±1.412	±418.7 ±182.6 ±25.5	-	0 ±0.289 ±0.867	±0.114 ±0.399 ±1.038	±3214.7 ±685.7 ±11.4	-
6	±0.262 ±0.802 ±1.412	0 ±0.490 ±1.005 ±1.593	-2557.6 1812.4 -599.1 65.2	1.411 4.269 39.23	±0.114 ±0.399 ±1.038	0 ±0.207 ±0.500 ±1.207	-46080 24459 -2880 21	1. 88 16.0 2194

* The form parameter $\varphi^{(n)} = \left|F_e^{(n)}\right|_{max} / F_e^{(n)}$.

It follows from Table 1.2-3 that:

1. The shift of the central minimum of the 2nd-order derivatives is approximately $-\tau$. Let us show that this result follows from Eq. (1.2-22). Since, near the minimum (y_e) y and τy are close to zero, $B_y \cong 1$, $F_{PMG}^{(1)} \cong -2y$, $F_{PMG} \cong 1$; for $\tau \ll 1$, $\tau^2 y \cong 0$, $\tau y^2 \cong 0$. By zeroing Eq. (1.2-22), we have

$F_{PMG}^{(3)} \cong 4[(y+3\tau)(1-3\tau y)+2y] \cong 4y+12\tau+8y=0. \quad (1.2-23)$

So, $y_e \approx -\tau$.

2. The shift of the central peak of the 4th-order derivatives is roughly two times larger than that of the 2nd-order derivative.

3. The left- and the right-side form parameters (the absolute ratios of the satellite intensity to the intensity of the central peak in the 2nd-order derivatives) are unequal.

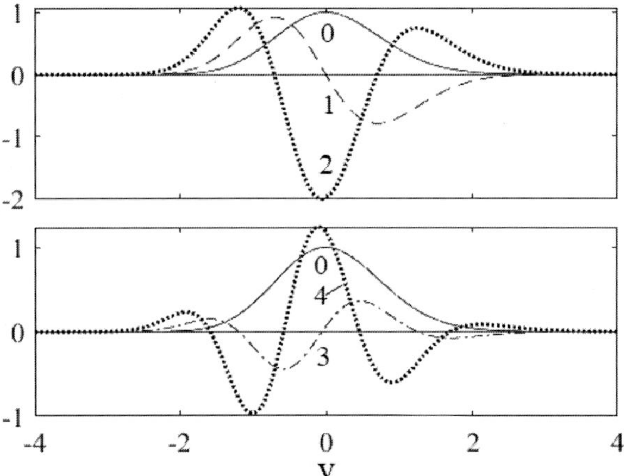

Figure 1.2-2. PMG peak and its derivatives. Numbers near curves designate the differentiation order. $\tau = 0.05$. $F_{PMG}^{(3)}$ and $F_{PMG}^{(4)}$ are multiplied by 0.1.

Asymmetrical Lorentzian (Dobosz) function:

$$F_D = exp\left(-\tau(1-tan^{-1}y)\right)F_L = Cexp\left(K_y\right) * K_y^{(1)}, \quad (1.2-24)$$

where $\tau \neq 0$; $F_L = 1/(1+y^2)$; $C = \exp(-\tau)\,/\,\tau$; $K_y = \tau tan^{-1}y$;

$K_y^{(1)} = \tau F_L$.

$$F_D^{(1)} = Cexp\left(K_y\right)\left[\left(K_y^{(1)}\right)^2 + K_y^{(2)}\right] = Cexp\left(K_y\right)T_y, \quad (1.2-25)$$

$K_y^{(2)} = -2\tau y F_L^2, T_y = \tau[(\tau-2y)]F_L^2,$

$$F_D^{(2)} = F_D T_y + Cexp\left(K_y\right)T_y^{(1)}, \quad (1.2-26)$$

$T_y^{(1)} = \tau(6y^2-4\tau y-2)F_L^3$.

Analytical expressions of the nth-order derivatives, for $n > 2$, are more cumbersome. Appendix B1 describes sequential high-order differentiation of the PMG. Tables 1.2-3 and 1.2-4 summarize the quantitative parameters of the Dobosz shape derivatives.

It follows from Table 1.2-4 that the Dobosz peak maximum is shifted by $\tau/2$. This result is obtained by zeroing the term T_y (Eq. (1.2-25)). The shift of the PMG peak is zero. The central peaks of the Dobosz shape 2nd- and the 4th-order derivatives are shifted approximately by $\tau/3$ and $\tau/4$, respectively (Table 1.2-4). Therefore, we suppose that impact of the asymmetry on the uncertainty in determining maximum of the polynomial modified Lorentzian asymmetrical peaks is significantly smaller than those of the asymmetrical Gaussians.

The form parameters of the 2nd-order derivatives of the PMG and Dobosz peaks (Table 1.2-5) are highly-sensitive to the asymmetry parameter τ. If τ increases, then the right-side satellite significantly decreases compared to the one on the left-side.

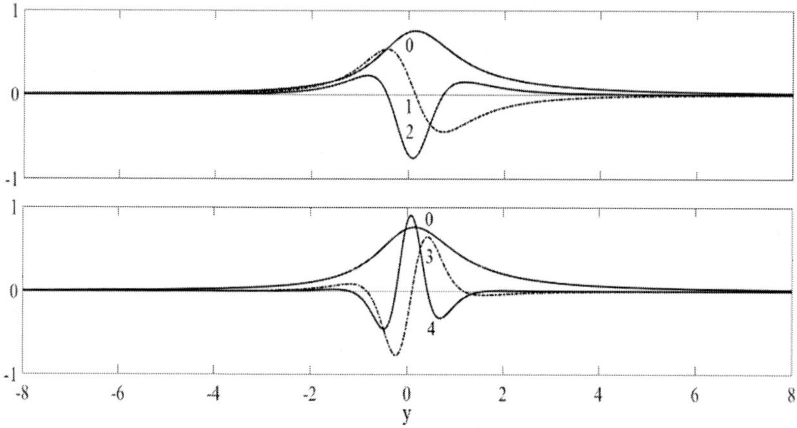

Figure 1.2-3. Dobosz peak and its derivatives. Numbers near curves designate the differentiation order. $\tau = 0.3$. $F_{PMG}^{(2)}$, $F_{PMG}^{(3)}$ and $F_{PMG}^{(4)}$ are multiplied by 0.5, 0.2 and 0.05, respectively.

Exercise 1.2-3

Readers are invited to study the dependencies of the positions and amplitudes of the extrema values of the PMG and Dobosz peak derivatives on the asymmetry parameter using Tables 1.2-3 and 1.2-4.

Table 1.2-3. Parameters of the PMG derivatives

n		τ				
		0	**0.05**	**0.10**	**0.15**	**0.20**
0	y_e	0	0	0	0	0
	$F_e^{(0)}$	1	1	1	1	1
1	y_r	0	0	0	0	0
	y_e	-0.7070 0. 7070	-0.7060 0.7060	-0.7020 0.7020	-0.6960 0.6950	-0.6890 0.6850
	$F_e^{(1)}$	0.8578 -0.8578	0.9206 -0.7992	0.9877 -0.7449	1.0593 -0.6947	1.1352 -0.6483
2	y_r	-0.7071 0.7071	-0.7058 0.7058	-0.7021 0.7016	-0.6963 0.6947	-0.6886 0.6850
	y_e	-1.2250 0 1.2250	-1.1970 -0.0500 1.2460	-1.16050 -0.0980 1.2610	-1.1290 -0.1430 1.2670	-1.0920 -0.1830 1.2650
	$F_e^{(2)}$	0.8925 -2.000 0.8925	1.0741 -2.0150 0.7444	1.2948 -2.0601 0.6247	1.5604 -2.1355 0.5287	1.8774 -2.2417 0.4522
3	y_r	-1.2247 0 1.2247	-1.1969 -0.0497 1.2465	-1.1645 -0.0977 1.2608	-1.1290 -0.1426 1.2670	-1.0917 -0.1834 1.2649
	y_e	-1.6510 - 0.5250 0.5200 1.6510	-1.5860 -0.5810 0.4590 1.7070	-1.5200 -0.6300 0.3900 1.7500	-1.4450 -0.6600 0.3150 1.7820	-1.3750 -0.6840 0.2430 1.7960
	$F_e^{(3)}$	3.9036 -1.0604 3.9036 -1.0604	1.4811 -4.4086 3.5780 -0.7670	2.0772 -5.1318 3.4025 -0.5643	2.9113 -6.1250 3.3562 -0.4250	4.0609 -7.505 3.4256 -0.3293
4	y_r	-1.6507 -0.5246 0.5246 1.6507	-1.5858 -0.5809 0.4593 1.7071	-1.5162 -0.6263 0.3882 1.7515	-1.4450 -0.6603 0.3149 1.7816	-1.3747 -0.6836 0.2429 1.7962
	y_e	-2.0200 -0.9600 0.9600 2.0200	-1.9130 -1.0030 -0.0990 0.8950 2.1180	-1.8020 -1.0270 -0.1910 0.8160 2.2010	-1.6940 -1.0350 -0.2720 0.7300 2.2620	-1.5910 -1.0300 -0.3400 0.6310 2.2990
	$F_e^{(4)}$	1.3949 - 7.4195 12.0000 - 7.4195 1.3949	2.3284 -9.6904 12.4240 -6.0209 0.8518	3.9120 -13.2583 13.7447 -5.2177 0.5375	6.5513 -18.7555 16.1121 -4.8363 0.3548	10.86068 -27.1010 19.7894 -4.7749 0.2470

y_r, y_e and $F_e^{(n)}$ are the roots, extrema and their values, respectively.

Table 1.2-4. Parameters of Dobosz derivatives

n		τ				
		0	**0.1**	**0.2**	**0.3**	**0.4**
0	y_e	0	0.0500	0.1000	0.1500	0.2000
	$F_e^{(0)}$	1	0.9071	0.8269	0.7576	0.6975
1	y_r	0	0.0500	0.1000	0.1500	0.2000
	y_e	-0.5770 0.5770	-0.5280 0.6280	-0.4800 0.6800	-0.4340 0.7340	-0.3890 0.7890
	$F_e^{(1)}$	0.6495 -0.6495	-0.6093 0.5690	-0.5736 0.5004	-0.5419 0.4419	-0.5136 0.3918
2	y_r	-0.5774 0.5774	-0.5281 0.6281	-0.4802 0.6802	-0.4338 0.7338	-0.3888 0.7888
	y_e	-1 0 1	-0.9430 0.0330 1.0590	-0.8880 0.0670 1.1210	-0.8350 0.1000 1.1850	-0.7850 0.1330 1.2520
	$F_e^{(2)}$	0.5000 -2.0000 0.5000	0.4866 -1.8127 0.4219	0.4748 -1.6484 0.3571	0.4644 -1.5039 0.3033	0.4552 -1.3766 0.2585
3	y_r	-1 0 1	-0.9428 0.0333 1.0595	-0.8879 0.0666 1.1213	-0.8352 0.0999 1.1853	-0.7847 0.1330 1.2516
	y_e	-1.3760 -0.3250 0.3250 1.3760	-1.3090 -0.2940 0.3560 1.4470	-1.2450 -0.2630 0.3880 1.5200	-1.1840 -0.2330 0.4210 1.5960	-1.1260 -0.2030 0.4530 1.6750
	$F_e^{(3)}$	0.4210 -4.6686 4.6686 -0.4210	0.4254 -4.3541 4.1106 -0.3418	0.4307 -4.0726 3.6305 -0.2782	0.4367 -3.8199 3.2165 -0.2272	0.4432 -3.5924 2.8588 -0.1861
4	y_r	-1.3764 -0.3249 0.3249 1.3764	-1.3093 -0.2938 0.3565 1.4465	-1.2451 -0.2630 0.3884 1.5197	-1.1839 -0.2328 0.4207 1.5960	-1.1256 -0.2030 0.4534 1.6752
	y_e	-1.7320 -0.5770 0 0.5770 1.7320	-1.6540 -0.5450 0.0260 0.6100 1.8140	-1.5800 -0.5140 0.0510 0.6440 1.8990	-1.5090 -0.4830 0.0770 0.6780 **1.9880**	-1.4420 -0.4530 0.1020 0.7120 2.0810
	$F_e^{(4)}$	0.3750 -10.1250 24.0000 -10.1250 0.3750	0.3937 -9.7008 21.7478 - 8.6753 0.2929	0.4137 - 9.3177 19.7646 -7.4541 0.2293	0.4350 -8.9704 18.0140 - 6.4239 0.1800	0.4575 -8.6546 16.4651 -5.5530 0.1417

y_r, y_e and $F_e^{(n)}$ are the roots, extrema and their values, respectively.

Table 1.2-5. Form parameters of the second-order derivatives

Peak									
PMG					Dobosz				
τ									
0	**0.05**	**0.10**	**0.15**	**0.20**	**0**	**0.10**	**0.20**	**0.30**	**0.40**
0.446	0.533	0.628	0.731	0.838	0.250	0.268	0.288	0.309	0.331
0.446	0.369	0.303	0.248	0.202	0.250	0.233	0.217	0.202	0.188

Left- and right-side parameters are in the upper and low rows, respectively.

Fractional derivatives

Fractional derivatives (FD) are the derivatives of non-integer orders [6]. The semi-differential analysis of irreversible voltammetric peaks was a pioneer FD method in analytical chemistry [7, 8]. There is a small study in which FDs were used in spectroscopy [9-17]. Possibly, this fact is explained by the lack of analytical formulas for estimating the FDs of typical peak shapes. Moreover, most of the studies [10, 12-15] involved the theoretical determination of the parameters of the overlapped Gaussian, Lorentzian, and Tsallis functions. The Tsallis statistical distribution ($T(z, x, \sigma)$ is based on nonextensive (Tsallis) entropy [18]. The authors of the article [10] defined this distribution as

$$T(z, x, \sigma) \sim \{[1 + (z-1)/(3-z)][(x-x_0)^2/\sigma^2]\}^{1/(z-1)}, \quad (1.2-27)$$

where σ is the standard deviation. According to the original study [18], if $z \to 1$, then $T(z, x, \sigma)$ is the Gaussian distribution. If $z = 2$, then Eq. (1.2-27) is equivalent to the Lorentzian distribution. The function $T(z, x, \sigma)$ as a peak shape has occasional applications in analytical spectroscopy [10, 13].

Different integral and finite-difference formulas of FDs [6] probably look scary for non-professional mathematicians. One of the definitions of the qth-order FD ($q > 0$) [9] is based on the Riemann-Liouville integral:

$$F^{(q)}[f(x)] = [1/\Gamma(q)] \int_0^x (x-t)^{-q-1} f(t) dt =$$

$$\lim_{N \to \infty} \left\{ [y^{-q}/\Gamma(-q)] \sum_{p=0}^{N-1} [\Gamma(p-q)/(p+1)] f(x-py) \right\}, (1.2-28)$$

where $\Gamma(q)$ is the gamma function; $y = x/N$.

FD is readily calculated using inverse FT [19] (see Appendix A1):

$$F^{(q)}[f(x)] = (1/2\pi)\int_{-\infty}^{\infty}(j\omega)^q\tilde{F}(\omega)\exp(j\omega x)\,d\omega. \quad (1.2-29)$$

The theoretical proof is given in the thesis [11].

We estimated Eq. (1.2-29) numerically using the Fast FT (Appendix Software SW2, function fractDeriv.m). The authors of the study [19] developed the fractional derivative differentiator.

The integer Savitzky-Golay (SG) differentiation (Chapter 2.3) was extended to the fractional case [16, 21]. Matrix $\boldsymbol{X}$ (Eq. (2.3-8)) was defined in the asymmetric interval $[1, 2m+1]$ instead of the symmetric one $[-m, m]$. Matrix $\boldsymbol{T} = (\boldsymbol{X}^T\boldsymbol{X})^{-1}\boldsymbol{X}^T$was changed to $\boldsymbol{d}^T\boldsymbol{T}$. The ith element of the vector $\boldsymbol{d}$ was estimated using the generalized derivative of the z^v [21]:

$$d_i = F_v^{(\alpha,\mu)}[z^v] = C_{\alpha,\mu}[\Gamma(\chi_v)/\Gamma(\chi_v-\alpha)]z^{\theta+v}, \quad (1.2-30)$$

where $i = 0, 1, .., 2t$; $2t$ is the polynomial power; α, μ are real constants, $C_{\alpha,\mu} = (\mu+1)^{\alpha-1}, \chi_v = v/(\mu+1)+1, \theta = (1-\alpha)\,(\mu+1)-1$.

The FDs were also calculated using the Haar wavelets [22].

Buslov [23] modified the integrand of Eq. (1.2.29); he used the absolute value of the factor: $|(j\omega)^q| = \omega^q$. If $q = 2$, then only the sign of a new derivative is changed. However, for $q = 1$ and 3, the modified derivatives look like those, obtained for $q = 2$ and 4, respectively, but they are wider.

The FD of a constant depends on the calculation method [Private communication. https://www.researchgate.net/profile/Anastasia_Gladkina]. E.g., the Riemann-Liouville derivative of a constant is not zero, however, in the Caputo method-zero [20]. The Liouville-Caputo and FD-FT derivatives are equivalent [20]. The FT of the constant is proportional to the delta function, which is zero everywhere except at the origin [1]. Therefore, from Eq. (1.2-29), we have $F^{(q)}[const]$=0. The FDs of the constant, estimated by the SG method (FD-SG), are also zeros. We checked these results numerically.

The FD-FTs of the linear and quadratic functions are not applicable since the high oscillations occur because of the jump discontinuity from zero at the endpoint (the Gibbs phenomenon) (Appendix A2). The FD-SG of these functions are constants and linearly increasing (Fig. 2.1-4, panels a and b, respectively), according to the 2nd-order differentiation rules, but they depend on μ.

Figure 1.2-5 compares the FD-FTs and FD-SGs of the Gaussian peaks. The parameters of Eq. (1.2-30) are $\alpha = 2$ (Eq. (1.2-29)); μ was chosen so that the FD-SG and FD-FT curves were close. The figure shows the difference between the curves. Also, parameter μ does not directly reflect the fractional order (q) of the FD-FTs. Therefore, below we will only study the FT fractional derivatives.

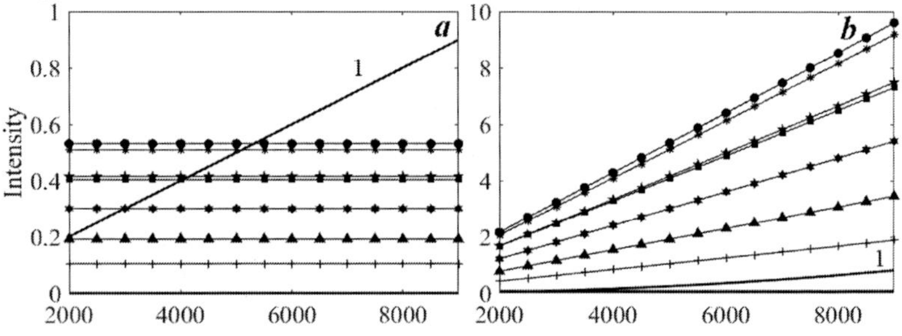

Figure 1.2-4. The FD-SGs of the linear ($y = x$) (a) and quadratic ($y = x^2$) (b) functions (curves 1). $\alpha = 2$. μ=0 (zero (a) and horizontal lines near zero (b)). μ= 0.1, 0.2,…, 0.7 (squares, circles, stars, pentagons, hexagons, triangles and pluses, respectively).

FDs have similar properties of the linearity as the integer-order derivatives [20].

Also, FD is the product of the integer n-order derivative and the fractional derivative of the order $q - n$:

$$F^{(q)} = F^{(n)}F^{(q-n)}. \quad (1.2-31)$$

Figure 1.2-6 demonstrates the normalized FD curves:

$$F^{(q)}_{norm} = F^{(q)}/\left[F^{(q)}_{max} - F^{(q)}_{min}\right]. \quad (1.2-32)$$

The denominator of Eq. (1.2-32) is the amplitude sweep, that is, it is the maximum distance between the extrema points (compare with Eq. (1.2-15)). The normalization allows better visualization of the curves since the impacts of the maximum peak intensity (F_0) and the width (we suppose that $\sim w^{-q}$ (Eq. (1.2-12)) are eliminated.

The FDs of the noisy peak were smoothed in the Fourier domain by slightly modified filter: $1/(1 + \alpha q^{2t}\omega^{2t})$ (Chapter 2.4).

Figure 1.2-6c shows that the smoothed FDs suffer from substantial form distortions. These factors must be considered when determining the initial peak parameters using FDs (Chapter 3.3-3).

Analysis of the FDs plots (Fig. 1.2-6) shows that the ratio of the left maximum to the right minimum, which characterizes curve asymmetry, decreases from a value more than 1(for $q = 0.6$ and 0.8) to 1 (for $q = 1$, the symmetrical curve) and further continues to fall while the FDs approach the 2nd-order derivative ($q = 2$). If $q > 1$, then the second (right) maximum appears, which may be called "a satellite" similar to the 2nd- order derivative.

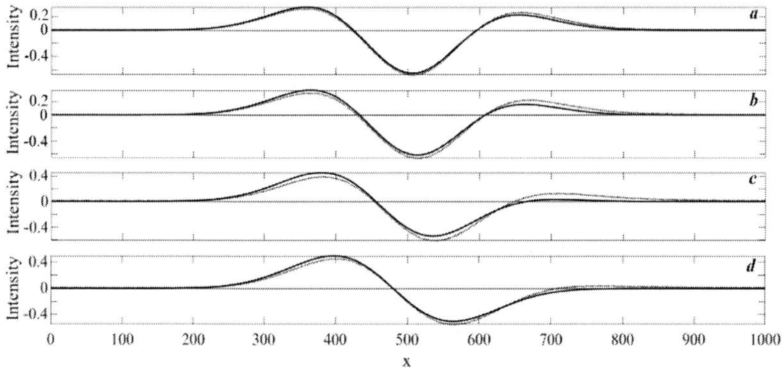

Figure 1.2-5. SG and FT normalized fractional derivatives of the Gaussian peak ($w = 200$) (continuous and dotted lines, respectively). $\alpha = 2$, $[q, \mu]$= (a) [1.9, -0.0085], (b) [1.8, -0.017], (c)[1.5, -0.05], and (d)[1.2, -0.15].

The authors of the study [10] established the linear dependence:

$$(x_0 - x_{max})/\sigma = a(1 - q), \qquad (1.2 - 33)$$

where $\sigma = w/(2\sqrt{ln2})$ and $w/2$ are the standard deviations for the Gaussian and Lorentzian distributions (peaks), respectively. However, according to our findings, this dependence is only linear for the Gaussian peak. For the Lorentzian shape, the linearity is rough (Fig. 1.2-7, a). The slopes (Eq. (1.2-33)) are $a \approx 0.200$ and 0.171 for the Gaussian and Lorentzian peaks, respectively. These values are not similar to those estimated in the paper [10] because of different FD calculation methods.

We obtained the following linear dependencies of the FD maxima (Fig. 2.1-7, b) and minima (Fig. 2.1-7, c):

$$ln\left\{F_{max}^{(q)}(w/1000)^q\right\} = a_{max}q + b_{max}, \qquad (1.2 - 34)$$

$$ln\left\{\left|F_{min}^{(q)}\right|(w/1000)^q\right\} = a_{min}q + b_{min}, \qquad (1.2 - 35)$$

where $a_{max}, b_{max}, a_{min},$ and b_{min} are given in Table 1.2-6.

The terms $F_{max}^{(q)}$ and $F_{min}^{(q)}$ are functions of q and w. The factor $(w/1000)^q$ almost eliminates the dependence on the width ($\sim w^{-q}$). The coefficients in Eqs. (1.2-34) and (1.2-35) hardly change when w decreases from 1000 to 250 (Table 1.2-6). The scaling factor 1000 underlines the small variations along the y-axis.

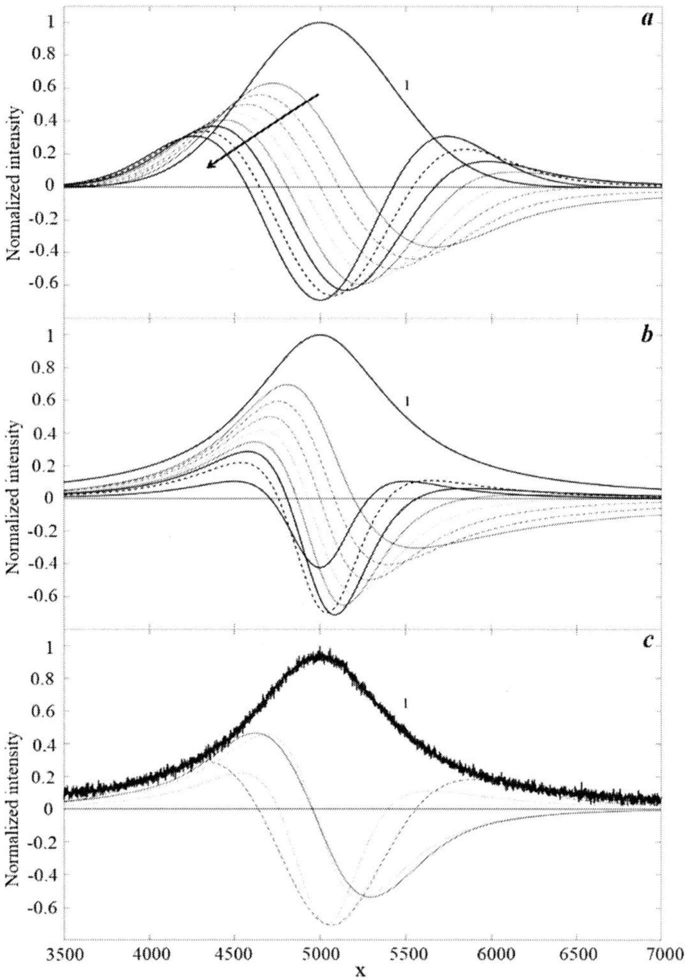

Figure 1.2-6. Fractional derivatives of the Gaussian (a) and Lorentzian (b) peaks ($x_{max} = 5000$, $w = 1000$) (curve 1). $q = 0.6, 0.8, \ldots, 2$ in accordance with the arrow. (c) Superposition of the noise-free and noisy (the signal-to-noise ratio is 50)- plots "b" smoothed by the FT filter ($\alpha = 2 * 10^{-5}, t = 2$). $q = 1.1$ and 1.8 (solid and dashed lines, respectively). The dotted lines correspond to the noisy-free data.

For the non-normalized FDs of the Gaussian and Lorentzian peaks, the absolute value of the minimum/maximum ratio depends on the differentiation order linearly and non-linearly, respectively (Fig. 1.2-8).

These dependencies are the FT form parameters since they are independent of the peak width.

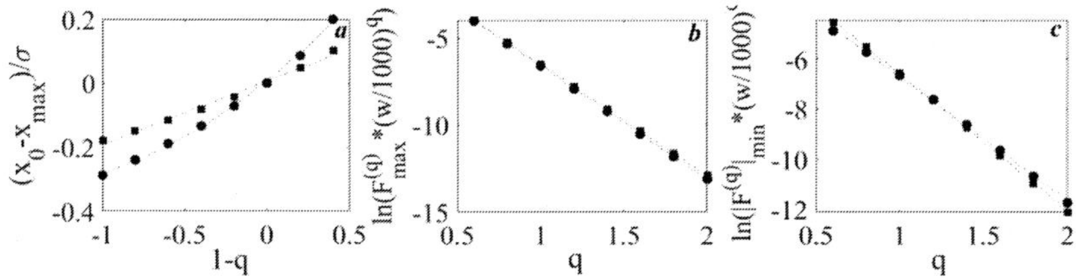

Figure 1.2-7. Dependencies of the FD zero-crossing points (a) and normalized maxima (b) and minima (c) values on the differentiation order for the Gaussian (squares) and Lorentzian (circles) peaks.

Table 1.2-6. Constants in Eqs. (1.2-34) and (1.2-35)

Parameters	Peak			
	Gaussian		Lorentzian	
	w=1000	w=250	w=1000	w=250
a_{max}	-6.3841	-6.3909	-6.4913	-6.5016
b_{max}	-0.1603	-0.1499	-0.1515	-0.1360
a_{min}	-5.4418	-5.4318	-4.9340	-4.9149
b_{min}	-1.1556	-1.1713	-1.7765	-1.8070

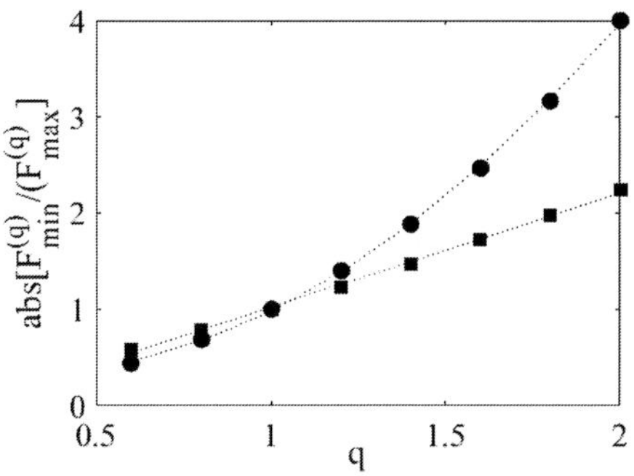

Figure 1.2-8. Dependencies of the absolute value of the minimum/ maximum ratio on the differentiation order for the non-normalized FDs of the Gaussian (squares) and Lorentzian (circles) peaks.

The positions of the extrema points in the 2nd-order derivatives allow us to acquire useful information about the parameters of the overlapped peaks, especially their maxima (Chapter 3.3). In this connection, we estimated the following relative shifts of the FD extrema for the derivatives of the orders q =1.6, 1.7,..., 2.4 (Table 1.2-7):

$$\delta_{x_{min}}^{(q)} = \left(x_{min}^{(q)} - x_{max}\right)/w, \quad (1.2-36)$$

$$\Delta RD_{x_{maxST}}^{(q)} = \left\{\left[x_{min}^{(q)} - x_{max}^{(q)}(ST)\right] - \left[x_{min}^{(2)} - x_{max}^{(2)}(ST)\right]\right\}/w, \quad (1.2-37)$$

where RD is the relative distance between the FD minimum and maximum, ST stands for left (L) or right (R) satellite. The table data weakly depend on the peak width.

Table 1.2-7 shows that only the relative distance between the FD minimum and the left satellite weakly depends on the variation of the derivative order near 2. This fact is the most noticeable for Lorentzians. The relative shift of the central negative FD peak ($\delta_{x_{min}}^{(q)}$) is approximately the linear function of the derivative order (Fig. 1.2-9). This behavior is similar to that of the zero-crossing points (x_0) (Eq. (1.2-33)) (Fig. 1.2-7, a). However, the accurate measurements of x_0 are error-prone due to the non-zero slope of the derivatives in the vicinity of the zero-crossing points, and the impact of the unsuppressed background. The positions of well-resolved FD minima are measured with small errors (Chapter 3.3). These positions are proportional to the peak widths. For the well-resolved doublet, when the first peak is three times narrower than the second one (Fig. 1.2-10), $\Delta x_2/\Delta x_1 \approx 3$. The corresponding shifts of the zero-points are also proportional to the peak widths (Eq. (1.2-33).

The ratio of the peak widths in a spectrum may be estimated using the slope of the dependencies $\Delta x(q)$:

$$w_2/w_1 = [\Delta x_2(q)]^{(1)}/[\Delta x_1(q)]^{(1)}. \quad (1.2-38)$$

However, noise in spectra and peaks overlap may significantly worsen the accuracy of this estimation (Chapter 3.3).

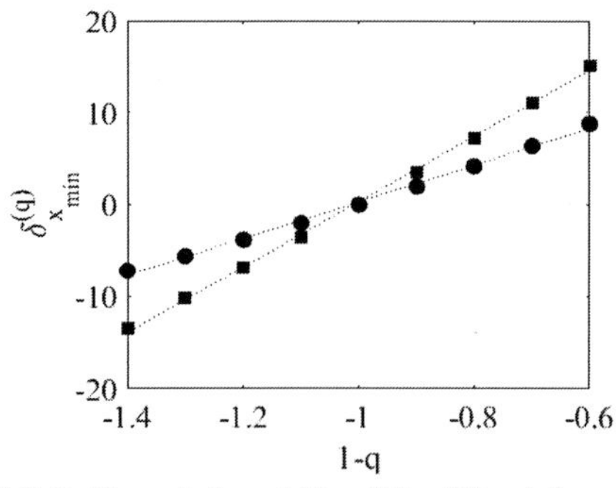

Figure 1.2-9. The relative shifts of the FD minima positions For the Gaussian (squares) and Lorentzian (circles) peaks.

Table 1.2-7. Relative shifts of the FD extrema

Peak	Shifts	*q*								
		1.6	**1.7**	**1.8**	**1.9**	**2.0**	**2.1**	**2.2**	**2.3**	**2.4**
Gaussian	$\delta^{(q)}_{x_{min}}$	15.1	11.1	7.3	3.6	0	-3.5	-6.9	-10.2	-13.5
	$\Delta RD^{(q)}_{x_{maxL'}}$ %	**3.5**	**2.5**	**1.6**	**0.7**	**0**	**-0.8**	**-1.5**	**-2.1**	**-2.8**
	$\Delta RD^{(q)}_{x_{maxR'}}$ %	-8.5	-5.9	-3.6	-1.7	0	1.6	3.0	4.2	5.3
Lorentzian	$\delta^{(q)}_{x_{min}}$	8.8	6.4	4.4	2.0	0	-2.0	-3.8	-5.6	-7.2
	$\Delta RD^{(q)}_{x_{maxL'}}$ %	**0.6**	**0.4**	**0.2**	**0**	**0**	**-0.2**	**0**	**0**	**0.4**
	$\Delta RD^{(q)}_{x_{maxR'}}$ %	-27.6	-17.4	-10.0	-4.4	0	3.6	6.4	8.8	10.8

w=1000.

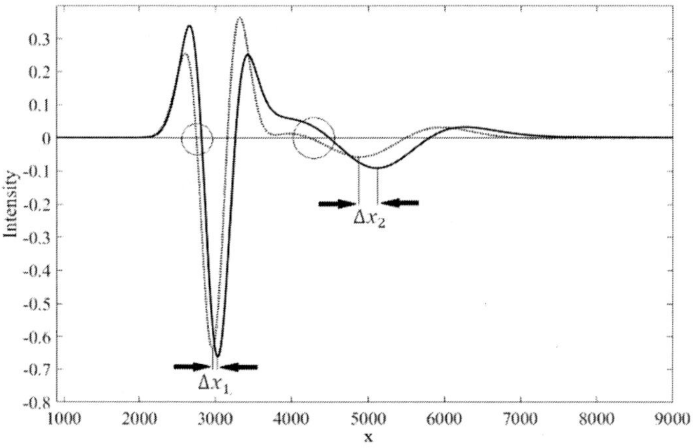

Figure 1.2-10. The shifts of the minima and zero-points (circles) in the FD (the order 1.8 and 2.2, solid and dotted lines, respectively) of the Gaussian doublet. The first peak is 3 times narrower than the second one.

X-scale transformations

The IR spectra are usually recorded as a function of energy (wavenumbers, $[cm^{-1}]$), while the UV-VIS spectra are recorded linearly with respect to wavelength ([nm]). The transformation of the wavelength to the wavenumber-scale may distort the peak shapes.

Antonov pointed out that "it is not crucial for conventional spectral analysis, such as absorption maxima estimation, molar absorptivity calculation, investigation of defined chemical reactions and equilibria, etc. At the same time, computational band decomposition is a mathematical procedure that crucially depends on the x scale on which the spectrum is presented." [24].

The linearity of the spectrum relative to its abscissa influences the differentiation algorithm [25-32]. Antonov et al. demonstrated that the broad peaks recorded linearly in the wavelength (λ) scale become asymmetric for the wavenumber's abscissa (ν) since

$$\lambda\,[nm] = k/(\nu[cm^{-1}]), \qquad (1.2-39)$$

where $k = 10^7$. Therefore, the derivatives of these peaks are distorted.

Let us consider the nature of this problem in the first order of approximation. Suppose that we have two Gaussian peaks (Chapter 1.1) defined in ν and λ scales, respectively:

$$F_\nu(\nu) = exp(-x_\nu^2), \qquad (1.2-40)$$

$$F_\lambda(\lambda) = exp(-x_\lambda^2), \qquad (1.2-41)$$

where $x_\nu = \beta\Delta\nu/w_\nu$; $\beta=2\sqrt{ln2}$; $\Delta\nu = \nu - \nu_0$; $x_\lambda = \beta\Delta\lambda/w_\lambda$; $\Delta\lambda = \lambda - \lambda_0$. The argument x_ν, transformed into the λ scale, according to Eq. (1.2-39), is

$$x_\nu(\lambda) = -\beta\Delta\lambda/(\lambda\lambda_0 w_\nu) = -\beta\Delta\lambda/[\lambda_0^2 w_\nu(1+\Delta\lambda/\lambda_0)], \qquad (1.2-42)$$

where $\lambda_0 = k/\nu_0$.

If $\Delta\lambda/\lambda_0 \ll 1$, then $x_\nu(\lambda) \approx x_\lambda$ and $w_\lambda = \lambda_0^2 w$.

Finally, from Eqs. (1.2-40) and (1.2-42), we have for $\Delta\lambda/\lambda_0 \ll 1$:

$$F_\nu(\lambda) = exp\{-\beta^2(\Delta\lambda)^2/[\lambda_0^4 w_\nu^2(1+\Delta\lambda/\lambda_0)^2]\} \approx$$

$$exp\{-\beta^2(\Delta\lambda)^2/[\lambda_0^4 w_\nu^2] * [1-2\Delta\lambda/\lambda_0]\} =$$

$$F_\lambda exp\{-\beta^2(\Delta\lambda)^3/[\lambda_0 w_\lambda^2]\}. \qquad (1.2-43)$$

Exercise 1.2-4

Readers are invited to derive equation $F_\lambda(\nu)$ symmetric to Eq. (1.2-43).

The exponential factor of Eq. (1.2-43) causes the asymmetry of the Gaussian peak (Fig. 1.2-11). The asymmetry of the broad peaks increases. The 2nd-order derivatives of these peaks reflect this effect (Fig. 1.2-12b).

Figure 1.2-13 demonstrates that the $F_\lambda(\lambda) \rightarrow F_\lambda(\nu)$ transformation of the four equal Gaussian peaks shifted along the wavelength axis (a), produces four Gaussian curves which widths are different (b). Therefore, the four derivatives $d^n F_\lambda(\lambda)/d\lambda^n$ of the peaks in Figure 1.2-13a are similar, but the corresponding derivatives $d^n F_\nu(\nu)/d\nu^n$ (Fig. 1.2-13b) are different.

Numeric differentiation of the peaks, whose widths are different (Fig. 1.2-13b), have technical problems since the equidistant points of the wavelength scale are non-linearly transformed to the wavenumbers (Eq. (1.2-39)). The straightforward solution was obtained using the differential calculus [25].

Given spectrum $F_\lambda(\lambda_i)$, defined at equidistant points ($i = 1,2,\dots,N$).

The 1st and the 2nd-order derivatives of $F_\lambda(\lambda_i)$ over ν are

$$dF_\lambda(\lambda_i)/d\nu = [dF_\lambda/d\lambda][d\lambda/d\nu] = -F_\lambda'\, k/\nu^2 = -F_\lambda'\, \lambda^2/k. \quad (1.2-44)$$

$$d^2F_\lambda(\lambda_i)/d\nu^2 = (-F_\lambda'\, \lambda^2/k)' = -F_\lambda''[d\lambda/d\nu][\lambda^2/k] + 2F_\lambda'\lambda^3/k^2 =$$

$$F_\lambda''\, \lambda^4/k^2 + 2F_\lambda'\lambda^3/k^2. \quad (1.2-45)$$

Figure 1.2-12b represents $d^2F_\lambda(\lambda_i)/d\nu^2$. The derivatives F_λ' and F_λ'' over λ (Eq. (1.2-45)) were estimated numerically using Savitzky-Goley filters (Chapter 2.3).

The equation of the 4th-order derivative was given in the article [27].

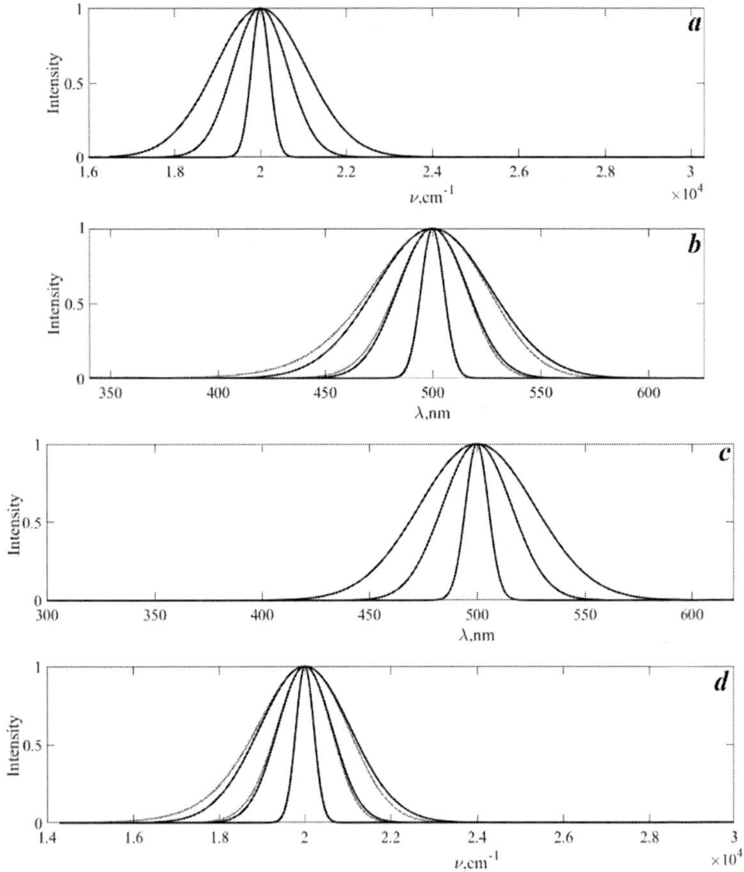

Figure 1.2-11. The x-scale transformations of the Gaussian peaks: $F_\nu(\nu) \to F_\nu(\lambda)$ $(a \to b)$ and $F_\lambda(\lambda) \to F_\lambda(\nu)$ $(c \to d)$ $\nu_0 = 20{,}000 cm^{-1}$, $w_\nu = 500, 1500, 2500 cm^{-1}$, $\lambda_0 = 500, w_\lambda = 12.5, 37.5, 62.5$. Dotted curves take the exponential factors, that cause the asymmetry into account.

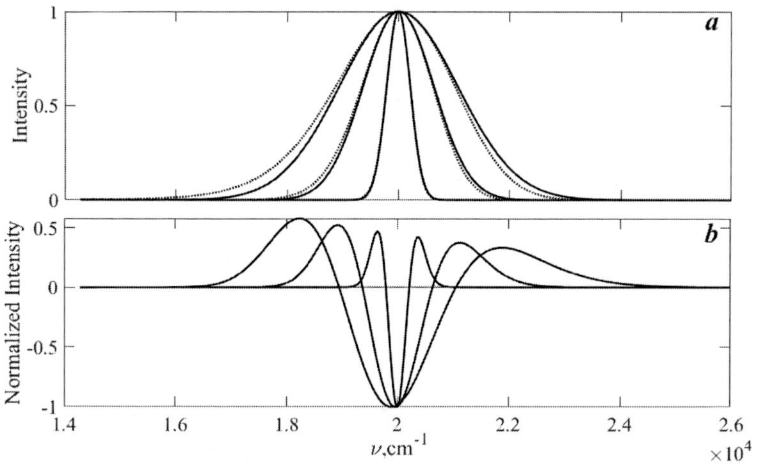

Figure 1.2-12. The x-scale transformation of the Gaussian peaks: $F_\lambda(\lambda) \rightarrow F_\lambda(\nu)$.
(a) ($\lambda_0 = 500, w_\lambda = 12.5, 37.5, 62.5$). Dotted curves take the exponential factors, that cause the asymmetry into account.
(b) The 2nd-order derivatives of "a".

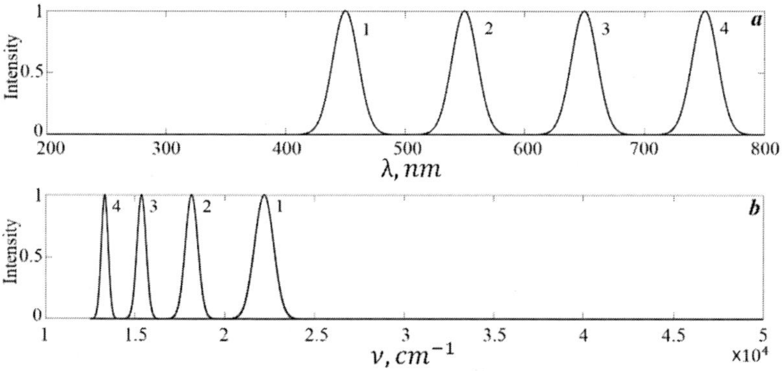

Figure 1.2-13. The $F_\lambda(\lambda) \rightarrow F_\lambda(\nu)$ transformations of the Gaussian peaks. $w_\lambda = 25$. Dotted curves take the exponential factors, that cause the asymmetry into account.

CHAPTER THREE

ANALYSIS OF NOISE IN SPECTRAL MEASUREMENTS*

Measurements of any physical value are influenced by random and systematic errors. Systematic errors may arise from incorrect measurements (a typical example is the improper preparation of the blank cuvette in the spectrophotometer) and from the imperfection of the instrument (e.g., spurious reflection and radiation). Often, systematic errors can be decreased until they are negligible by improving the apparatus and the measurement process (e.g., by correct calibration). However, random errors cannot be eliminated in principle; they only can be decreased by improving the measurement procedure (e.g., by an expansion of the measurement scale) and by the analog or numerical processing of obtained results (e.g., by smoothing). It is important to emphasize that reducing the random errors can cause unpredictable, significant distortions of the actual value of the quantity to be measured (systematic errors).

Random errors vary randomly with time. They are due both to the errors in the analog-to-digital conversion of the measured value and to the impact of different external factors on the measurement process (e.g., a variation of the sample temperature, vibrations). In spectroscopy, the sources of the random errors are the random noises arising in different parts of the spectrometer, mainly, in the radiation detector. The origins of these noises are very different and can be approximately described in every case by the particular mathematical model based on probability theory. The following classification was given in a series of articles which were summarized by Mark and Workman [1].

---

*Major parts of this chapter were taken from the following book: Dubrovkin, J. Mathematical Processing of Spectral Data in Analytical Chemistry: A Guide to Error Analysis. Cambridge Scholars Publishing. 2018.

The sources of noise

Detector-dependent noise

The thermal noise (IR, NIR spectrometers) is the noise in the thermal detectors. This noise does not depend on the intensity of the electromagnetic radiation that falls on the sensor. In the time domain, this noise has a Gaussian (normal) intensity distribution. (We hope that readers remember this from their university courses in probability and statistics). This noise is often called "white noise" because it has the constant power density of the Fourier spectrum (Appendix A1) in the vast range (Fig. 1.3-1). Noise power is proportional to the width of an interval of Fourier frequencies (bandwidth) of the recording device. For more details on noise, see Appendix C.

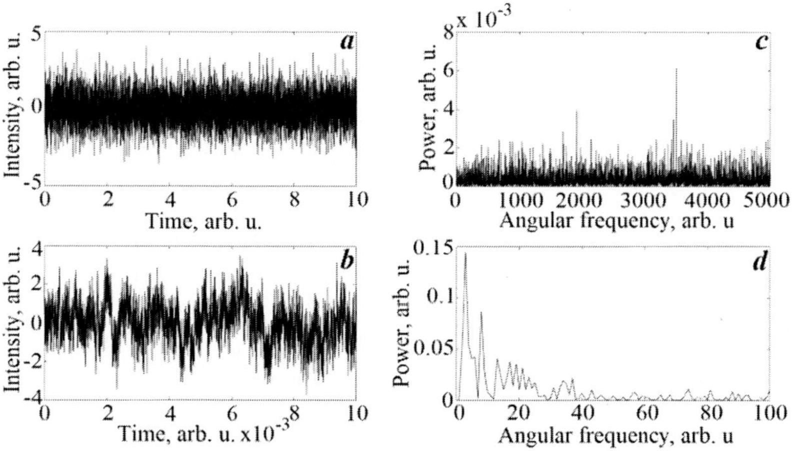

Figure 1.3-1. White and pink noises in the time (a, b) and the Fourier (c, d) domain. The noise has a zero mean value; the standard deviation is one.

For numerical experiments, analytical signals are usually simulated by the composition of the peaks of known shapes. This model includes a noise generated by a computer program. For example, in the MATLAB package, the function “randn” generates pseudo- normal noise with zero mean and unit standard deviations. Numerical experiments show that these noise parameters are only accurate if the noise array contains hundreds of points. Figure 1.3-2 demonstrates that, the noise parameters are different in short intervals. Moreover, the noise is well approximated by a parabola, and not by a zero line as expected. Therefore, the research may observe a false structure in the analytical signal under processing.

However, the mean noisy data obtained in a large number of repetitions, lack this flaw. We will discuss this issue below.

The standard deviation of the absorbance (A) measurements distorted by a low-intensity thermal noise [2]:

$$\sigma_{Term} = \{k_1/ln(10)\}\sqrt{1 + 100^A}, \quad (1.3-1)$$

where k_1 is the constant indicating precision of the spectrophotometer.

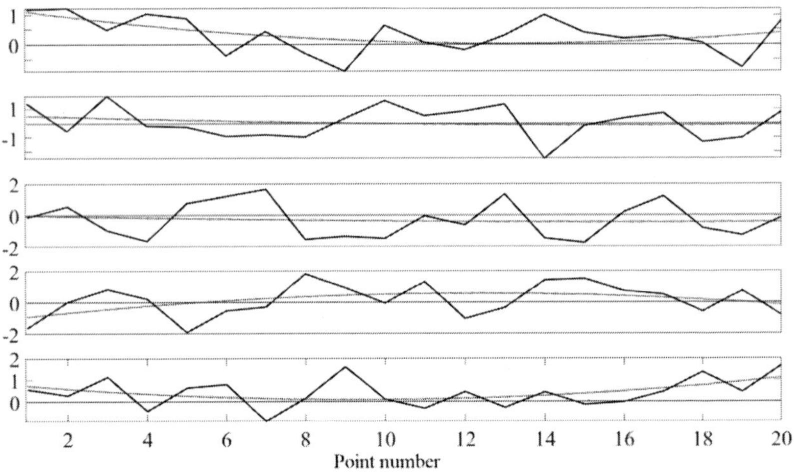

Figure 1.3-2. Normal noise, generated by MATLAB function “randn” with zero mean and unit standard deviation (solid line), was approximated by a parabola (dotted line). The estimated means and the standard deviations from the top to the bottom subplots: [0.3168, 0.6111; 0.0649, 1.1256; -0.3137, 1.1218; 0.1367 1.0445; 0.4027 0.6751].

Plot $[\sigma_{Term}/A](A)$ (Fig. 1.3-3) shows that the absorbance region 0.2-1.2 is the most suitable for spectroscopic measurements since the relative standard deviation is approximately constant in this range.

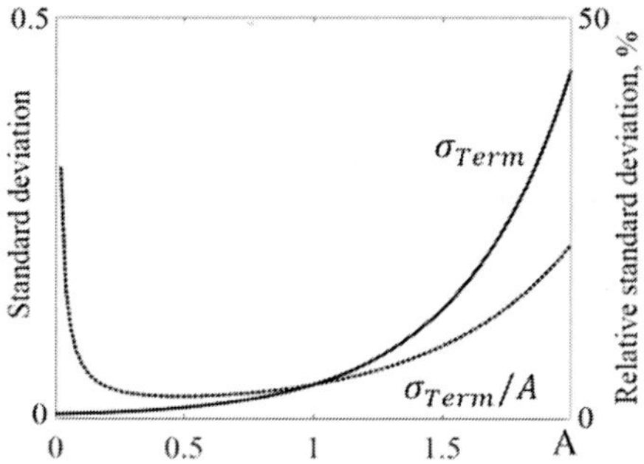

Figure 1.3-3. Eq. (1.3-1). $k_1 = 0.01$.

The shot-noise (UV-VIS, X-ray, and gamma-ray spectrometry) in the photon-counting detectors has a Poisson intensity distribution in the time domain. Its intensity increases with the square root of the signal. This noise is also white.

The standard deviation of the absorbance, which is distorted by a low-intensity short-noise, is

$$\sigma_{Shot} = \{k_2/(ln(10))\}\sqrt{1+10^A}, \qquad (1.3-2)$$

where k_2 is a constant similar to k_1 in Eq. (1.3-1).

Exercise 1.3-1

The readers are invited to represent the noise data obtained by their laboratory instruments, like in Figure 1.3-4.

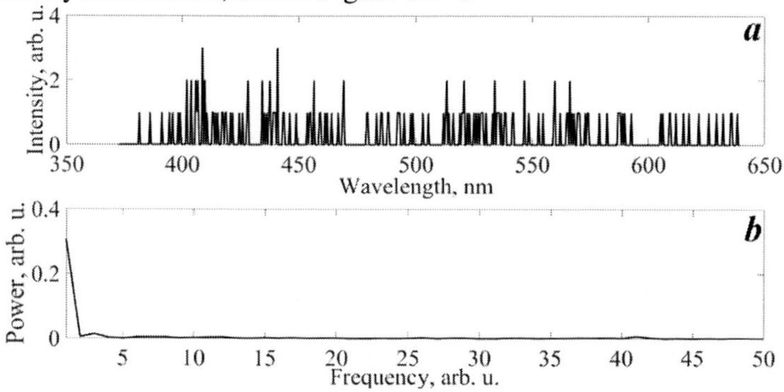

Figure 1.3-4. Noise measured on Bruker Optics 2501S spectrograph by D. V. Ushakou (Pomeranian University in Slupsk, Poland) and the noise power spectrum (panels a and b, respectively).

Variations in energy, which are incident on the detector due to the vibrations of the source, and the changing geometry of the radiation, cause the flicker (pink) noise ($1/f$-noise). The noise intensity is proportional to the signal energy. The noise power spectrum depends on the frequency: $f^{-\alpha}$, where $\alpha \simeq 1$ (Fig. 1.3-1) (Appendix C). If the noise is small, then the standard deviation of absorbance σ_f is approximately constant.

Detector-independent noise

The noise sources include the mechanical vibrations of the optical instruments and different kinds of instabilities:

- The variation of the pathlength in the absorption spectroscopy is due to the changes in the sample position [2].
- The variability of the sample properties that cannot be measured. However, these properties influence the measurement property (e.g., the changing of light reflectance in the transmittance measurements, the inhomogeneous of the sample, and the artefacts of the blood motion in the blood analysis using Functional NIR Spectroscopy (Optical Topography) [3, 4]).
- Random drifts of optical and electronic devices (the flicker noise) caused by slow changes of their parameters due to the temperature variations and other factors.

The mathematical analysis of these sources can only be performed in particular cases using appropriate assumptions that simplify the solution of the problem.

Computer modelling of correlated noise

Let us consider the noise intensity distribution in the time domain [5]. As was pointed out above, the Gaussian (normal) and Poisson noise distributions are usually used to model noise in spectroscopic measurements. In the time domain, noise is characterized mathematically by the error covariance matrix $\boldsymbol{COV}$, in which a diagonal element

$$COV_{ii} = \sigma_i^2 \qquad (1.3-3)$$

represents the noise variance (dispersion) in the ith point. If no two points of the noise in the time domain are correlated with each other, the non-diagonal elements $COV_{ij} = 0$. If all variances σ_i^2 (Eq. (1.3-3)) are the same, the noise is called *homoscedastic*, otherwise it is *heteroscedastic*. The heteroscedastic noise source, in particular, is a randomly varying baseline (background) [6, 7], due to both instrumental and physical-chemical factors. Total baseline compensation is practically impossible. Suppose that the baseline is approximated by the 2nd-order polynomial, whose coefficients are random numbers. This baseline is added to some spectrum. The procedure is repeated r times. Then the intensity of the tth spectrum in the point i is

$$y_{it} = y_i + \eta_i + a_{0t} + a_{1t}i + a_{2t}i^2 \,, \qquad (1.3-4)$$

where y_i is the undistorted value; η_i is the normal noise with zero mean and dispersion σ_y^2; a_{qt} are the constant coefficients ($q = 0, 1,$ and 2), which change randomly for each spectrum. The estimation of the covariance matrix element is [8]

$$COV(y_k, y_l) = [1/(r-1)]\, \textstyle\sum_{t=1}^{r} (y_{kt} - \bar{y}_k)\,(y_{lt} - \bar{y}_l), \qquad (1.3-5)$$

where the bar is the average symbol. According to Eq. (1.3-4):

$$\bar{y}_i = y_i + \bar{a}_0 + \overline{(a_1)}i + \overline{(a_2)}i^2. \qquad (1.3-6)$$

Substituting Eqs. (1.3-4) and (1.3-6) into Eq. (1.3-5), we have

$$COV(y_k, y_l) = \sigma_y^2 \xi_{kl} + BS(y_k, y_l), \qquad (1.3-7)$$

where $BS(y_k, y_l) = [1/(r-1)] \sum_{t=1}^{r} \Phi(k)\,\Phi(l)$;

$\Phi(v) = \delta_{a_{0t}} + \delta_{a_{1t}} v + \delta_{a_{2t}} v^2$;

$\xi_{kl} = \begin{Bmatrix} 1, k = l \\ 0, k \neq l \end{Bmatrix}$ is the Kronecker symbol; $\delta_{a_{qt}} = a_{qt} - \bar{a}_q$.

Suppose that $\delta_{a_{qt}}$ is a normal random variable with a zero mean and covariance matrix $\sigma_{a_q}^2 \boldsymbol{I}$ (***I*** is the identity matrix). Then, by neglecting the small contributions to the sum of the cross members for sufficiently large r, we obtain from Eq. (1.3-7):

$$COV(y_k, y_l) = \sigma_y^2 \xi_{kl} + \sigma_{a_0}^2 + \sigma_{a_1}^2 kl + \sigma_{a_2}^2 k^2 l^2 \,. \qquad (1.3-8)$$

Correlations between analytical points may also be due to the pre-processing, e.g., digital smoothing [6]. Since a digital filter is usually much shorter than a spectrum, the covariance matrix will be sparse (populated primarily with zeros) and, therefore, singular. If the matrix is near to singular, then a correctly computed inverse is impossible.

CHAPTER FOUR

RESOLUTION LIMIT IN DERIVATIVE SPECTROSCOPY

One of the central concepts of optics and spectroscopy is the resolution limit of optical instruments, whose estimation involves specific criteria [1-10]. The most common of these is the historically first suggested diffraction-related Rayleigh criterion [4], and the mathematical Sparrow criterion [5]. With the advent of computer technology, more elaborate mathematical criteria have been developed to estimate the resolution limit: the relative depth of the dip in the center of a symmetrical doublet [6], the absolute value of the minimum in the Fourier transform of this doublet [7], and the information measure of the distinctiveness between two spectra [8]. However, the comparative studies of the above methods were only conducted for symmetrical doublets [9]. Further, the resolution of asymmetrical doublets was also studied [10, 11].

In the recent article [12], the 2nd-order derivatives of symmetrical peaks, which were modelled by a sum of Lorentzian and Gaussian functions, were obtained by the SG method. The resolution enhancement was empirically studied in the wide range of the SG filter parameters, sampling interval, and the relative combinations of the Lorentzian and Gaussian components in the peak profile. The practical recommendations [12] are not new. They were summarized many years ago [13].

Below, we discuss the resolution problem of the derivatives of the doublets consisting of the symmetrical and asymmetrical peaks, whose intensities and widths are varied.

Resolution limit of doublets

Consider the following doublet:

$$D(y) = F(y + y_0) + RF((y - y_0)/r), \qquad (1.4 - 1)$$

where $F(y)$ is the doublet peak; $y = \beta i/w$; β is the shape parameter; $\beta = 2\sqrt{ln2}$ and 2 for the Gaussian and Lorentzian shapes, respectively;

$i = -m, -m+1, \dots, 0, \dots, m-1, m$; w is the FWHM; $y_0 = \beta\delta/2$; $\delta = \Delta/w$ is the relative separation of the doublet components; $\Delta = 2i_0$ is the absolute separation; $\pm i_0$ is the positions of the doublet peaks; R and r are the relative intensity and the relative width of the second doublet peak, respectively.

We studied pairs of symmetrical and asymmetrical peak shapes: Gaussian-PMG and Lorentzian–Dobosz (Chapter 1.2). The asymmetry coefficients were normalized to the constant value of the asymmetry parameter [14].

Resolution criteria

Sparrow criterion

According to the Sparrow (SP) resolution limit (δ_{SP}) [2, 5], the doublet consisting of equal symmetrical peaks (Fig. 1.4-1, panel a) is resolved when the 1st- and the 2nd-order derivatives in the center of the doublet (Eq. (1.4-1)) are equal to zero (Fig. 1.4-1, panel b):

$$D^{(1)}(0) = D^{(2)}(0) = 0. \qquad (1.4-2)$$

The mathematical expressions for the derivatives of the typical peak shapes can be found in Chapter 1.2.

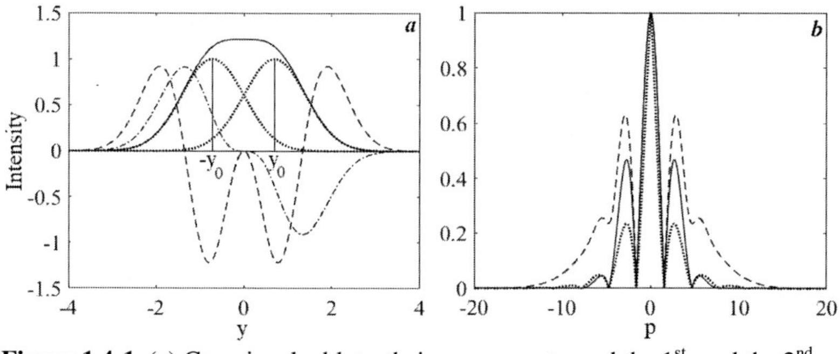

Figure 1.4-1. (a) Gaussian doublets, their components, and the 1st- and the 2nd-order derivatives (solid, dotted, dash-dotted and dashed curves respectively). $R = r = 1$. $y_0 = 0.7071$ (resolution limit). (b) Absolute value of FT of the symmetrical Gaussian ($y_0 = 1$) and Lorentzian (solid and dotted curves, respcctively), and asymmetrical Gaussian ($y_0 = 1, R = 2, r = 0.5$) (dashed curves) doublets. $Abs(\tilde{s}_2)$ (Eq. (1.4-8)) is normalized to the unit maximum.

Exercise 1.4-1

We ask the readers to estimate the SP resolution limit:

1. For symmetrical doublets using the roots given in Table 1.2.2.

2. By generating the 1st- and the 2nd-order derivatives of the doublet with predefined peak shapes and parameters: Δ, R, r, and τ (using formulas of Chapter 1.2 or MATLAB programs). The conditions, represented by Eq. (1.4-4), may be visually satisfied by varying the doublet separation Δ (Fig. 1.4-1, b).

The SP criterion does not take the impact of the noise on the central hollow of the symmetrical doublet into account. Suppose that the doublet is distorted by the normal noise with zero mean and the standard deviation σ. Then the minimum hollow that can be detected with the probability 90% is equal to the width of the 90% noise confidence interval $[-1.65\,, 1.65\,] * \sigma$ [6]. The relative hollow:

$$\rho_{min} = 3.30\sigma/F_{max} \geq 3.30/\xi, \qquad (1.4-3)$$

where $\xi = F_{max}/\sigma$ is the signal-to-noise ratio.

If $\xi \to \infty$, then $\rho_{min} \to 0$, that is, the *SP* criterion is valid. Figure 1.4-2 shows that the minimum hollow increases significantly if the noise increases. Smoothing of the noisy doublets reduces the noise, but the doublet peaks broaden. A trade-off between these factors, in principle, may decrease the resolution limit. The use of the even-order derivatives further reduces the minimally distinguishable doublet separation.

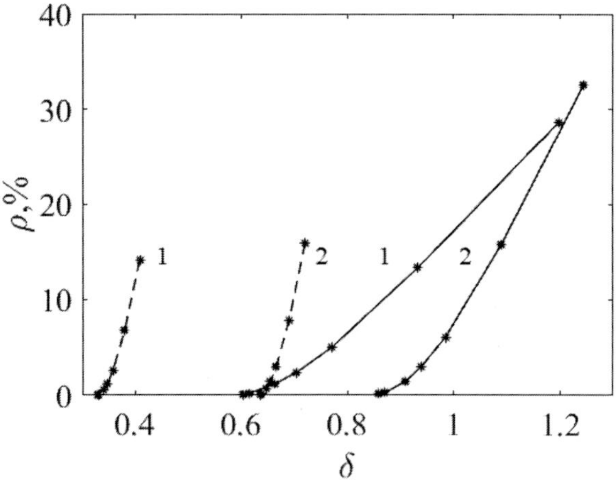

Figure 1.4-2. Dependencies of the relative hollow ρ_{min} on the relative separation (δ) for the Lorentzian (1) and Gaussian (2) doublet peaks and their 2nd-order derivatives (solid and dashed lines, respectively).

The dependence of the SP resolution limit (δ) on the signal-to-noise ratio (ξ) of doublets and their 2nd-order derivatives (Table 1.4-1) shows that $\delta(\xi)$ is the decreasing function. This result follows from Eq. (1.4-3). When ξ increases, then the minimum relative hollow decreases. Therefore, the peaks shift to each other, thus reducing δ.

Table 1.4-1. The SP resolution limit

ξ_0	Doublet				The 2nd-order derivative			
	Gauss		Lorentz		Gauss		Lorentz	
	δ	ξ	δ	ξ	δ	ξ	δ	ξ
10	1.2433	10.1	1.1966	11.5	0.7208	20.7	0.4083	23.3
20	1.0895	20.9	0.9334	24.7	0.6902	42.3	0.3801	48.7
50	0.9844	54.4	0.7714	65.9	0.6660	109.3	0.3579	127.9
100	0.9388	111.9	0.7048	136.8	0.6548	223.0	0.3477	261.9
200	0.9094	229.2	0.6617	281.2	0.6472	452.1	0.3407	532.3
1000	0.8704	1185	0.6051	1466	0.6374	2357	0.3316	2750
2000	0.8589	2400	0.5854	2979	0.6350	4691	0.3294	5553

ξ_0 is the signal-to-noise ratio of a single peak.

In the asymmetrical case, the SP criterion (Fig. 1.4-3) has a new form:

$$D^{(1)}(Y_0) = D^{(2)}(Y_0) = 0. \quad (1.4-4)$$

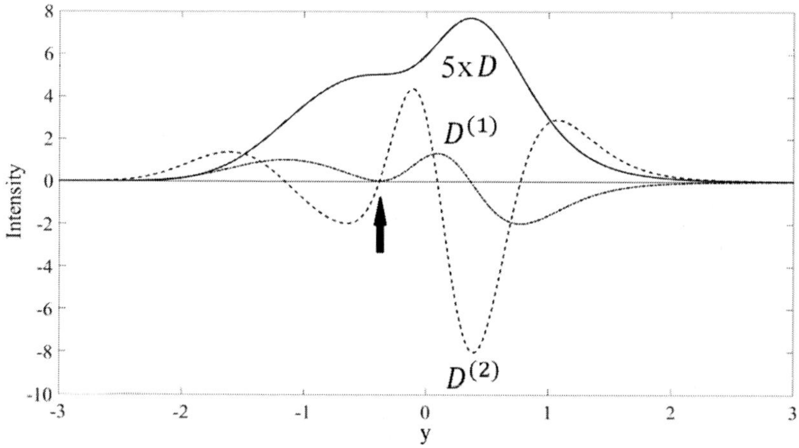

Figure 1.4-3. Resolution of the asymmetrical PMG doublet ($R = 1$, $r = 0.5, \tau_1 = \tau_2 = 0.1142, \delta_y = 0.9089, Y_0 = -0.384$ (arrow)) (solid curve) and its 1st- and 2nd-order derivatives.

Information measure

The information measure to distinguish between spectra s_1 and s_2 is [8]

$$\mu(s_1, s_2) = (1/2\pi) \int_{-\infty}^{+\infty} |(\tilde{s}_1 - \tilde{s}_2)|^2 / G(\omega)\, d\omega, \qquad (1.4-5)$$

where the tilde character is the symbol of the Fourier transform (FT), G is the spectral noise power (Appendix C), and ω is the angular frequency. The probability that two spectra can be distinguished from one another is

$$P(\mu) = 0.5 + \Phi\left(\sqrt{\mu/8}\right), \qquad (1.4-6)$$

where $\Phi(z) = \sqrt{(1/2\pi} \int_0^z e^{-t^2/2}\, dt$ is the cumulative distribution function of the standard normal distribution. $P(\mu) = 90, 95$ and 99% for $\mu \approx 13.1, 21.8$ and 43.4, respectively.

The FT of the Gaussian doublet with coinciding maxima:

$$\tilde{s}_1 = \tilde{F}_1 + \tilde{F}_2, \qquad (1.4-7)$$

and with maxima separated by δ:

$$\tilde{s}_2 = \tilde{F}_1\, exp(-jp\delta/2) + \tilde{F}_2\, exp(jp\delta/2), \qquad (1.4-8)$$

where $j = \sqrt{-1}$ and $p = \omega w$.

According to Eqs. (1.1-6) and (1.1-7),

$$\tilde{F}_1 = Cw\Theta(p), \qquad (1.4-9)$$

$$\tilde{F}_2 = CwRr\Theta(rp), \qquad (1.4-10)$$

where $C = \sqrt{\pi}/\beta_G$ and π/β_L ; $\Theta(p) = \exp(-p^2/4\beta_G^2)$ and $exp(-|p|/\beta_L)$ for Gaussians and Lorentzians, respectively. For these peaks, $\Theta(p)$ is a real function, while for asymmetrical peaks, it is a complex function.

The information measure μ was estimated for the Gaussian and Lorentzian peaks in the general case in Appendix D. Here, we only consider a particular case of symmetrical doublets, when $R = r = 1$, that is, $\tilde{F}_1 = \tilde{F}_2$.

$$\tilde{s}_1 = 2\tilde{F}_1, \tilde{s}_2 = 2\tilde{F}_1 \cos(p\delta/2). \qquad (1.4-11)$$

From Eqs. (1.4-5), (1.4-9) and (1.4-11), we have

$$\mu(s_1, s_2) = (2/\pi) \int_{-\infty}^{+\infty} \left|\tilde{F}_1\right|^2 [1 - \cos(p\delta/2)]^2 / G(\omega)\, d\omega. \qquad (1.4-12)$$

The theoretical power of the white noise with the variance σ^2 is $G(\omega) = \pi\sigma^2/\omega_0$ (Appendix C). Physically, if $|\omega| > \omega_0$, then the integrand of Eq. (1.4-5) approaches zero; therefore, the integrals in the boundaries $\pm\infty$ and $\pm\omega_0$ are equals.

Finally, we have the following for the Gaussian and Lorentzian doublets, respectively (Appendix D):

$$\mu_G = T_G p_0 \xi^2 [3 - 4exp(\delta^2 ln2/2) + exp(\delta^2 2ln2)], \qquad (1.4-13)$$

$$\mu_L = T_L p_0 \xi^2 [3 - 4/(1 + \delta^2/4) + 1/(1 + \delta^2)], \qquad (1.4-14)$$

where $T_G = 1/\sqrt{2\pi ln2}$; $T_L = 0.5$; $p_0 = w\omega_0$; $\xi = F_0/\sigma$ is the signal-to-noise ratio; F_0 is the peak amplitude.

For the strongly overlapping doublets ($\delta < 0.5$):

$$\mu = Z_\mu p_0 \xi^2 \delta^4, \qquad (1.4-15)$$

where Z_μ=0.3453 and 0.3750 for the Gaussian and Lorentzian doublets, respectively.

It follows from Eq. (1.4-15) that the signal-to-noise ratio, required to identify the internal doublet structure with the probability no less 95%, is

$$\xi \geq Z_\xi/(\delta^2\sqrt{p_0}), \qquad (1.4-16)$$

where Z_ξ=7.95 and 7.62 for the Gaussian and Lorentzian doublets, respectively.

Consider the following example:

$\delta = 0.5$; $w = 1$; $p_0 = \omega_0 = 1/\tau = 10$, where $\tau = 0.01$s is the time constant of the spectrometer. Then, from Eq. (1.4-16) we have $\xi_{min} \approx 3$.

Exercise 1.4-2

We offer the readers to estimate the information measure for the strongly overlapped doublets (similar to Eq. (1.4-15)) when R and r are not units.

Petrash criterion

The Petrash resolution criterion (PT) [7] is based on Michelson studies in the interferometry [15].

According to the PT, the resolution measure equals to the amplitude of the second lobe in the modulus of the FT of the doublet. If $R = 1, r = 1$, then this measure, obtained from Eq. (1.4-8), is

$$\tilde{s}_\delta(p_{max}, \delta) = 2Cw\Theta(p_{max})cos(\delta p_{max}/2), \qquad (1.4-17)$$

where p_{max} is the point located next to the global maximum at $p = 0$ (Fig. 1.4-1b). p_{max} lays between two zero points of $cos(\delta p/2)$ in the interval $[1, 3] * \pi/\delta$. The maximum values were estimated numerically.

The PT criterion may show infinite resolution since it does not take a noise that influences the second lobe intensity into account. To study this issue, we performed a numerical experiment. The Gaussian doublet ($R = r = w = 1, \delta = 0.5$) was distorted by the normal noise with zero mean and standard deviation σ. The FTs of the noise-free and noisy doublets were obtained using the fast algorithm (FFT) (Fig. 1.4-4, panel a and b, respectively). Since the doublet was defined in the wide interval of abscissa [-100, 100], the discrete FT was very close to the continuous FT.

The signal-to-noise ratio in the maximum of the second lobe ($\tilde{F}_{LM}$), normalized to the maximum of the central lobe (Fig. 1.4-4 a), is

$$\xi_{Four} = \tilde{F}_{LM}/\sigma_{Four} \sim \tilde{F}_{LM}/\sigma\sqrt{h} = \xi\tilde{F}_{LM}/\sqrt{h}, \qquad (1.4-18)$$

where σ_{Four} is the standard deviation of the noise in the Fourier domain estimated in the high frequency FT region (Fig. 1.4-4 b), and h is the sampling interval.

For a given h, σ was adjusted to $\xi_{Four} \approx 3$, similar to the SP criterion. The corresponding values of the maximum signal-to-noise ratio of the doublet were approximately 12.5, 40, 125, and 400 for $h = 10^{-5}$, $10^{-4}, 10^{-3}$, and 10^{-2}, respectively.

Since, as a rule [13], $h/\tau < 10$, the signal-to-noise ratio, estimated by the PT criterion using one point in the Fourier spectrum, is more significant than the corresponding value required by the information measure. This measure uses all the information contained in the doublet.

Generally, the PT resolution measure, obtained from the modulus of Eq. (1.4-8), is

$$|\tilde{s}_\delta(p,\delta)| = Cw\sqrt{\Psi}, \qquad (1.4-19)$$

where $\Psi = \Theta^2(p) + R^2r^2\Theta^2(rp) + 2Rr\Theta(p)\Theta(rp)cos(\delta p)$.

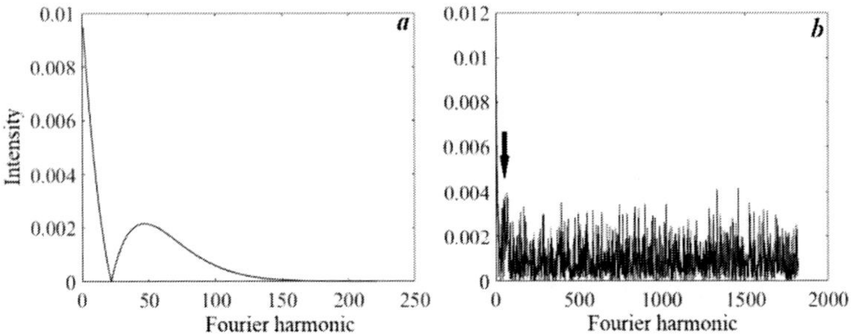

Figure 1.4-4. FFT of the noise-free (a) and noisy (b) Gaussian doublets ($R = r = w = 1, \delta = 0.5$). The arrow points to the lobe. (b) $h = 10^{-5}$, $\sigma = 0.08, \tilde{F}_{LM} = 0.0022, \sigma_{Four}$ =0.0073.

Figure 1.4-1 (d) represents the plots of the normalized values $|\tilde{s}_\delta(p,\delta)|/|\tilde{s}_\delta(0,\delta)|$ versus p (Eq. (1.4-19)). The plots show that the first and the second lobes of the FT of the asymmetrical doublet, consisting of the unequal width peak ($r = 0.5$) and amplitudes ($R = 2$), strongly overlap. However, the separation of the doublet peaks ($\delta = 2$) is larger than the Sparrow limit ($\delta_{SP} = 1.149$). For asymmetrical doublets, it is impossible to measure the second lobe amplitude directly. Therefore, the PT resolution criterion appears to be of no practical use.

Information measure (continued)

For the asymmetrical Gaussian, Lorentzian, and Voigt doublets, analytical calculations appeared to be very complicated; therefore, μ was estimated numerically.

To compare the information resolution limits (δ_{inf}) of the different doublets under similar conditions (i.e., the same information measure), we estimated δ_{inf} for specific doublets from the equation

$$\mu(\delta_{inf}, R, r) = \mu(\delta_{SP}, R = 1, r = 1). \qquad (1.4-20)$$

The solution of Eq. (1.4-20) is only of theoretical interest. The numerical results are discussed below.

Consider two anti-symmetrical doublets composed of equal-width components ($r = 1$) with relative intensities of the second doublet peak: $R_1 = 2$ and $R_2 = 1/R_1 = 0.5$ (Fig. 1.4-5, panel a, plots I and II, respectively). The second doublet was obtained by dividing the first doublet's intensities by 2 (plot III) and inverting. According to Eq. (1.4-14), the informational measures μ (Eq. (1.4-5)) calculated for these doublets must be the same. This condition is satisfied for $R > 1$ and any r if $\chi(\delta, R, r)$ (Eq. (1.4-14)) is divided by R^2. Then, for $r = 1$, doublets I and II coincide. However, in the general case ($r \neq 1$); after division and inverting, two anti-symmetrical doublets having $R < 1$ and $R > 1$, do not coincide (Fig. 1.4-5, panel b, plot III) and, therefore, the μ values are also different.

We compared the resolution limits of doublets belonging to the group, which consisted of the same peak shapes but which are different in the peak parameters, using the Sparrow and the information criteria separately.

Results and Discussion

Doublets consisting of equal-width components ($r = 1$)

Tables 1.4-1and 1.4-2 show that

1. Peaks with relative peak intensities R and $1/R$ have the same Sparrow resolution limit (δ_{SP}).

2. In all cases, δ_{SP} has minimum value (the resolution is maximum) if $R = 1$ increases. The information resolution limit δ_{inf} of the Gaussian doublets ($\tau = 0$) slightly decreases while R increases (Fig. 1.4-6). In the case of the Lorentzian peaks ($\tau = 0$), the dependence $\delta_{inf}(R)$ has a convex form. For more detailed information, we refer readers to the book [11].
3. Asymmetry of the Gaussians and Dobosz peaks may improve or worsen the resolution.

Table 1.4-1. The Sparrow resolution limit for PMG and Dobosz doublets

R	r	Doublets					
		PMG			Dobosz		
		τ					
		0	**0.0519**	**0.1142**	**0**	**0.2**	**0.5**
0.5	**0.5**	1.0610	0.8904	0.8539	0.7932	0.8892	1.0153
	1.0	**1.8580**	1.8769	1.8434	**1.8613**	1.8137	1.7287
	2.0	2.2984	2.4423	2.5895	2.8200	2.7457	2.6410
1.0	**0.5**	1.0432	0.9933	0.9029	1.0998	1.1609	1.2517
	1.0	1.4142	1.5263	1.5300	1.1540	1.3404	1.4864
	2.0	2.0865	2.1660	2.2146	2.1996	2.0760	1.8854
2.0	**0.5**	1.1492	1.0745	0.9871	1.4100	1.4487	1.5092
	1.0	**1.8580**	1.8084	1.7238	**1.8613**	1.9038	1.9610
	2.0	2.1210	2.5756	2.6959	1.7290	2.1355	2.3194

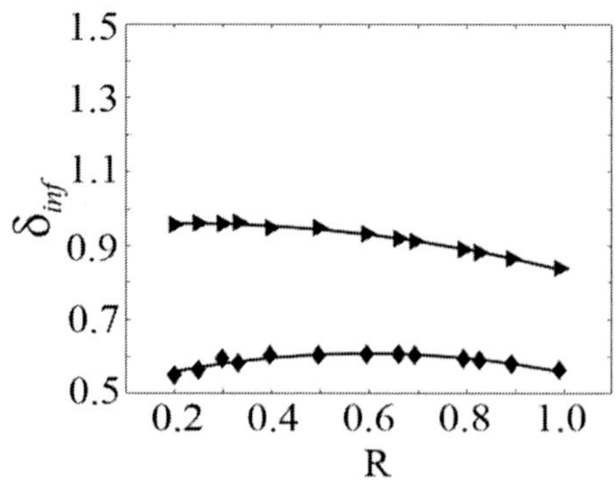

Figure 1.4-6 [10]. The information resolution limit for the Gaussian (triangles) and Lorentzian (diamonds) peaks.

Doublets consisting of unequal-width components

The minima of the dependencies of the information resolution limit on R $\left(\delta_{inf}^{min}(R)\right)$ are located at $R = 1$ (Fig. 1.4-6). In contrast to this, the dependencies $\delta_{SP}(R)$ reach their minima, δ_{SP}^{min} (corresponding to the best resolution), when $R \approx r$ ($r = 1 \div 0.5$ and $\tau = 0$) (Fig. 1.4-6). However, if $r = 0.3$, then, for the Gaussian doublet, the minimum value $\delta_{SP}^{min} = 0.3277$ is located in the neighbourhood of $R = 0.19$ (Fig. 1.4-8, panel b). We found that the doublet shoulder shifts from the left to the middle and then to the right position, when the function $\delta_{SP}(R)$ passes the point of its minimum (δ_{SP}^{min}) (Fig. 1.4-9). Plot a of this figure shows that the transition to the left wing causes a steep increase of the dependence $\delta_{SP}(R)$ when $r < 1$. This effect is very noticeable if $r = 0.3$ (Fig. 1.4-9, plots c and d).

Consider two pairs of the overlapping peaks: [weak, narrow; strong, broad] (Fig. 1.4-8, plots a and b) and [weak, broad; strong, narrow] (Fig. 1.4-9, plot d). The Sparrow resolution limit of the first pair is less than that of the second. However, this is not always the case (Fig. 1.4-9, plot c). The data presented in Figure 1.4-3 show that the information criterion is $\delta_{inf}(R) < \delta_{inf}(1/R)$ if $R > 1$ and $r \neq 1$. We explain the observed result as follows. The weak, narrow peak is immersed in a broad contour (Fig. 1.4-8, plot b), in contrast to the weak, broad peak, whose left wing becomes resolved (Fig. 1.4-9, plot d). Since the information measure takes this wing into account, the resolution is better in the last case.

The dependencies $\delta_{SP}^{min}(r)$ and $\delta_{inf}^{min}(r)$ are decreasing functions. The central part of the dependence $\delta_{lim}(R)$ has a V-form. The wings of the dependence $\delta_{SP}(R)$ are increasing functions, while the wings of the dependence $\delta_{inf}(R)$ reach maximum values and then slowly decrease. The evenness of the function $\delta_{inf}(R)$, relative to the point $R = 1$, is impaired when $r < 1$. The Voigt profiles (Fig. 1.4-10) represent complex combinations of the dependencies $\delta_{inf}(R)$ for the Gaussian and Lorentzian peaks. As a rule, a small asymmetry of the Gaussians increases the δ_{SP}values, but small anomalies are observed if $R \leq 0.3$. For the Lorentzians, dependence $\delta_{SP}(\tau)$ is a decreasing function in the range of $R = 5 - 0.7$. For $R \leq 0.7$, the shoulder is shifted from the left to the middle and then to the right position (Fig. 1.4-11). Therefore, the behaviour of the function $\delta_{SP}(\tau)$ becomes irregular. For the Gaussians, the dependence $\delta_{inf}(\tau)$ is an increasing function, while for the Lorentzian peaks, it usually decreases.

In conclusion, we note that the Sparrow criterion is useful to visually evaluate the resolution limits of overlapping peaks. The main drawbacks of this criterion are the non-smooth character of the dependence $\delta_{SP}(R)$ for unequal-widths doublet components, and a possible non-unique solution (Fig. 1.4-12). The plots show that the SP criterion is valid for the left and the right shoulders.

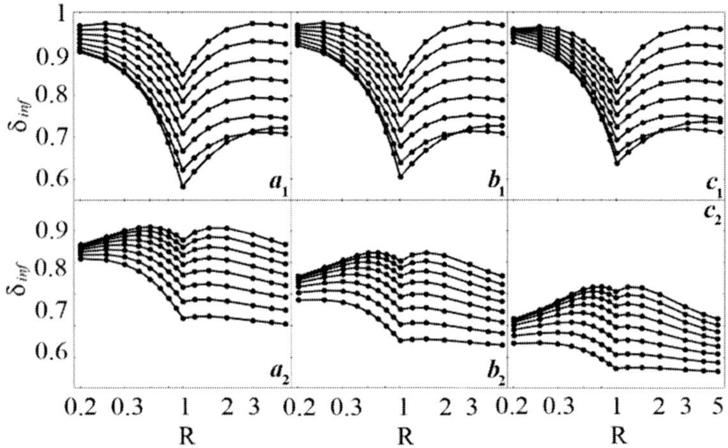

Figure 1.4-7 [10]. The informational resolution limit for the asymmetrical Gaussian ($\tau = 0$ (a_1), 0.0519 (b_1), and 0.1142 (c_1)) and Lorentzian ($\tau = 0$ (a_2), 0.2 (b_2), and 0.5 (c_2)) doublets. $r = 1, 0.9, \dots, 0.3$ (from top to bottom).

Table 1.4-2 [10]. The information resolution limit for symmetrical peaks

R	Gaussians		Lorentzians	
	r=0.3	r=0.5	r=0.3	r=0.5
0.2	0.9036	0.9148	0.5187	0.5403
5	0.7228	0.7448	0.3121	0.3961
0.5	0.8209	0.8443	0.4798	0.5357
2	0.6879	0.7404	0.3322	0.4313

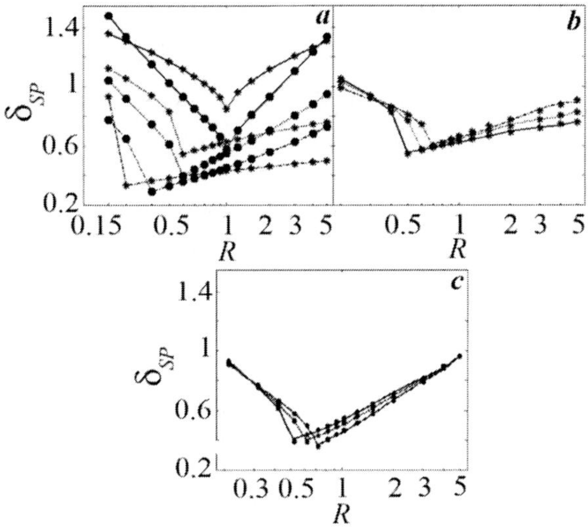

Figure 1.4-8 [10]. The SP resolution limits for the Gaussian (*) and Lorentzian (•) doublets: (a) $r = 1, \tau = \{0, 0.5, 0.3\}$; (b) $r = 0.5$, $\tau = \{0, 519, 1142\}$ x10^{-4} ;(c) $r = 0.5, \tau = \{0, 0.2, 0.5\}$. Each τ-triple is represented by solid, dotted and dash-dotted curves.

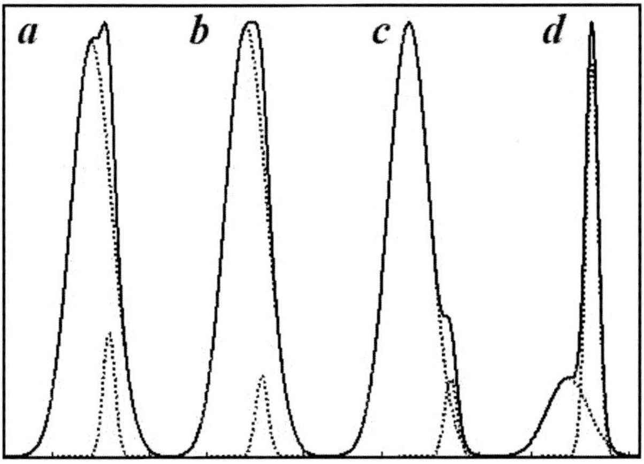

Figure 1.4-9 [10]. Gaussian doublets. $r = 0.3$. (a) $R = 0.3, \delta_{SP} = 0.3632$; (b) $R = 0.19, \delta_{SP} = 0.3277$; (c) $R = 0.18, \delta_{SP} = 0.8733$; (d) $R = 5, \delta_{SP} = 0.4975$.

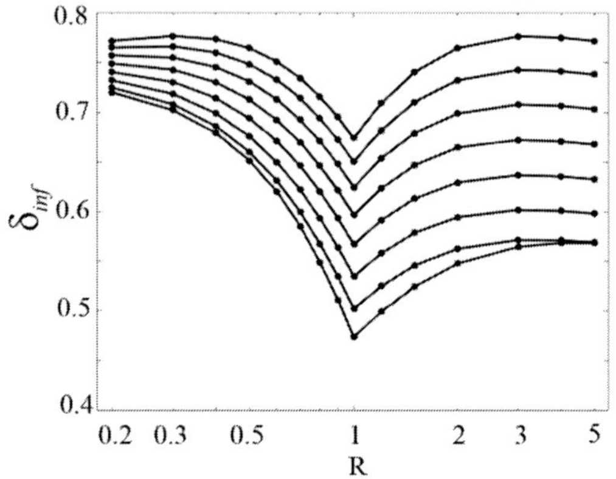

Figure 1.4-10 [10]. Dependencies of the informational resolution limit for the Voigt doublets. $t_G = t_L = 0.5$. For the curves from top to bottom: $r = 1, 0.9, \ldots, 0.3$.

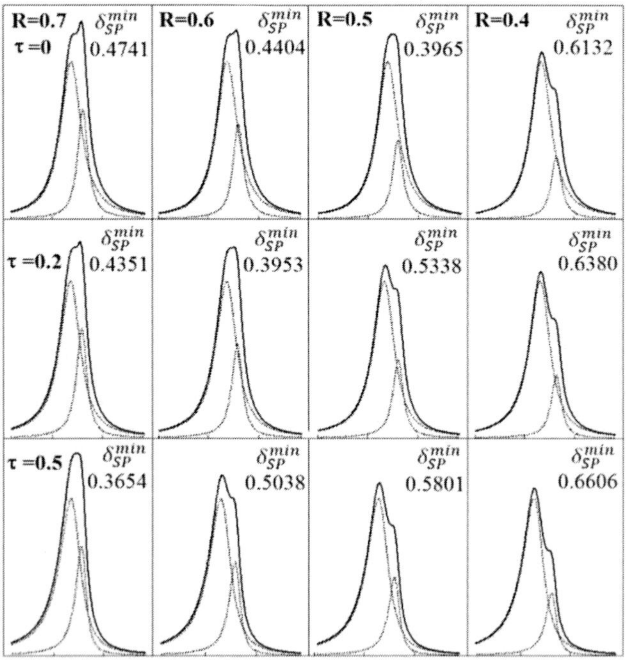

Figure 1.4-11 [10]. Anomalous behaviour of the SP resolution limit for the asymmetrical Lorentzian doublets.

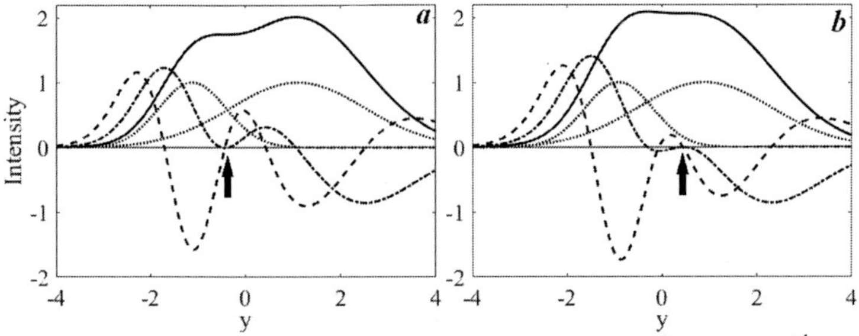

Figure 1.4-12. Gauss doublets, their components and the 1st- and the 2nd-order derivatives (solid, dotted, dash-dotted and dashed curves, respectively). $R = r = 2$. The SP criterion is valid for $\Delta_y = 2.121$ (a) and 1.810 (b).

Resolution limit of the even-order derivatives of doublets

The SP resolution limits of the 2nd- and 4th-order derivatives of the symmetrical doublets are estimated similarly to Eq. (1.4-2)):

$$D^{(3)}(0) = D^{(4)}(0) = 0, \qquad (1.4-21)$$

$$D^{(5)}(0) = D^{(6)}(0) = 0. \qquad (1.4-22)$$

The minimum relative separations in the 2nd- and 4th-order derivatives (Table 1.2-2) for the Gaussian and Lorentzian peaks are, respectively:

$$\delta_{min} = \Delta/w = 0.6302, 0.5238, \text{and } 0.3250, 0.2282. \qquad (1.4-23)$$

Exercise 1.4-3

We ask the readers to do following:

1. To estimate the SP resolution limit (Eq. (1.4-23)) using the roots given in Table 1.2.2.
2. To show that the 2nd- and 4th-order derivatives improve the maximum resolution by factors 1.3475, 1.7766 and 16212, 2.5302 for the Gaussian and Lorentzian peaks, respectively.

Table 1.4-4 shows the resolution limit of the 2nd-order derivatives of the doublets composed of the asymmetric peaks with equal widths and intensities ($R = 1, r = 1$). The limit depends on the asymmetry coefficient τ, but differently for the PMG and Dobosz peaks. Generally, resolution studies using the SP criterion in the derivatives of the asymmetric doublets are difficult due to the redistribution of intensities in favor of narrow peaks. It follows from the inverse dependence of the intensity of the derivative peak on its width in a power equal to the derivative order.

Narrowing the even-order derivatives' central peak is accompanied by the appearance of additional petals (Chapter 1.2). The petals of adjacent peaks may overlap, thereby producing a false structure and/or demonstrating "super-resolution".

Table 1.4-4. Resolution limit of the 2nd-order of the doublets ($R = 1, r = 1$) composed of the asymmetric peaks

Peak	τ	δ_y	Y_0
PMG	0	1.049296	0
	0.0519	1.10691	-0.4415
	0.1142	1.04358	-0.6204
Dobosz	0	0.64984	0
	0.2	0.71268	-0.0860
	0.5	0.75657	-0.0546

In the simplest cases, the SP resolution limit of the asymmetrical doublets indicates the traces of the resolved weak maxima in the 2nd-order derivatives. These maxima, which are strongly shifted from their accurate locations, may be false peaks (Fig. 1.4-13) (see Chapter 3.2).

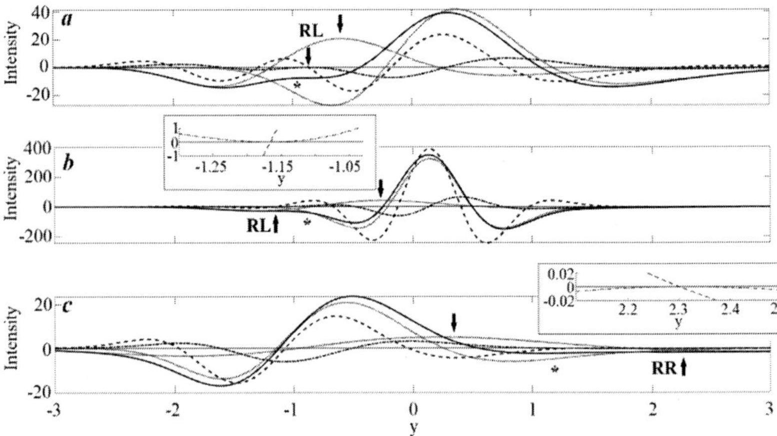

Figure 1.4-13. The SP resolution limits for the negative 2nd-order derivative of the PMG asymmetric doublets (solid line). RL and RR stand for the left and right-side cases, respectively. The 2nd-order derivative doublet components, the 3rd- and the 4th-order derivatives are designated by dotted, dash-dotted, and dashed lines, respectively. Asterisks point out the traces of weak maxima. Curve intensities are adjusted to improve visibility. $R, r, \Delta_y, \tau_1, \tau_2 = \{[2, 1, 0.9636, 0.1142, 0.1142]$, $[2, 0.5, 0.2887, 0.1142, 0]$, and $\{[1, 2, 0.8890, 0.1142, 0.1142]\}$ for panel a-c, respectively.

The modified SP resolution criterion, which is based on the minimum relative hollow $\rho_{min} = (F_{max} - F_{min})/F_{max}$ in the doublet derivative center, takes the noise into account (Eq. (1.4-3)).

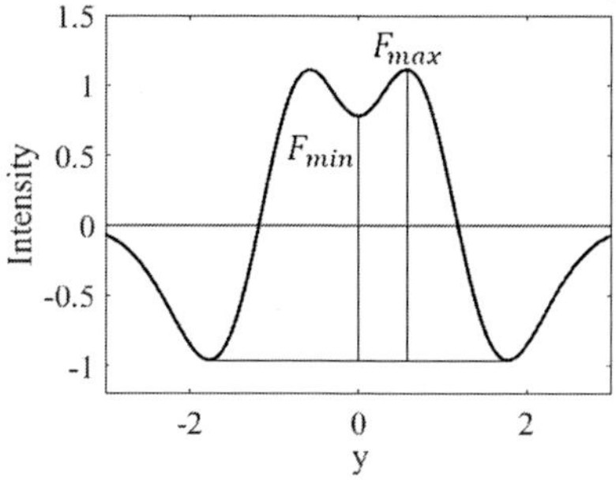

Figure 1.4-14. The hollow in the center of the negative 2nd-order derivative of a doublet.

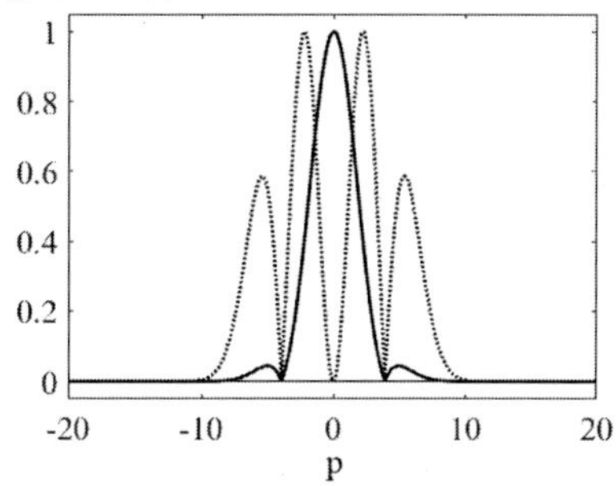

Figure 1.4-15. Normalized FT of the Gaussian doublet ($\delta = 0.8, w = 1$) and its 2nd-order derivative (solid and dashed curves, respectively).

The dependencies of ρ_{min} on the relative separation (δ) for the 2nd-order derivatives of the symmetrical Lorentzian and Gaussian doublets (Fig. 1.4-2) and Table 1.4-1 show that if $\delta \rightarrow \delta_{min}$, then the signal-to-noise ratio becomes more than 1000.

The information measure μ (Eq. (1.4-5)) is invariant to the differentiation since the nominator and denominator of the integrand are multiplied by the same Fourier operator.

The PT criterion demonstrates that the 2nd-order differentiation improves resolution (Fig. 1.4-15). Multiplication of the FT spectra by the differentiation operator p^2 significantly intensifies the lobe. However, the noise also dramatically increases in the derivatives. Therefore, as for the SP criterion (Fig. 1.4-2), the resolution improvement, due to the differentiation, may only be achieved at the cost of reducing the signal-to-noise ratio in derivatives.

Conclusion

The Sparrow criterion is useful for the resolution limit estimation of the even-order derivatives of symmetrical doublets. However, in asymmetrical cases, there are some technical problems in terms of the accurate resolution limit identification.

The information criterion is useful to evaluate the probability of the distinctiveness between the Fourier transforms of two overlapping spectra. The advantage of this criterion is the smoothness of the dependence $\delta_{inf}(R)$. Here, similar to the use of the Sparrow criterion, increasing the signal-to-noise ratio can improve the distinctiveness. We see future perspectives in analytical applications of the information criterion for determining the reliability of the spectral curve deconvolution.

CHAPTER FIVE

INFORMATION CONTENT OF THE DERIVATIVE SPECTRUM

Using information theory (IT) in analytical chemistry and, notably, in analytical spectrometry allowed researchers to reveal new peculiarities of the objects under study [1]. A sample of the object may be characterized by some unknown quantity of intrinsic analytical information (I_{intr}) [2] (Fig. 1.5-1). The analytical instrument extracts from I_{intr} only a small part I_{instr} contained in the analytical signal (AS). Finally, the information content of output data $I_{data} \leq I_{instr}$ due to data processing.

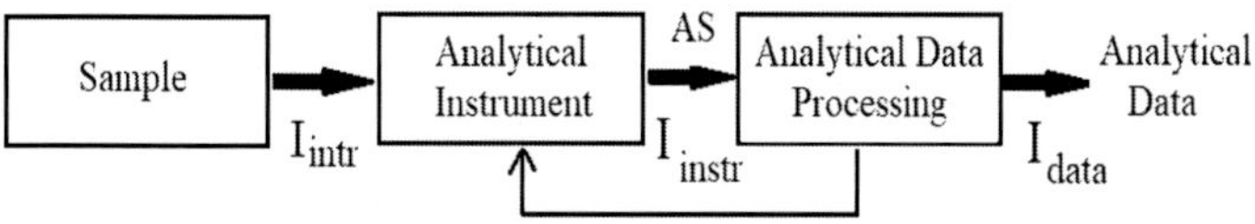

Figure 1.5-1. Information flow in analytical system (adapted from [2]).

Two right rectangles (Fig. 1.5-1) represent the information flow of the derivative spectrometry methods. Unfortunately, in some publications, the success of the derivative spectrometry methods is associated, in a veiled form, with gains achieved by the differentiation. However, according to the general principles of information theory, any linear transformation cannot increase the data information content [3]. The theoretical study [4] discussed this problem in spectroscopy.

We will use the Savitzky-Golay (SG) digital differentiation filters to demonstrate the information changes occurring in derivatives. This less formal method does not require specific knowledge in spectroscopy instrumentation [4].

According to information theory, the maximum information content of a given spectrum depends on the maximum number of the possible combinations of the independent data points, which can be measured [5]. This number is

$$N_{comb} = (N_y)^{N_x}, \quad (1.5-1)$$

where N_y and N_x are the numbers of sampling intervals along y and x axis of the spectra, respectively. N_y is an increasing function of the signal-to-noise ratio. N_x increases if the spectral resolution becomes higher.

Since the information is additive [3], the logarithm of N_{comb} must be used. The base of the logarithm defines the information units. These units are not essential since we will study the ratio of the logarithms below.

The maximum information content of a spectrum [5] is

$$I_{max}^{(0)} = N_{\Delta}^{(0)} ln(1 + \xi^{(0)}), \quad (1.5-2)$$

where $N_{\Delta}^{(0)} = (\lambda_2 - \lambda_1)/\Delta\lambda^{(0)}$ is the number of distinguishable spectral intervals in the range $\lambda_2 - \lambda_1$; $\xi^{(0)}$ is the signal-to-noise ratio.

Suppose that $\Delta\lambda^{(0)}$ is the resolution limit of the symmetrical doublet composed of the narrowest peaks $(w_{min}^{(0)})$, measured by the spectral instrument. Therefore,

$$\Delta\lambda^{(0)} = w_{min}^{(0)} \delta_{\xi^{(0)}}^{(0)}, \quad (1.5-3)$$

where $\delta_{\xi^{(0)}}^{(0)} = (\Delta\lambda^{(0)}/w_{min}^{(0)})_{\xi^{(0)}}$ is the function of the relative resolution limit of the signal-to-noise ratio.

The maximum information content of the 2nd-order derivative of a spectrum is

$$I_{max}^{(2)} = N_{\Delta}^{(2)} ln(1 + \xi^{(2)}), \quad (1.5-4)$$

where $N_{\Delta}^{(2)} = (\lambda_2 - \lambda_1)/\Delta\lambda^{(2)}$; $\Delta\lambda^{(2)}$ is the resolution limit of the 2nd-order derivatives of the symmetrical doublet, whose peaks have a width of $w_{min}^{(2)}$;

$$\Delta\lambda^{(2)} = w_{min}^{(2)} \delta_{\xi^{(2)}}^{(2)}, \quad (1.5-5)$$

where $\delta_{\xi^{(2)}}^{(2)} = (\Delta\lambda^{(2)}/w_{min}^{(2)})_{\xi^{(2)}}$.

Functions $\delta_{\xi^{(0)}}^{(0)}$ and $\delta_{\xi^{(2)}}^{(2)}$ are estimated using the dependences of the minimum relative hollow ($\rho_{min} = 3.30/\xi$) on the relative separation ($\delta = 2\Delta\lambda/w$) for the doublets and their derivatives (Chapter 1.4).

Consider the ratio of the information contents of the spectrum and its 2nd-order derivative, using Eqs. (1.5-2)-(1.5-5):

$$\mathcal{R}_{inf} = \frac{I_{max}^{(0)}}{I_{max}^{(2)}} = \frac{N_{\Delta}^{(0)} ln(1+\xi^{(0)})}{N_{\Delta}^{(2)} ln(1+\xi^{(2)})} = \frac{\delta_{\xi^{(2)}}^{(2)} ln(1+\xi^{(0)})}{\delta_{\xi^{(0)}}^{(0)} ln(1+\xi^{(2)})} = \mathcal{R}_{\lambda}\mathcal{R}_{w}\mathcal{R}_{\xi}, \quad (1.5-6)$$

where $\mathcal{R}_{\lambda} = \Delta\lambda^{(2)}/\Delta\lambda^{(0)}$; $\mathcal{R}_{w} = w_{min}^{(0)}/w_{min}^{(2)}$; $\mathcal{R}_{\xi} = ln(1+\xi^{(0)})/ln(1+\xi^{(2)})$.

The factors are $\mathcal{R}_{\lambda} < 1, \mathcal{R}_{w} > 1$. However, $\mathcal{R}_{\xi}$ may be larger or smaller than 1.These factors depend on the accuracy of the numerical differentiation. Therefore, we estimated Eq. (1.5-5) numerically using the SG differentiation.

Suppose that the narrowest peak ($F(x)$) measured by the instrument is the Gaussian with a unit width and intensity ($w_{min}^{(0)} = 1, F_{max} = 1$). The absolute value of the width is not principal since the readers can imagine that the range of the peak abscissa is large enough to decrease the relative width to the small value that you need. We calculated the full width ($w_{min}^{(2)}$) at the half amplitude ($I_{max}/2$) between the central part and the satellite of the 2nd-order SG derivative of the peak. The polynomial power was 2. Table 1.5-1 represents the widths (m) of the SG digital filters. Since the central peak is very steep, a small sampling interval guarantees a high calculation accuracy for the width.

The standard deviation of the noise in the derivatives is

$$\sigma^{(2)} = \sigma^{(0)} \sum_{\alpha=-m}^{m} W_{\alpha}^{2}, \quad (1.5-7)$$

where W_{α} is the normalized coefficient of the SG filter.

Table 1.4-1 gives the parameters $\xi^{(0)}$ and $\delta_{\xi^{(0)}}^{(0)}$.

The 2nd-order SG derivative of the Gaussian doublet $F(\lambda+\Delta\lambda) + F(\lambda-\Delta\lambda)$ was obtained using the practical sampling interval ($h = 10^{-2}$). The peak separation ($2\Delta\lambda$) was varied to satisfy the resolution criterion (Eq. (1.4-3)). The hollow (H_c) in the doublet center is

$$H_c \cong 3.3\sigma^{(2)}. \quad (1.5-8)$$

The signal-to-noise ratio is

$$\xi^{(2)} = I_{max}/\sigma^{(2)}. \quad (1.5-9)$$

Figure 1.5-2 demonstrates some of the derivatives' plots. The central hollow (H_c) and the maximum amplitude (I_{max}) decrease while the filter broadens (m increases). The standard deviation $\sigma^{(2)}$ decreases because of smoothing. The combine effect of these changes improves the signal-to-noise ratio $\xi^{(2)}$ (Eq. (1.5-9)) (Table 1.5-1).

The table's data shows that

1. The following factors increase the ratio of the information ($\mathcal{R}_{inf} = I_{max}^{(0)}/I_{max}^{(2)}$), contained in the spectrum and its derivative: better accuracy

of differentiation (less the filter width or the sampling interval for $m = const$), and worse derivatives' signal-to-noise ratio (lesser smoothing).

Since for each column of the table $I_{max}^{(0)} = const$, $I_{max}^{(2)}$ decreases. That is, the loss of information is the price that need to be paid to increase the resolution in the derivatives.

2. The relative loss of the information due to the differentiation decreases if the signal-to-noise ratio in the spectrum (ξ_0) increases; we suppose that m and h are constants. The information loss is especially noticeable when $m = 10$.

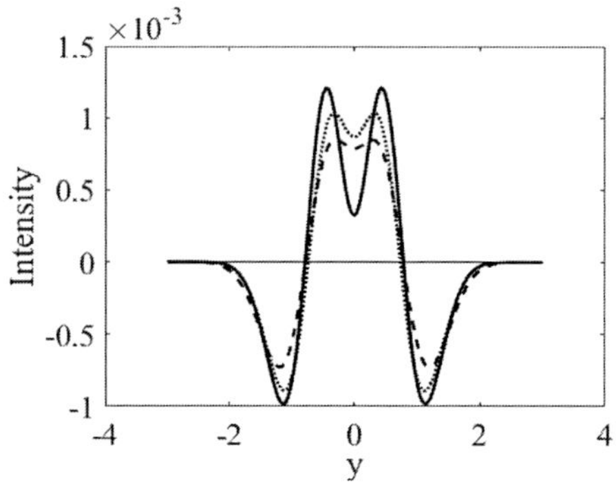

Figure 1-5.2. Negative 2nd-order derivative of the Gaussian doublets. The parameters correspond to the column $\xi_0 = 50, h = 0.02$ of Table 1.5-1. $m = 10$, 20, and 30 (solid, dotted, and dashed lines, respectively).

In conclusion, in this chapter, we have addressed the question of the information content of spectra and its 2nd-order derivatives from a mathematical point of view. This question should not be confused with the issue of the amount of independent information present in spectra (or their derivatives) for the prediction of the chemical properties of the compounds under study (e.g., [6]).

The information content of the analytical signal should not also be confused with the amount of information extracted from the signal in qualitative and quantitative analysis.

According to Eckschlager and Danzer [1], the quantity of the information, obtained by the analytical procedure, equals

$$I = ln\left[(c_2 - c_1)/\left(\sigma_c\sqrt{2\pi e}\right)\right] - 0.5(\delta_c/\sigma_c)^2, \quad (1.5-10)$$

where $c_2 - c_1$ is a priory range of the analyte concentrations; and δ_c and σ_c are their systematic and random errors.

In Chapters 3.2 and 4.4, we show that differentiation may increase this information. From an informational point of view, signal processing is similar to gold mining. The amount of ore minerals decreases due to the processing, but the useful product yield increases.

Table 1.5-1. Parameters of Eq. (1.5-6). Case study: the 2nd-order SG derivatives of the Gaussian doublets

Parameters							
m	ξ_0	50		100		200	
	$\xi^{(0)}$	54.4		111.9		229.2	
	$\delta^{(0)}_{\xi^{(0)}}$	0.9844		0.9388		0.9094	
10	$\mathcal{R}_{inf}$	8.8620	3.9299	7.0332	3.4887	5.5734	3.1793
	$\xi^{(2)}$	3.3	8.2	4.6	16.2	8.5	33.1
	$\delta^{(2)}$	1.2023	0.8110	0.9078	0.7486	0.7957	0.7171
20	$\mathcal{R}_{inf}$	3.3860	2.1689	3.0943	2.1807	2.8858	2.1854
	$\xi^{(2)}$	10.9	38.5	21.9	79.0	44.8	161.6
	$\delta^{(2)}$	0.7794	0.7419	0.7299	0.7201	0.6999	0.7058
30	$\mathcal{R}_{inf}$	2.3376	1.8842	2.3046	1.9703	2.2762	2.0329
	$\xi^{(2)}$	28.1	85.2	57.6	174.0	118.0	354.2
	$\delta^{(2)}$	0.7326	0.7812	0.7066	0.7669	0.6898	0.7573

The SG filter power is 2. The sampling interval $h = 0.01$ and 0.02 in the left and the right columns, respectively.

PART II:

DERIVATIVE SPECTROSCOPY TECHNIQUES

INTRODUCTION

The design of differentiators for analytical instruments as, for any technical devices, is not a simple task since the signal-to-noise ratio in derivatives significantly worsens because of the increase of the higher Fourier frequencies (Appendix A). Also, sometimes the measurement process of the output signals is slow. Therefore, one problem is the differentiation of the slowly varying signals which derivatives over time have low intensity. In this case, cumbersome electronic amplifiers or digital devices are required.

There are two principal groups of mathematical differentiation methods used in spectroscopy that differ in the kind of the signal used in the processing technique. In the first group, analog or discrete devices differentiate the output signal of the electronic unit of the spectrometer. In the second group, the optical module carries out the differentiation. This so-called modulation spectroscopy is a useful tool in physics. Analysts use numerical differentiation performed by computerized devices.

The Fourier Transform (FT) (Appendix A) is a standard method for studying differentiation procedures.

CHAPTER ONE

ANALOG DIFFERENTIATION

Analog electronic systems transform a continuously varying input signal, which is represented by a voltage (or current) in accurate proportion, to the output signal, which is also continuous in time. Their working principle is modelling the dependence between the signals by some function required for signal processing, e.g., a derivative over time. The main electronic elements of analog devices are capacitors and resistors.

At an early stage of development of the derivative spectrometry (1960s), electro-mechanical analog differentiators were developed [1]. These instruments only generated 1st-order derivatives. Their technical specifications were low.

Analog differentiators are no longer used for analytical applications. Therefore, we will only provide a brief description of these devices, including a historical excursion.

Figure 2.1-1 shows the simplest passive differentiation RC circuits. Readers familiar with the basic electricity course will quickly understand the following simple formulas. Others may omit mathematical expressions and only pay attention to the final result.

Consider the 1st-order differentiator (Fig. 2.1-1). The current is the derivative of the electric charge (Q) on the capacitor C over time:

$$i = dQ/dt = C\, dU_C/dt. \qquad (2.1-1)$$

According to Kirchhoff low the output voltage

$$U_{out} = U_{in} - U_C. \qquad (2.1-2)$$

According to Ohm's law, using Eq. (2.1-1), we obtain

$$U_{out} = Ri = \tau\, dU_C/dt, \qquad (2.1-3)$$

where $\tau = RC$ is the time constant of the circuit.

Substituting U_C from Eq. (2.1-2) into Eq. (2.1-3), we have

$$U_{out} = \tau\, dU_{in}/dt - \tau\, dU_{out}/dt. \qquad (2.1-4)$$

If τ is small enough so that $\tau\, dU_{out}/dt \ll U_{out}$, then

$$U_{out} \approx \tau\, dU_{in}/dt. \qquad (2.1-5)$$

So, the output signal is proportional to the derivative of the input signal.

The solution for the Eq. (2.1-4) readily demonstrates the inertial distortions of 1st-order RC derivatives. For example, for the Gaussian input $U_{in} = exp[-(t-t_0)^2]$ (Appendix E1):

$$U_{out} = -2exp(t/\tau)\int_0^t (t-t_0)exp(-(t-t_0)^2 + t/\tau)dt. \qquad (2.1-6)$$

Since this integral has no solution in the elemental functions, we used numerical integration.

Figure 2.1-2 shows that with an increase of the time constant, the 1st-order derivative is shifted towards more time, and its shape is distorted.

A combination of two RC circuits, which separated by a buffer (amplifier) (Fig. 2.1-1b), performs the 2nd-order differentiation. In the smoothing (integration) RC circuit, the resistor, and the capacitor swap their places (Fig. 2.1-1c).

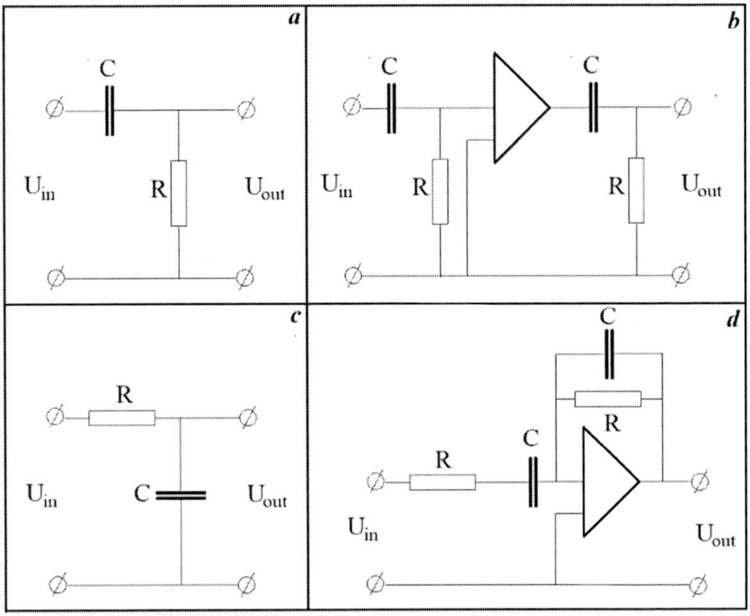

Figure 2.1-1. RC circuits. The 1st-(a, d) and 2nd -order (b) differentiators, and integrator (c). The buffer (b) and operational (d) amplifiers are designated by triangles. Operators set the parameters of the capacitors and resistors.

It is common to study electronical devices in the Fourier domain using the Fourier Transform (FT). So, the module of the frequency characteristic of the RC differentiator (Fig. 2.1-1a) is as follows:

$$\left|G^{(1)}(\omega)\right| = x/\sqrt{1+x^2}, \qquad (2.1-7)$$

where $x = \tau\omega$, ω is the angular (Fourier) frequency.

For the 2nd-order RC differentiator (Fig. 2.1-1b):

$$\left|G^{(2)}(\omega)\right| = K_B\, x^2/(1+x^2), \qquad (2.1-8)$$

where K_B is the amplifier gain.

If $x \ll 1$, then the RC differentiators produce the quasi-ideal n^{th}-order derivatives:

$$\left|G^{(n)}(\omega)\right| \cong x^n. \qquad (2.1-9)$$

(For simplicity, $K_B = 1$).

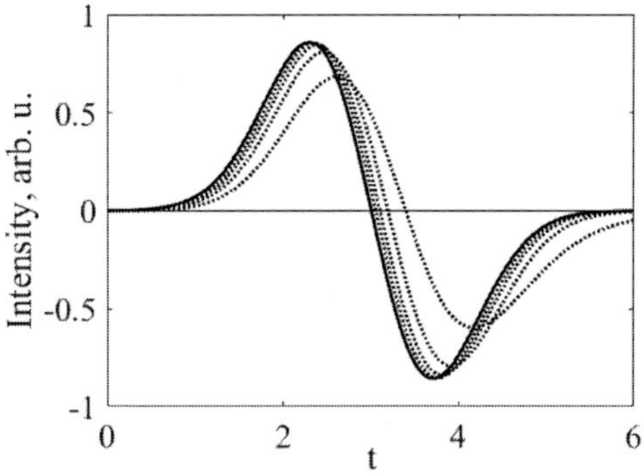

Figure 2.1-2. Inertial distortions of the 1st-order *RC* derivatives of the Gaussian: $exp[(t-2)^2]$ ($\tau = 0, 0.05, 0.1, 0.2$, and 0.5 from the upper to the bottom curves, respectively). The curves are multiplied by $1/\tau$, $\tau \neq 0$.

The signal-to-noise ratio of the nth-order RC derivative of a single peak, the full width of which at the half maximum (FWHM) is w, is [1]

$$\xi^{(n)} \sim \sqrt{2n+1}(\tau/w)\xi^{(0)}, \qquad (2.1-10)$$

where $\xi^{(0)}$ is the signal-to-noise ratio of the peak.

To decrease the inertial distortions, $\tau/w \ll 1$. Therefore, for the inertial distortions and noise in derivatives to be small, you need to increase $\xi^{(0)}$.

We studied the inertial distortions of the 2nd-order RC differentiator combined with the RC integrator circuit (2.1-1c) of the Gaussian peak [1]. In this case (Appendix E2), neglecting $x^2 \ll 1$ and using the exponential form of a complex number [2], we have

$$G^{(2)}(j\omega)\sim\frac{x^2}{(1+xj)^2}*\frac{1}{1+x_sj}=\frac{x^2(1-xj)^2}{(1+x^2)^2}*\frac{(1-x_sj)}{1+x_s^2}\cong$$

$$\frac{x^2(1+x_{comb}j)}{(1+x^2)^2(1+x_s^2)}=\frac{x^2\sqrt{1+x_{comb}{}^2}Ex}{(1+x^2)^2(1+x_s^2)}, \qquad (2.1-11)$$

where $j=\sqrt{-1}$; $x_s=\tau_s\omega$; τ_s is the time constant of the smoothing circuit; $x_{comb}=(2\tau+\tau_s)\omega$; $Ex=exp[-j*atan(x_{comb})]$.

According to Eq. (1.1-9), factor Ex means shifts of the derivatives towards more time. If $2\tau+\tau_s$ is small ($x_{comb}\ll 1$), then the factor approximately equals $exp[-j(2\tau+\tau_s)\omega]$. That is, the shift is $2\tau+\tau_s$. Also, this complex term causes the asymmetry of the derivatives [3].

We estimated [1] the integral, which describes the 2nd-order derivative of the Gaussian peak (Appendix A1) to be

$$F_G^{(2)}(\lambda)=(1/2\pi)\int_{-\infty}^{\infty}G^{(2)}(j\omega)\tilde{F}_G(\omega)\exp(j\omega\lambda)\,d\omega, \qquad (2.1-12)$$

where $\tilde{F}_G$ is the FT of the Gaussian peak (Eq. (1.1-7)). It was shown that, when the time constant $2\tau+\tau_s$ is increased, the central peak broadens, and the satellites are deformed: the first grows in the direction of the time-scanning and becomes narrower, while the second decreases and expands.

If the time constant $2\tau+\tau_s$ increases, then the resolution of the RC derivative peaks decreases [1]. The conditions, $\tau/\tau_s\approx 0.1$ and $(2\tau+\tau_s)/w\approx 0.1$, are appropriate for suitable signal-to-noise ratio and resolution.

If $x\ll 1$ and $x_s\ll 1$, then from Eq. (2.1-11) we have

$$G^{(2)}(j\omega)\sim x^2-x^2x_e^2+\cdots, \qquad (2.1-13)$$

where $x_e=\sqrt{2\tau^2+\tau_s^2}\,\omega$.

The term $x^2x_e^2$, which is proportional to ω^4, means a small contribution of the 4th-order derivative into the output signal. Subsequent members correspond to the higher order derivatives. However, if the input signal U_{in} is proportional to the analyte concentration, then this proportion is valid for all terms of Eq. (2.1-13) and, therefore, for the differentiator output: that is, the RC derivatives are suitable for quantitative analysis.

Also, one must take into account that

$$d^{(n)}U_{out}/d\lambda^n=v^n d^{(n)}U_{out}/dt^n, \qquad (2.1-14)$$

where $v=d\lambda/dt$ is the scanning speed of the instrument. Therefore, to obtain intensive output, high scanning speed is required. But the time constants must be reduced to keep the inertial distortions at the same level. In conclusion, the question is raised: "Is high accuracy of differentiation needed for quantitation estimation in analytical spectroscopy?" For example, the active circuit (Fig. 2.1-1d) significantly improves the accuracy at the cost of deterioration of the signal-to-noise ratio. Our experience showed that the answer is "No," but differentiation devices must have good reproducibility of the output signal. Also, the time scanning of the spectral instrument must be stable.

RC differentiators were widely used in the 1960 - 1970s for analytical researches. We resolved strongly overlapping IR bands of the complexes of salts with acetonitrile molecules in 1965 using a home-made attachment to the single-beam IR spectrometer [1].

Later, a quantitative analysis of industrial products was carried out by this device (Chapter 4.2): determination of amount of epoxy resin in chloroparaffins and analysis of the ternary mixtures of dichlorophenols, control on the content of the 2.4-dichlorophenol in herbicides, of the phenol and chlorinated phenoxyacetic acids in wastewater.

Now analytical instruments perform numerical differentiation; their essential advantage over RC devices is the lack of inertial distortions.

CHAPTER TWO

DIGITAL DIFFERENTIATION: FINITE DIFFERENCES

Digital processing of analytical signals, including differentiation, is a standard method used in qualitative and quantitative analyses. The discrete derivatives of spectra are usually obtained by the finite difference method, polynomial Savitzky-Golay filters, the spline method, and Fourier Transform (FT). To understand the intricacies of these techniques, readers need a sound mathematical background. The author has tried to provide a simple explanation. The appendices will also explain some additional minor details.

According to the definition of the derivative [1],

$$F^{(1)}(\lambda_i) = \lim_{\lambda\to 0}(\Delta F/\Delta\lambda) = \lim_{h\to 0}\{[F(\lambda_i + h/2) - F(\lambda_i - h/2)]/h\}, \quad (2.2-1)$$

where $h = \Delta\lambda$ is the sampling interval.

So, if h is small, then the increment ΔF is approximately proportional to the 1st-order derivative $F^{(1)}(\lambda_i)$. Similarly, the 2nd-, 3rd-, and 4th-order derivatives, which are obtained by the central differences method, are [2]

n	Equations (2.2-2) - (2.2-4): $F^{(n)}(\lambda_i)h^n \cong \Delta^n F$
2	$F(\lambda_i - h) - 2F(\lambda_i) + F(\lambda_i + h)$
3	$F(\lambda_i - 1.5h) + 3F(\lambda_i - 0.5h) - 3F(\lambda_i + 0.5h) + F(\lambda_i + 1.5h)$
4	$F(\lambda_i - 2h) - 4F(\lambda_i - h) + 6F(\lambda_i) - 4F(\lambda_i + h) + F(\lambda_i + 2h)$

Figure 2.2-1 illustrates the calculation for the coefficients of the central differences. The terms of the 1st-order operator of the central difference [-1, 0, 1] are multiplied by -1, 0, and 1. The second and the third products are shifted to the right side by $h/2$ and h, respectively. The sum is the coefficients of the 2nd-order operator.

Exercise 2.2-1

Readers are invited to check the calculation of the coefficients of the

3rd-order operator's coefficients (Eq. (2.2-4)).

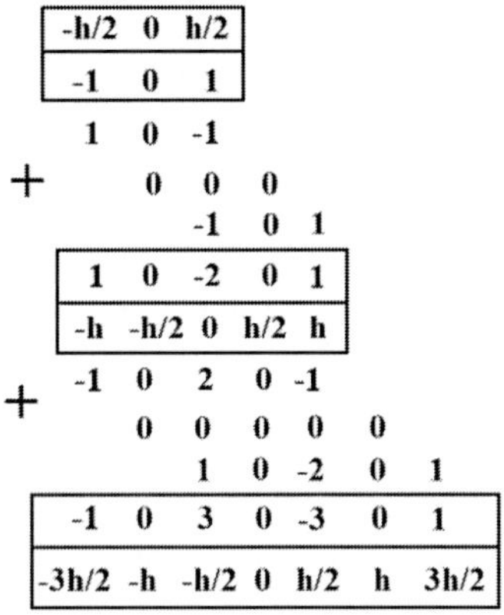

Figure 2.2-1. Calculation of the central differences' coefficients.

General equation of the nth-order operator of the central differences [2]:

$$\Delta^n F_k = \sum_{p=0}^{n} (-1)^p \binom{n}{p} F_{k+n-p}, \qquad (2.2-5)$$

where

$$n = 2,3,\ldots, k = \begin{cases} \pm 0.5, \pm 1.5, \ldots \; for\; n\; odd \\ 0, \pm 1, \pm 2, \ldots \; for\; n\; even \end{cases}, \binom{n}{p} = \frac{n!}{p!\,(n-p)!}.$$

Using Eq. (1.2-15), we obtain the maximum signal-to-noise ratio of the nth-order derivative of a single peak disturbed by white noise with a standard deviation σ:

$$\xi_{max}^{(n)} = F_e^{(n)}/\sigma^{(n)} = \mathcal{D}^{(n)} F_0 H^n / \left(\sigma\sqrt{K_d}\right) = \mathcal{D}^{(n)} \xi_0 H^n / \sqrt{K_d}, \quad (2.2-6)$$

where $\sigma^{(n)} = \sigma\sqrt{K_d}/h^n$; $K_d = \sum_{\alpha=1}^{n+1} W_\alpha^2$, W_a is the coefficient of the nth-order operator (Eq. (2.2-5)); $H = h/w$ is the relative sampling interval; w is the full peak width at half maximum (FWHM); and $\xi_0 = F_0/\sigma$ is the signal-to-noise ratio of the peak. $K_d = 2, 6, 20$, and 70 for $n = 1, 2, 3$, and 4, respectively.

In reality, the signal-to-noise ratio is less than $\xi_{max}^{(n)}$ due to the disturbance of the derivatives.

The frequency characteristic of the nth-order operator (Eq. (2.2-5)) equals the nth-power of the characteristic of the 1st-order operator (digital filter) (Fig. 2.2-1):

$$G^{(n)}(jx) = \left\{\frac{1}{h}\left[-exp\left(-\frac{jx}{2}\right) + exp\left(\frac{jx}{2}\right)\right]\right\}^n = \left[\frac{2j}{h}sin\left(\frac{x}{2}\right)\right]^n, (2.2-7)$$

where $= \sqrt{-1}$; $x = \omega h$; ω is the angular (Fourier) frequency.

Compare the frequency characteristics of the 2nd-order RC differentiator (Eq. (2.1-8)) and the operator, given by Eq. (2.2-7), for $x << 1$:

$G_{RC}^{(2)}(\omega){\sim}\omega^2(1 - \tau^2\omega^2)$ and $G_{opr}^{(2)}(\omega){\sim}\omega^2(1 - h^2\omega^2/12)$. (2.2 − 8)

The second terms in the parentheses define the systematic errors of the differentiators. If $h \approx \tau v$, where v is the scanning speed, then the numerical differentiation is more accurate than analog.

According to Eq. (2.2-6), the only way to improve the signal-to-noise ratio of derivative (for $\xi_0 = const$), is to increase interval H. However, the distortions will also increase and the resolution will be poorer.

Similar to the RC method, the combination of differentiation with smoothing may improve the signal-to-noise ratio. The most straightforward smoothing procedures are the moving average and averaging of data intervals.

In the first case, the digital filter

$$\hat{F}_i = (1/N)\sum_{\alpha=-m}^{m} F_{i+\alpha}, \qquad (2.2-9)$$

where $i = m + 1, m + 2, \ldots, L - m$; L is the number of data points; $N = 2m + 1$.

Using the formula of the sum of cosines [4] and simple trigonometry, we obtain the frequency characteristic of the filter (Eq. (2.2-9)):

$$G_{mva}(x) = \frac{1}{N}\sum_{\alpha=-m}^{m} exp(j\alpha x) = \frac{2}{N}\sum_{\alpha=0}^{m} \cos(\alpha x) = \frac{1}{N}\left\{\frac{2sin\frac{(m+1)x}{2}cos\frac{mx}{2}}{sin\frac{x}{2}} - 1\right\} = \frac{sin\frac{Nx}{2}}{Nsin\frac{x}{2}}. \qquad (2.2-10)$$

In the Digital Signal Processing (DSP), Eq. (2.2-10) is the FT of the rectangular pulse [5].

If $x \ll 1$, then using the first two terms of the Taylor series, we have:

$$G_{mva}(x) \approx \frac{\frac{Nx}{2} - \frac{(Nx)^3}{48}}{N\left(\frac{x}{2} - \frac{x^3}{48}\right)} \approx 1 - \frac{m(m+1)}{6x^2}. \qquad (2.2-11)$$

A combination of Eq. (2.2-8) and the moving average filter gives a new differentiation filter, which has the width Nh. For example, $N = 5$:

$$F_{comb}^{(1)}(\lambda_i) = \frac{F(\lambda_i + h/2) - F(\lambda_i - h/2)}{h} = \frac{(F_{k+1} - F_k)}{h} = \qquad (2.2-12)$$

$$\frac{[F_{k-1} + F_k + F_{k+1} + F_{k+2} + F_{k+3}] - [F_{k-2} + F_{k-1} + F_k + F_{k+1} + F_{k+2}]}{5h}$$

In the next point:

$$F_{comb}^{(1)}(\lambda_{i+1}) = (F_{k+2} - F_{k+1})/h = (F_{k+4} - F_{k-1})/5h. \qquad (2.2-13)$$

If the moving average is absent, then the derivative in the first point is equal to Eq. (2.2-12):

$$F^{(1)}(\lambda_i) = (F_{k+3} - F_{k-2})/5h. \qquad (2.2-14)$$

But the next point is shifted by $5h$:

$$F^{(1)}(\lambda_{i+5}) = (F_{k+8} - F_{k+3})/5h. \qquad (2.2-15)$$

Figure 2.2-2 illustrates the difference between the two filters.

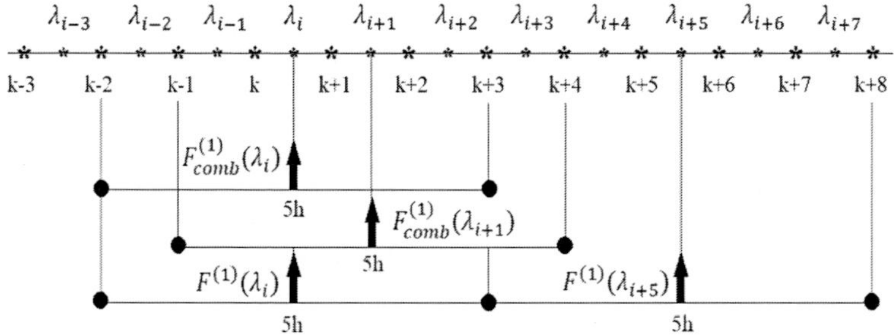

Figure 2.2-2. Diagram of the calculations of the 1st-order derivative in combination with and without the moving average.

Consider the combination of the smoothing filter with the 2nd-order differentiation (Eq. (2.2-2)). For example, $N = 5$:

$$\begin{array}{lccccccc} & 1 & 1 & 1 & 1 & 1 & & \\ + & & -2 & -2 & -2 & -2 & -2 & \\ & & & 1 & 1 & 1 & 1 & 1 \\ \hline & 1 & -1 & 0 & 0 & 0 & -1 & 1 \end{array}$$

This filter is readily obtained from Eqs. (2.2-12) and (2.2-13).

The noise coefficient of such a filter is

$$K_{combdif} = (1 + 1 + 1 + 1)/N^2 = 4/N^2. \qquad (2.2-16)$$

The combination of Eq. (2.2-8) and the multi-pass moving average filter, repeated r times, has the frequency characteristic for $x << 1$:

$$G^{(2)}_{comb}(\omega) = G^{(2)}_{opr}(\omega)[G_{mva}(x)]^r \sim \omega^2(1 - \{[2rm(m+1)+1]/12\}\, x^2). \qquad (2.2-17)$$

Eq. (2.2-13) shows that, in the first approximation, the systematic error of the combined filter depends on

$$B^{(2)}_{comb}(r) = [2rm(m+1)+1]/12. \qquad (2.2-18)$$

If $r = N$ then the noise coefficient (Appendix F, Eq. (F1-4)):

$$K_{combTrdif} = 6/N^4. \qquad (2.2-19)$$

Let $N_{r=1}$ and $N_{r=N}$ be the widths of the two filters, which have equal systematic errors. Then from Eq. (2.2-18), we obtain $2m_{r=1}(m_{r=1}+1)+1 = 2N_{r=N}m_{r=N}(m_{r=N}+1)+1$. If $N^2 >> 1$, then $N^2_{r=1} = N^3_{r=N}$. So,

$$N_{r=N} = \sqrt[3]{N^2_{r=1}}. \qquad (2.2-20)$$

By substituting Eq. (2.2-20) into Eq. (2.2-19), we have

$$K_{combTrdif} = 6/\left(N^2_{r=1}\sqrt[3]{N^2_{r=1}}\right) < K_{combdif} = 4/N^2_{r=1} \qquad (2.2-21)$$

since $N > 2$. For example, for $N_{r=1} = 9$ the combination filter decreases white noise by the factor $\sqrt{K_{combTrdif}} \approx 2$, but the systematic error does not change.

The term (Eq. (2.2-11))

$$A_{mv} = [m(m+1)/6]x^2 = [(N^2-1)/24]\omega^2 h^2 \qquad (2.2-22)$$

defines the systematic error of the moving average. Since $-\omega^2$ is the Fourier operator of the 2nd-order differentiation, the 1st-order approximation of the smoothed signal in the time domain is

$$F_{mv}(\lambda) \approx F(\lambda) - F^{(2)}(\lambda). \qquad (2.2-23)$$

For example, in the maximum of the Gaussian peak from (Eqs. (1.2-5), (2.2-22), and (2.2-23), we have the following systematic error:

$$\Delta F_{mv}(\lambda_0) = F(\lambda_0) - F_{mv}(\lambda_0) = 8ln2H^2[(N^2-1)/24]. \qquad (2.2-24)$$

In the averaging of data intervals (the sequential groups of N points):

$$\bar{F}_i = \sum\nolimits_{\alpha=-m}^{m} F_{i+\alpha}/N, \qquad (2.2-25)$$

where $i = m + Nq$, $q = 0, 1, \ldots$

To estimate the systematic errors of the averaging of data intervals, we need to calculate its frequency characteristic. However, the frequency properties of the method depend on i. Earlier, the properties were studied for $i = const$ [3] using harmonic analysis [6]. Here, we estimate the

systematic errors in the maximum of the Gaussian peak directly (Fig. 2.2-3).
If $H = h/w \ll 1$, then near the Gaussian peak maximum ($y = 0$):
$$F(y) = \exp(-y^2) \cong 1 - y^2 = 1 - Ki^2, \qquad (2.2-26)$$
where $y = 2\sqrt{ln2}x$; $x = Hi$; $K = 4ln2H^2$; $i = 0, \pm1, \pm2, \dots, \pm m$.
The accurate mean is $\bar{F}(0) = 0.9348$. The approximate value is
$$\bar{F}_{appr}(0) = 1 - 2K\sum_{i=1}^{m}\frac{i^2}{N} = 1 - 8ln2H^2\left[\frac{N^2-1}{24}\right]. \qquad (2.2-27)$$
Figure 2.2-3 confirms these calculations.

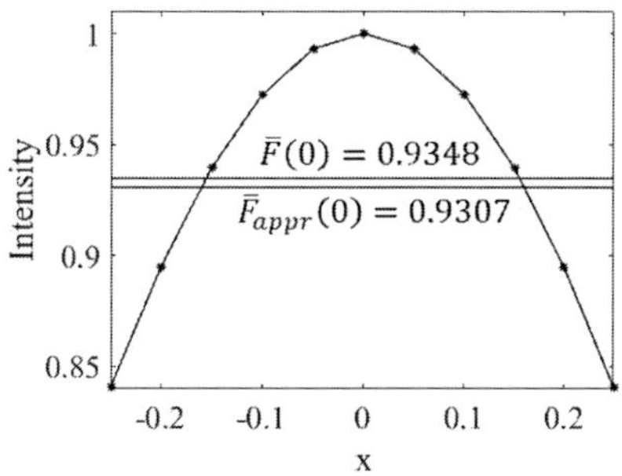

Figure 2.2-3. Estimation of the systematic error in the maximum of the Gaussian peak ($w = 1, N = 11, h = 0.05$).

The second term of Eq. (2.2-27), which defines the systematic error of the averaging of data intervals, is equals to the systematic error of the moving average (Eq. (2.2-24)). This conclusion is intuitively clear.

The 1st-order differentiation, combined with the averaging of data intervals, produces the derivatives with the sampling interval h (Fig. 2.2-2), but the signal-to-noise ratio increases because of the averaging.

In spectroscopic literature, the averaging of data intervals, combined with differentiation, is known as a Norris-Williams derivation [7, 8].

We proved theoretically that the moving average filter is optimal for constant and lincar signals (Chapter 2.5). In other cases, polynomial Savitzky-Golay (SG) smoothing filters are more effective. Similarly, the difference method is inferior to polynomial SG differentiation since the first involves the linear approximation of the input data. This approximation does not allow a more flexible selection of the digital SG filters' optimal parameters by varying their length and the power.

CHAPTER THREE

DIGITAL DIFFERENTIATION: SAVITZKY-GOLAY FILTERS

Introduction

Savitzky-Golay filters (SGF) have been the most popular digital filters used in chemometrics since the publication of a famous article in 1964 [1]. The journal Analytical Chemistry published some articles [2-6], concerning the technical details of SGF. During the last six decades, there have been numerous theoretical and experimental studies of these filters, which have been carried out by experts in analytical chemistry and signal processing. These studies are available in online resources.

Remark-1

In our opinion, the following translation from Russian to English could compete with the Savitzky-Golay paper and with the subsequent articles concerned with this issue [Grabar, L. P. (1967). Numerical differentiation by means of Chebyshev polynomials orthonormalized on a system of equidistant points. USSR Comput. Math. Math. Physics, 7, 215-220].

Remark-2

Curious readers can find simple methods for the SGF calculation in some particular cases described in the remarkable book by Cornelius Lanczos [7]. The author brilliantly illustrated the least-squares methods with practical calculations that non-professional readers can quickly reproduce. This book also showed the historical development of different applications of numerical analysis.

Remark-3

Many articles, written in Russian [9-11], escaped the attention of the English-speaking reader. This chapter represents modified old findings (sometimes reinvented in English publications). We have provided simple mathematical details, which are usually avoided so that the readers are forced to blindly believe in the correctness of the given mathematics.

To facilitate reading, the formulas from previous chapters will be repeated.

Models

The models in the signal (time) domain (ν) (e.g., time, wavelength) are the Gaussian $(F_{G0}(\nu))$ and Lorentzian $(F_{L0}(\nu))$ peaks with a unit maximum intensity,

$$F_{G0}(\nu) = exp\left\{-4ln2\left[\frac{\nu-\nu_m}{w}\right]^2\right\}, \quad (2.3-1)$$

$$F_{L0}(\nu) = 1/\{1 + 4[\nu - \nu_m/w]^2\}, \quad (2.3-2)$$

where ν_m is the maximum of the peak position; w is the full peak width at the half maximum (FWHM).

Integral Fourier transforms (FT) of these peaks in the frequency domain (ω):

$$\tilde{F}_{G0}(\omega) = \int_{-\infty}^{\infty} F_{G0}(\nu)\, exp(-j\omega\nu)\, d\nu =$$

$$(w/2)\sqrt{\pi/ln2}\, exp(-(\omega w)^2/16ln2), \quad (2.3-3)$$

$$\tilde{F}_{L0}(\omega) = (w\pi/2)\, exp(-0.5|\omega w|), \quad (2.3-4)$$

where $j = \sqrt{-1}$.

It is common to perform signal processing and analysis in the frequency domain using Fourier operators $(\mathcal{F}(\omega))$, which represent mathematical procedures in this domain [7].

The peak restored from the FT using Inverse FT (IFT) is

$$F_{0pr}(\nu) = (1/2\pi)\int_{-\infty}^{\infty} \tilde{F}_0(\omega)\mathcal{F}(\omega)\, \exp(j\omega\nu)\, d\omega. \quad (2.3-5)$$

SGF Calculations

Let $W_\alpha^{(m,2t,n)}$ be a normalized coefficient of the nth-order differentiation SGF (if $n = 0$, then it is a smoothing filter) [1]:

$$W_\alpha^{(m,2t,n)} = \{W_{-m}, W_{-m+1}, \dots, W_0, W_1, \dots, W_m\}. \quad (2.3-6)$$

$N = 2m + 1$ is the width of the SGF. The filter is applied to the evenly sampled data with a sampling interval (step) h: $\nu_i (i = 1, 2, \dots, I)$. The N data points are approximated by the polynomial of the power $2t \geq m$:

$$Pol_{m,2t}(\alpha) = \sum_{i=0}^{2t} a_i \alpha^i, \quad (2.3-7)$$

where the constants a_i are estimated using the least-squares fitting of the polynomial to the data points.

Remark-4

The SGF was motivated by the Weierstrass approximation theorem [12].

According to this theorem, "every continuous function defined in a closed interval can be uniformly approximated, as closely as desired, by a polynomial function."

In the central point $\alpha = 0$, the nth-order derivative of the polynomial (Eq. (2.3-7)) $(n \leq 2t)$:

$$Pol_{m,2t}^{(n)}(0) = n!\,a_n, \qquad (2.3-8)$$

where $a_n = \sum_{\alpha=-m}^{m} T_{n+1,\alpha} F_0(\mathrm{v}_{i+\alpha})$, $T_{n+1,\alpha}$ is the element of the matrix $\boldsymbol{T} = (\boldsymbol{X}^T\boldsymbol{X})^{-1}\boldsymbol{X}^T$; superscripts -1 and T stand for the matrix transposition and inversion, respectively,

$$\boldsymbol{X} = \begin{vmatrix} (-m)^0 & \dots & (-m)^k & \dots & (-m)^{2t} \\ \vdots & \dots & \vdots & \dots & \vdots \\ 1 & \dots & 0 & \dots & 0 \\ \vdots & \dots & \vdots & \dots & \vdots \\ m^0 & \dots & m^k & \dots & m^{2t} \end{vmatrix}.$$

If the input signal (for example, a peak (Eqs. (2.3-1) and (2.3-2))) has passed through the SGF, then the filter output is the derivative of the peak:

$$F_{df}^{(n)}(\mathrm{v}_i) = h^{-n} \sum_{\alpha=-m}^{m} W_{\alpha}^{(m,2t,n)} F_0(\mathrm{v}_{i+\alpha}), \qquad (2.3-9)$$

where $i = m+1, m+2, \dots, I-m$;

$$W_{\alpha}^{(m,2t,n)} = n!\,T_{n+1,\alpha}. \qquad (2.3-10)$$

For $n = 0, 2, 4, \dots$, the FSG coefficients are symmetrical relative to the central point $\alpha = 0$. For the odd n, they are asymmetrical, and $W_0 = 0$.

Eq. (2.3-10) shows that the standard SGF does not process the first ($\boldsymbol{P}_b = 1,2,\dots,m$) and the last ($\boldsymbol{P}_e = I-m+1, I-m+2,..,I$) (total $2m$) points. To overcome this drawback, the data may be extended by adding a copy of the reverse vector $\boldsymbol{P}_b$ at the beginning, and the $\boldsymbol{P}_e$ at the end. Another way to do this is using the same polynomial. However, in both cases, the accuracy of the differentiation reduces. For more details, see [4, 5].

The SG method's remarkable property is the independence of the filter coefficients of the processed data [7]: that is, the SGF coefficients may be calculated and tabulated in advance before processing [1, 2]. This fact explains the widespread use of SGF in analytical chemistry in the early years of non-computerized measurement techniques.

In their original paper, Savitzky and Golay estimated the filter coefficients using Eq. (2.3-8), similar to the Matlab function, sgolay.m. This is the most straightforward code of the SGF generation.

```
function W = savitzkyGolayModif(m, power, n)
x=- m: m;
N=2*m+1;
x=x(:);
X=ones(N,1);
for k=1:2t
 X=[X, x.^k];
end
matrixW = (X'*X)\X';
W=matrixW(n+1,:)*factorial(n);
end
```

The main drawback of this direct method is the inversion $(\boldsymbol{X}^T\boldsymbol{X})^{-1}$. If matrix $\boldsymbol{X}$ has a large condition number (a collinearity problem [7]), then the inversion may cause unpredictable numerical errors. This phenomenon is observed for a large m. For example,
savitzkyGolayModif (100, 6,0)
Warning: Matrix is close to singular or badly scaled. Results may be inaccurate. RCOND = 2.557404e-24

Remark 5

The cumbersome mathematical expressions of the smoothing and differentiation SGF coefficients [1-3] avoid the inversion problem. However, they correspond to the limited number of differentiation orders and polynomial powers.

To simplify the SGF synthesis, we performed the orthogonal transform of the variables in Eq. (2.3-7) [11]:

$$Pol_{m,2t}(\alpha) = \sum_{i=0}^{2t} \theta_i \varphi_i(\alpha), \qquad (2.3-11)$$

where θ_i is a constant; $\varphi_i(\alpha)$ is the ith-order orthogonal polynomial defined on the interval $\alpha \in [-m, m]$.

The polynomials obey an orthogonality relation over the range $[-m, m]$:

$$\sum_{\alpha=-m}^{m} \varphi_i(\alpha)\varphi_j(\alpha) = 0, \qquad (2.3-12)$$

where $i \neq j$.

The recursive formula for the fast calculation of the polynomials is as follows: $\varphi_0(\alpha) = 1$; $\varphi_1(\alpha) = \alpha$. If $i \geq 2$, then

$$\varphi_{i+1}(\alpha) = \alpha\varphi_i(\alpha) - L_i\varphi_{i-1}(\alpha), \qquad (2.3-13)$$

where $i = 0, 1, \ldots, 2m$; $L_i = i^2(N^2 - i^2)/[4(4i^2 - 1)]$.
For example,

k	$\varphi_k(\alpha)$
2	$\alpha^2 - L_1$
3	$\alpha^3 - \alpha \sum_{i=1}^{2} L_i$
4	$\alpha^4 - \alpha^2 \sum_{i=1}^{3} L_i + L_1 L_3$
5	$\alpha^5 - \alpha^3 \sum_{i=1}^{4} L_i + \alpha(L_1 L_3 + L_4 \sum_{i=1}^{2} L_i)$
6	$\alpha^6 - \alpha^4 \sum_{i=1}^{5} L_i + \alpha^2(L_1 L_3 + L_4 \sum_{i=1}^{2} L_i + L_5 \sum_{i=1}^{3} L_i) - L_1 L_3 L_5$

The last terms are

$$\varphi_{2t-1}(0) = 0,$$

$$\varphi_{2t}(0) = (-1)^t \prod_{k=1}^{t} L_{2k-1}. \quad (2.3-14)$$

The odd-degree polynomials are asymmetrical, the even-degree polynomials are symmetrical relative to the central point $\alpha = 0$.

From the differentiation of Eq. (2.3-13) we have

$$\varphi_i(\alpha) = \sum_{k=0}^{i} \varphi_i^{(k)}(0)\alpha^k/k! \quad (2.3-15)$$

$$\varphi_i^{(k)}(0) = \begin{cases} 0, \begin{cases} k > i; \\ i = 2p, k = 2q-1; \\ i = 2p-1, k = 2q; \\ p, q = 0, 1, 2, \ldots, t; \end{cases} \\ k!, \ k = i; \end{cases} \quad (2.3-16)$$

Therefore,

$$\varphi_{2p+1}(\alpha) = \sum_{k=0}^{p} \varphi_{2p+1}^{(2k+1)}(0)\alpha^{2k+1}/(2k+1)!$$

$$\varphi_{2p}(\alpha) = \sum_{k=0}^{p} \varphi_{2p}^{(2k)}(0)\alpha^{2k}/(2k)! \quad (2.3-17)$$

$$\varphi_{2p-1}(\alpha) = \sum_{k=0}^{p} \varphi_{2p-1}^{(2k-1)}(0)\alpha^{2k-1}/(2k-1)!$$

By substituting Eqs. (2.3-17) into Eq. (2.3-13), we obtain the nth-order derivative over α at $\alpha = 0$:

$$\varphi_i^{(n)}(0) = n\varphi_{i-1}^{(n-1)}(0) - L_{i-1}\varphi_{i-2}^{(n)}(0). \quad (2.3-18)$$

Readers can check Eq. (2.3 −18) using the above expressions for the polynomials (Eq. (2.3-13)).

From Eqs. (2.3-8) and (2.3-11) we have

$$Pol_{m,2t}^{(n)}(0) = \sum_{i=0}^{2t} \theta_i \varphi_i^{(n)}(0). \quad (2.3-19)$$

The modified matrix $\boldsymbol{X}$ (Eq. (2.3-8)):

$$\boldsymbol{X} = \begin{vmatrix} \varphi_0(-m) & \dots & \varphi_k(-m) & \dots & \varphi_{2t}(-m) \\ \vdots & \dots & \vdots & \dots & \vdots \\ \varphi_0(0) & \dots & \varphi_k(0) & \dots & \varphi_{2t}(0) \\ \vdots & \dots & \vdots & \dots & \vdots \\ \varphi_0(m) & \dots & \varphi_k(m) & \dots & \varphi_{2t}(m) \end{vmatrix}. \qquad (2.3-20)$$

The calculation of the term $T_{n+1,\alpha}$ does not require the matrix inversion:

$$T_{n+1,\alpha} = \varphi_i(\alpha)/\widetilde{M}_i, \qquad (2.3-21)$$

where $\widetilde{M}_i = \sum_{\alpha=-m}^{m} \varphi_i^2(\alpha) = \dfrac{(i!)^2 N \prod_{k=1}^{i}(N^2 - i^2)}{2^{2i}\left[\prod_{k=1}^{i}(2i-1)\right]^2 (2i+1)}$.

So, θ_i is evaluated using Eq. (2.3-21). Substituting Eq. (2.3-21) into Eq. (2.3-19), we obtain the SGF coefficients:

$$W_\alpha^{(m,2t,n)} = Pol_{m,2t}^{(n)}(0) = \sum_{i=0}^{2t}(\varphi_i(\alpha)/\widetilde{M}_i)\varphi_i^{(n)}(0). \qquad (2.3-22)$$

The SGF coefficients can be calculated recursively from Eq. (2.3 −22), taking Eq. (2.3-16) into account:

$$W_\alpha^{(m,2t,n)} = \begin{cases} W_\alpha^{(m,2t,n-2)} + (\varphi_k(\alpha)/\widetilde{M}_k)\varphi_k^{(n)}(0), & k = \begin{cases} 2t, & n = 2q; \\ 2t-1, n = 2q-1; \\ q = 0, 1, \dots, t-1. \end{cases} \\ n!\,(\varphi_n(\alpha)/\widetilde{M}_n), & 2t = n; \\ 0, & 2t < n. \end{cases} \qquad (2.3-23)$$

Remark 6

In the middle of the 1980s, we generated SGFs on programmable micro-calculators using the above methods. Personal computers were not yet available for regular users in Russia.

The SGF coefficients were calculated using Gram [6] and Legendre [13] polynomials.

The obtained mathematical expressions for the SGF coefficients are beneficial when determining the optimal parameters of these filters.

Frequency properties of the SGF

The frequency properties of the SGF were studied using Harmonic (or Fourier) Analysis (HA) [7] (Appendix A). The HA originated in the middle 18th century in the solution to the string oscillation problem. Mathematicians use HA to represent periodical and non-periodical functions as a sum of harmonics (Fourier Transform-FT). Fourier spectrum is the transformation of AS into a set of harmonics. The amplitudes of these harmonics depend on their frequencies. In the beginning, the Fourier spectrum of the analytical signal contains the most informative intensive low-frequency harmonics. However, the high-frequency harmonics reflect the negative impact of noise on the AS. The signal-to-noise ratio may be significantly improved by an appropriate choice of the frequency characteristic of the filter, while the useful signal will only be slightly distorted.

In Digital Signal Processing, the HA is one of the most effective tools to study analogous and digital devices' properties. The frequency characteristic of the device (e.g., digital or analogous filter) is the dependence of its response to the input harmonic signal (sinusoidal or, in general, complex exponent) on the signal frequency.

It is common to describe instrumental distortions of the input signals in its native (time) scale (domain) by the convolution of the signal with the instrumental function (e.g., the function of a spectrometer). Analysis of instrumental distortions becomes significantly simpler using HA in the so-called Fourier domain since multiplication replaces the cumbersome convolution procedure. The HA's great success was achieved because of the Fast FT invention, which revolutionized HA. The SGF convolutes the input data with the filter coefficients. The filter frequency characteristic allows us to understand the filter's properties using simple mathematics.

Let us study the frequency characteristic of the SGF (Eq. (2.3-9)) using the input signal in the form of the complex exponent. Substituting the power series of the exponent $exp(y) = \sum_{k=0}^{\infty} y^k/k!$ [7] into Eq. (2.3-9), we have

$$G^{(n)}(j\omega) = h^{-n} \sum_{\alpha=-m}^{m} W_{\alpha}^{(m,2t,n)} exp(-j\alpha\omega h) = \sum_{k=0}^{\infty} (-j\omega)^k h^{k-n} M_{k,m,2t}^{(n)}/k!, \quad (2.3-24)$$

where $M_{k,m,2t}^{(n)} = \sum_{\alpha=-m}^{m} \alpha^k W_{\alpha}^{(m,2t,n)}$ is the kth-moment of the SGF impulse characteristic.

To determine the cases, when the moments are zeros, we take the symmetry (asymmetry) of the FSG coefficients relative to the central point $\alpha = 0$ for the even and the odd n, respectively, into account. In the last case, $W_0 = 0$. So, for the even n, the odd-order moments are zero. For the odd n, the even-order moments are zero. $M_{k,m,2t}^{(n)} = 0$, if $k \neq n < 2t$.

Finally, according to [9-11]

$$M_{k,m,2t}^{(n)} = \begin{cases} 0, \begin{cases} k < n; \\ k = 2t,\ n = 0, 2, \dots, 2t-2; \\ k = 2t-1, n = 1, 3, \dots, 2t-3; \\ k = 2p+1, n = 2q; \\ k = 2p, n = 2q+1, p, q = 0, 1, \dots \end{cases} \\ k!, k = n. \end{cases} \qquad (2.3-25)$$

Readers can calculate the moments using the Matlab function:

```
function Mom = momentsSGFun(power, m, n, k)
 rngAlpha=-m: m;
 W = savitzkyGolayModif(m, power ,n);
 Mom=sum(W.*rngAlpha.^k);
end
```

Smoothing SGF

The coefficients W_α of the smoothing SGF are symmetrical relative to the central point $\alpha = 0$. Therefore, substituting the Euler formula $exp(-j\alpha\omega h) = \cos(\alpha\omega h) - jsin(\alpha\omega h)$ into Eq. (2.3-24), we have

$$G^{(0)}(x) = W_0^{(m_1,2t_1,0)} + 2\sum_{\alpha=1}^{m} W_\alpha^{(m_1,2t_1,0)} \cos(\alpha x), \qquad (2.3-26)$$

where $x = \omega h$; subscript 1 stands for the parameters of the smoothing SGF. By using the power series: $\cos(y) = \sum_{k=1}^{\infty}(-1)^k y^{2k}/(2k)!$, and the properties of the moments: $\left|M_{k,m_1,2t_1}^{(0)}\right| = \begin{cases} 0, k < 2t_1 \\ k!, k = 2t_1 \end{cases}$, for $x \ll 1$ we obtained [8-11]

$$G^{(0)}(x) \cong 1 - A_{m_1,2t1} x^{2t_1+2}, \qquad (2.3-27)$$

where $A_{m_1,2t1} = \left|M_{2+2t_1,m_1,2t_1}^{(0)}\right|/(2+2t_1)! = C_{m_1,2t1}\, N_1^{2t_1+2}$; $C_{m_1,2t1}$ is a function of the filter parameters . $C_{m_1,2t_1} \to C_{2t_1}$, if $m_1 \to \infty$ (Fig. 2.3-1, Table 2.3-2).

We found precise approximations (e.g., Fig.2.3-2):

$$M_{2+2t_1,m_1,2t_1}^{(0)}/N^{2t_1+1} \cong sl * N + inc, \qquad (2.3-28)$$

where the slope (sl) and the intercept (inc) are $\{-0.5369, 0.2051\} * 10^{-2}$, $\{0.3407, -0.4452\} * 10^{-3}$, and $\{-0.2160, 0.6431\} * 10^{-4}$, for $2t_1 = 2, 4$, and 6, respectively. In the following example, we demonstrate two analytical estimations of C_2, using the direct SGF formula and Eq. (2.3-22).

$M_{4,m_1,2}^{(0)} = \sum_{\alpha=-m}^{m} W_\alpha \alpha^4$, where

$W_\alpha = [3(3m^2 + 3m - 1) - 15\alpha^2]/[(2m+3)(2m+1)(2m-1)]$.

The sums:

$\sum_{-m}^{m} \alpha^4 = 2m(m^4/5 + m^3/2 + m^2/2 - 1/30);$

$\sum_{-m}^{m} \alpha^6 = 2m(m^6/7 + m^5/2 + m^4/2 - m^2/6 - 1/42).$

If $m \to \infty$, then

$\sum_{-m}^{m} \alpha^4 = 2m^5/5\,; \sum_{-m}^{m} \alpha^6 = 2m^7/5\,;$

$C_2 = |2m^5/5 * \; 9m^2 - 15 * 2\,m^7/7|/\{8m^3 * 4! * (2m)^4\} = 2.2321 * 10^{-4}.$

Using Eq. (2.3-22), we have $M_{4,m_1,2}^{(0)} = \sum_{m}^{-m} \alpha^4 \sum_{i=0}^{2} (\varphi_i(\alpha)/\widetilde{M}_i)\varphi_i^{(0)}(0).$

Finally,

$\varphi_0(\alpha) = 1;\ \varphi_1(\alpha) = 0;\ \varphi_2(\alpha) = \alpha^2 - L_1;\ L_1 = N^2/12;\ \widetilde{M}_0 = N; \widetilde{M}_2 = N^5/180.$

If $m \to \infty$, then $C_2 = abs\ (2m^5/5 * [1/2m + (1/2m) * 180/144] - 2\,m^7/7 * 180/(12 * (2m)^3)/[4! * (2m)^4]) = 2.2321 * 10^{-4}.$

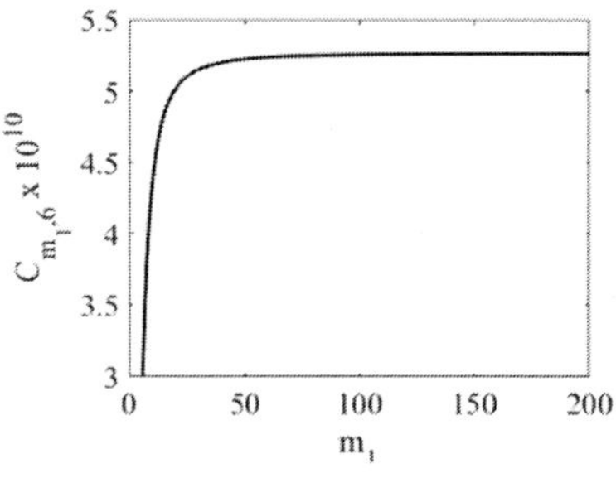

Figure 2.3-1. The dependence $C_{m_1,6}(m_1)$ (Eq. (2.3-27)).

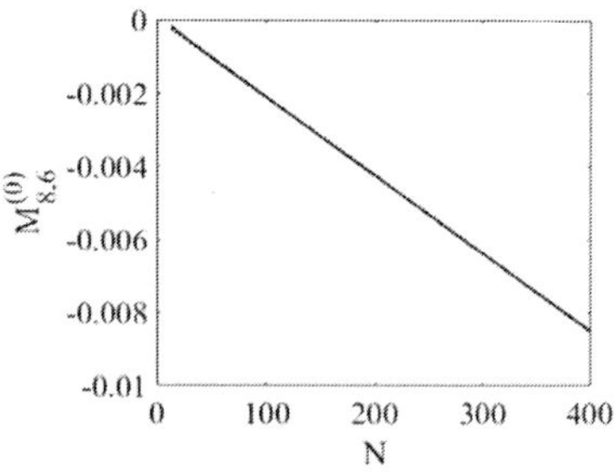

Figure 2.3-2. Approximation of the 8th-order moment of the smoothing SGF (dotted line).

Remark 7

Generally, if $m \to \infty$, then $\sum_{-m}^{m} \alpha^z = 2m^{z+1}/(z+1).$

The main drawback of the 1th-order approximation (Eq. (2.3-27)) is the sharp reduction to the negative values when x increases. We modified this equation by introducing the denominator in the second term and by an appropriate choice of the power in Eq. (2.3-28). This method is partly similar to the use of Tikhonov regularization [14]. For example, Figure 2.3-3 demonstrates the best approximation:

$$G^{(0)}(x) \cong 1 - C_{m_1,6}\,(N_1 x)^{7.62}/(1 + C_{m_1,6}\,(N_1 x)^{7.62}). \qquad (2.3-29)$$

The second term of Eq. (2.3-27) approximates the systematic error of the smoothing [8-11].

Figure 2.3-3 demonstrates how the smoothing filter decreases the high-frequency Fourier region but conserves the informative low-frequency part ($x << 1$). In this region, the amplitudes of the filter characteristic are close to the unit.

Eq. (2.3-29) describes the smoothly fading filter, which damped the wavy part of the SGF filter (Fig.2.3-3). Unfortunately, similar to the SGF, this filter cannot be built in the time domain. It can only be used in the frequency domain. Below, we study the impact of this filter on the signal in the time domain using the inverse FT.

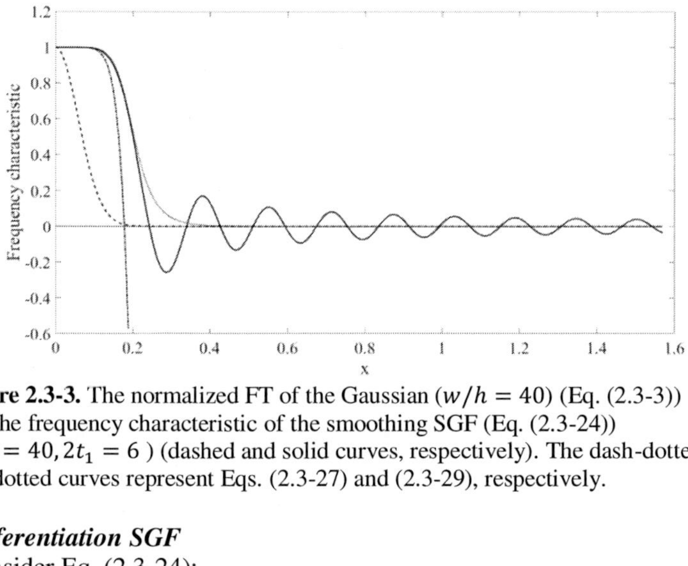

Figure 2.3-3. The normalized FT of the Gaussian ($w/h = 40$) (Eq. (2.3-3)) and the frequency characteristic of the smoothing SGF (Eq. (2.3-24)) ($m_1 = 40, 2t_1 = 6$) (dashed and solid curves, respectively). The dash-dotted and dotted curves represent Eqs. (2.3-27) and (2.3-29), respectively.

Differentiation SGF

Consider Eq. (2.3-24):

$$G^{(n)}(j\omega) = \sum_{k=0}^{\infty}(-j\omega)^k\, h^{k-n}M_{k,m,2t}^{(n)}/k! = h^{-n}\, M_{0,m,2t}^{(n)} -$$

$$j\omega\, h^{1-n}M_{1,m,2t}^{(n)} - \omega^2\, h^{2-n}M_{2,m,2t}^{(n)}/2 + j\omega^3\, h^{3-n}M_{3,m,2t}^{(n)}/3! + \quad (2.3-30)$$

$$\omega^4\, h^{4-n}M_{4,m,2t}^{(n)}/4! - j\omega^5\, h^{5-n}M_{5,m,2t}^{(n)}/5! + \omega^6\, h^{6-n}M_{6,m,2t}^{(n)}/6! \ldots$$

Since $M_{0,m,2t}^{(n)} = 0$, $M_{k,m,2t}^{(k)} = k!$, and the odd and the even-odd moments are zeros for the even and the odd-order differentiation SGFs, respectively, we have

$$G^{(1)}(j\omega) = -j\omega + j\omega^3 h^2 M^{(1)}_{3,m,2t}/3! - j\omega^5 h^4 M^{(1)}_{5,m,2t}/5! + \cdots =$$
$$-(j\omega)\left\{1 - \omega^2 h^2 M^{(1)}_{3,m,2t}/3! - \omega^4 h^4 M^{(1)}_{5,m,2t}/5! + \cdots \right\};$$
$$G^{(2)}(j\omega) = \omega^2 M^{(2)}_{2,m,2t}/2 + \omega^4 h^2 M^{(2)}_{4,m,2t}/4! + \omega^6 h^4 M^{(2)}_{6,m,2t}/6! + \cdots =$$
$$\omega^2\left\{1 + \omega^2 h^2 M^{(2)}_{4,m,2t}/4! + \omega^4 h^4 M^{(2)}_{6,m,2t}/6! + \cdots\right\}, \qquad (2.3-31)$$

and so on.

Let us transform two above equations:

$$G^{(1)}(x) \cong G^{(1)}_{id}\left\{1 - x^2 M^{(1)}_{3,m,2t}/3! - x^4 M^{(1)}_{5,m,2t}/5! + \cdots \right\}; \qquad (2.3-32)$$
$$G^{(2)}(x) \cong G^{(2)}_{id}\left\{1 + x^2 M^{(2)}_{4,m,2t}/4! + x^4 M^{(2)}_{6,m,2t}/6! + \cdots\right\}, \qquad (2.3-33)$$

where $x = \omega h$; $G^{(n)}_{id} = (j\omega)^n$ is the Fourier operator of the precise differentiation.

The terms in the braces play the role of smoothing filters. They reduce the amplitude of the precise differentiator frequency characteristic, which increases significantly in the high-frequency region. Therefore, the high-frequency noise distorts the input signal.

Figure 2.3-4 represents the amplitude-frequency responses of the 2nd-order differentiation FSG and the precise differentiator. If $x << 1$, then the FSG performs the precise differentiation. When x increases, the amplitude of the FSG characteristic decreases: that is, the filter begins the smoothing process. Mathematically, this process is explained by introducing the smoothing component into the differentiation FSG.

According to Eq. (2.3-25), the first moments in braces $M^{(n)}_{k,m,2t} \neq 0$, if $k = n + 2t$. Therefore, if $x << 1$, then, similar to Eqs. (2.3-32)-(2.3-33), we obtain the two-term approximation of Eq. (2.3-29):

$$G^{(n)}(x) \cong G^{(n)}_{id}\left\{1 - B^{(n)}_{m,2t} x^{2t}\right\}, \qquad (2.3-34)$$

where

$$B^{(n)}_{m,2t} = \left|M^{(n)}_{n+2t,m,2t}\right|/(n+2t)! = C^{(n)}_{m,2t} N^{2t}; \qquad (2.3-35)$$

$C^{(n)}_{m,2t}$ is a function of the filter parameter m. $C^{(n)}_{m,2t} \to C^{(n)}_{2t}$, if $m \to \infty$ (Table 2.3-2).

Remark 8

According to [15],

$$\left|C^{(0)}_{2t_1}\right| = \{(2t_1+1)(2t_1+1)!\ (2t_1)!\}/\{2^{(2t_1+1)}(2t_1+2)(4t_1+3)!\,(t_1)!^2\}.$$

Digital processing of the single peak in the signal domain (v)

Suppose that the SGF is applied to the single peak (e.g., Gaussian (Eq. (2.3-1)) and Lorentzian (Eq. (2.3-2))), which FT is $F_0(\omega)$. Then, the output signal in the time domain is the inverse FT ($\mathcal{F}$) of the FT data:

$$F_{df}^{(n)}(\nu) = \mathcal{F}\{F_0(\omega)G^{(n)}(j\omega)\}. \quad (2.3-36)$$

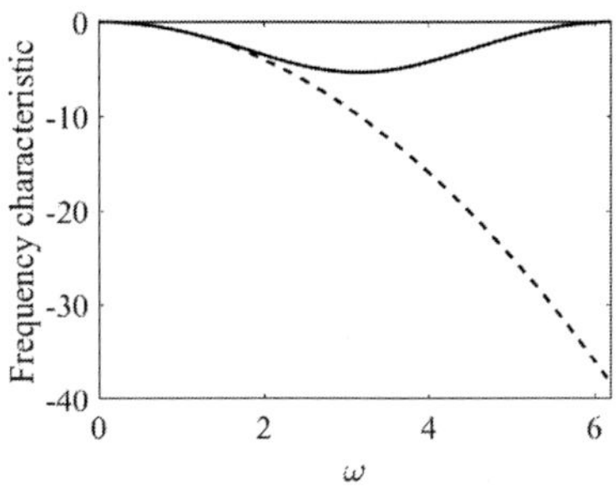

Figure 2.3-4. The frequency characteristic of the ideal and SGF 2nd-order differentiation (dashed and solid curves, respectively). $m = 2, 2t = 4, h = 1$.

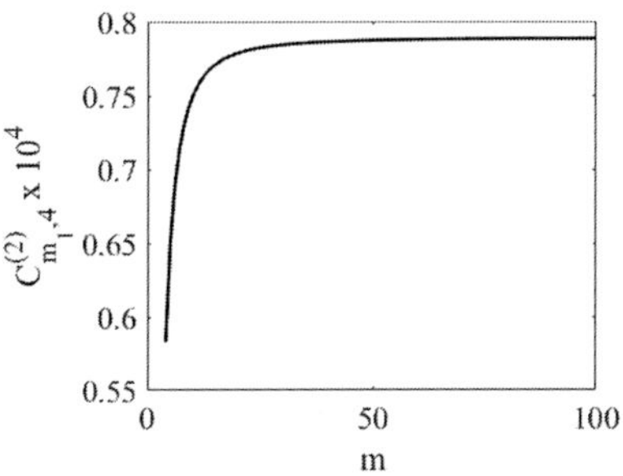

Figure 2.3-5. The dependence $C_{m.4}^{(2)}(m)$ (Eq. (2.3-35)).

By substituting Eq. (2.3-29) into Eq. (2.3-36), and, taking the fact that $(j\omega)^k$ is the Fourier operator of the kth-order differentiation into account, we have

$$F_{df}^{(n)}(\nu) = \sum_{k=0}^{\infty}(-1)^k\, h^{k-n}M_{k,m,2t}^{(n)}F_0^{(k)}(\nu)/k!. \quad (2.3-37)$$

For further study, we transform Eq. (2.3 −36), using Eq. (2.3-37) and the expressions of the peak derivatives at the peak maximum ν_m (Chapter 1.2) [11]. For example, if $n = 2t = 2$, then:

$$F_{df}^{(2)}(\nu_m) = F_0^{(2)}(\nu_m)\left\{1 - \left(\left|M_{4,m,2}^{(2)}\right|/4!\right)H^2\left|R^{(4)}/R^{(2)}\right| + \cdots\right\}, \quad (2.3-38)$$

where $H = h/w$ is the relative sampling interval; w is the FWHM; $R^{(2p)} = F_0^{(2p)}(\nu_m)w^{2p}$. For the Gaussian and Lorentzian peaks, $R_G^{(2p)} = (-4ln2)^p(2p)!/p!$ and $R_L^{(2p)} = (-4)^p(2p)!$, respectively (Chapter 1.2).

At the extremum of the first derivative ν_{e1} (for $2t = 2$):

$$F_{df}^{(1)}(\nu_{e1}) = F_0^{(1)}(\nu_{e1})\left\{1 - \left(\left|M_{3,m,2}^{(1)}\right|/3!\right)H^2\left|R^{(3)}/R^{(1)}\right| + \cdots\right\}, \quad (2.3-39)$$

where $R^{(2p-1)} = F_0^{(2p-1)}(\nu_{e1})w^{2p-1}$. The ratios $\left|R^{(n)}/R^{(1)}\right|$ are given in

Table 2.3-1.

Table 2.3-1. The ratios $|R^{(n)}/R^{(1)}|$

Peak	n			
	3	5	7	9
Gaussian	11.09	184.5	3410	276.1 x 10^3
Lorentzian	18	0	136.1 x 10^3	638.4 x 10^5

Since $H << 1$, then the two terms are a good approximation of Eqs. (2.3-38) and (2.3-39). The second terms in the braces of these equations define *the relative systematic errors* (bias) of the SGF differentiation, observed at the extrema of the 2nd- and 1st-order derivatives, respectively:

$$\varepsilon_s = \left(\left|M^{(n)}_{n+2t,m,2t}\right|/(n+2t)!\right)H^{2t}\left|R^{(2t+n)}/R^{(n)}\right|. \quad (2.3-40)$$

Since for the Lorentzian peak $R^{(5)} = 0$ (Table 2.3-1), we take $2t = 6$ and

$$\varepsilon_s = \left(\left|M^{(1)}_{7,m,6}\right|/7!\right)H^{6}\left|R^{(7)}/R^{(1)}\right|. \quad (2.3-41)$$

Suppose that the nth-order differentiation is performed using z repetitions of the $(n_0)^{th}$-order differentiation SGF, built on the polynomial of the power $2t$. That is, $n = zn_0$. In this case, Eq. (2.3- 40) is multiplied by z.

If the smoothing SGF, built on the polynomial of the power $2t_1$, is repeated r times then

$$\varepsilon_s^{sm} = r\left(\left|M^{(0)}_{2+2t_1,m_1,2t_1}\right|/(2+2t_1)!\right)H^{2t_1+2}\left|R^{(2+2t_1)}\right|. \quad (2.3-42)$$

The systematic error of the combined smoothing and differentiation SGFs:

$$\varepsilon_s^{comb} = \varepsilon_s^{sm} + \varepsilon_s. \quad (2.3-43)$$

As was stated above, parameters $A_{m_1,2t_1}$ (Eq. (2.3-27)) and $B^{(n)}_{m,2t}$ (Eq. (2.3-35)) define the systematic signal distortions in the frequency domain. For further study, we transform Eqs. (2.3-35) and (2.3-42) (for $r = 1$) using Eq. (2.3-27):

$$\varepsilon_s = P^{(n)}_{m,2t}(NH)^{2t}, \quad (2.3-44)$$

$$\varepsilon_s^{sm} = P_{m_1,2t1}(N_1H)^{2t_1+2}, \quad (2.3-45)$$

where $P^{(n)}_{m,2t} = C^{(n)}_{m,2t}\left|R^{(2t+n)}/R^{(n)}\right|$ and $P_{m_1,2t_1} = C_{m_1,2t_1}\left|R^{(2+2t_1)}\right|$. Table 2.3-2 represents the asymptotic values of these constants.

Suppose that the systematic errors are low. Then, the relative random error in the extremum point of the SG derivative of the single peak, distorted by the white noise, is

$$\varepsilon_r = \sqrt{K}/\left(\xi_0 \mathcal{R}^{(n)} H^n\right), \quad (2.3-46)$$

where $K = \sum_{\alpha=-m}^{m} W_\alpha^2$ is the SGF white noise factor; ξ_0 is the signal-to-noise ratio of the peak, and $\mathcal{R}^{(n)} = \left|F_{max}^{(n)} - F_{min}^{(n)}\right| w^n$ (the peak has the unit amplitude); $\mathcal{R}^{(1)} = 2.9$ and 2.6; $\mathcal{R}^{(2)} = 8$ and 10; $\mathcal{R}^{(4)} = 150$ and 546 for the Gaussian and Lorentzian peaks, respectively.

We found that

$$K_{m,2t}^{(n)} = \left(l_{m,2t}^{(n)}\right)^2 / N^{2n+1}, \quad (2.3-47)$$

where $l_{m,2t}^{(n)} \to l_{2t}^{(n)}$ when $m \to \infty$ (Table 2.3-2).

Figure 2.3-6 represents some dependencies $l_{m_1,2t_1}^{(0)}(m_1)$ for the smoothing FSGs. They quickly approach to the limits $l_{m_1}^{(0)}$.

Remark 9

According to [13]: $\left|l_{2t}^{(0)}\right| = \{2(2t+1)2t!/(2^{2t+1}/(t!)^2)\}$.

We could not obtain the general analytical expression for the combined filters' noise factor due to the correlations between their components. Therefore, only some essential particular cases were numerically studied. Instead of r repeated convolutions of the filter in the time domain, the frequency characteristic of the filter was raised to the power r. The noise factor was estimated in the Fourier domain.

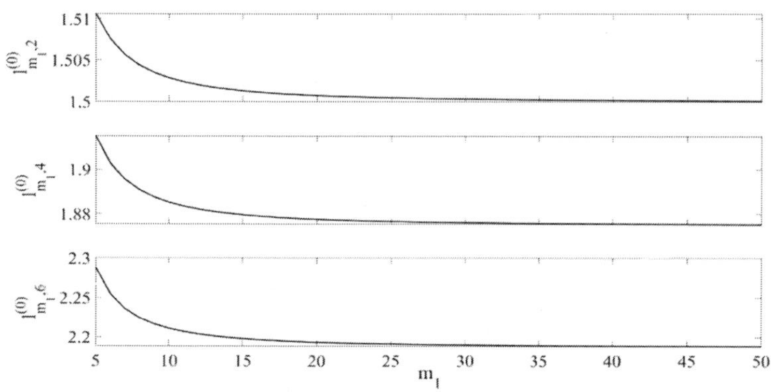

Figure 2.3-6. The dependences $l_{m_1,2t_1}^{(0)}(m_1)$ for the smoothing FSG.

For example, if the smoothing SGF was repeated r times, then

$$K^{sm}(r) = K^{sm}_{m_1,2t1} r^{-2/(4t1+3)}. \quad (2.3-48)$$

If the 1st-order differentiation SGF was repeated n times, then

$$K^{(n)} = \left(K^{(1)}_{m,2t}\right)^{a^{(n)}_{m,2t}}, \quad (2.3-49)$$

where $a^{(n)}_{m,2t} = n^{\varphi_{m,2t}}$. If $m >> 1$, then $\varphi_{m,2t} = \varphi_{2t} = 0.78, 0.73$, and 0.67 for $2t = 2, 4$, and 6, respectively.

In what follows, we estimate the optimal widths of SGF, when $m >> 1$. In this case, the SGF characteristic constants of the broad filters do not approximately depend on the filter width. The criterion of the optimal filter is the minimum of the relative error or the bias-variance tradeoff. That is,

$$N_{opt} = argmin\left\{\varepsilon = \sqrt{\varepsilon_s^2 + \varepsilon_r^2}\right\}(N). \quad (2.3-50)$$

Analysis of the smoothing SGF

By substituting Eqs. (2.3-45)-(2.3-47) into Eq. (2.3-50), we have for $m >> 1$:

$$\varepsilon = \sqrt{\varepsilon_s^2 + \varepsilon_r^2} = \sqrt{\left(P^{(0)}_{2t1}(N_1 H)^{2t_1+2}\right)^2 + \left(l^{(0)}_{2t}\right)^2/(N_1\xi_0^2)}. \quad (2.3-51)$$

Eq. (2.3-51) has the minimum value, if

$$N_{1,opt} = \xi_0^{-2/t_2} H^{-(t_2-1)/t_2} Z^{(0)}_{2t_1}, \quad (2.3-52)$$

where $t_2 = 4t_1 + 5$, $Z^{(0)}_{2t_1} = \left\{\left(l^{(0)}_{2t_1}/P^{(0)}_{2t_1}\right)^2/(t2-1)\right\}^{1/t_2}$.

The $N_{1,opt}$, estimated by Eq. (2.3-52), must be odd value in practice.

It follows from Eq. (2.3-52) that, for $m >> 1$, the optimal width of the smoothing SGF, expressed as a part of the FWHM, is

$$\Xi^{(0)}_{opt}(H, \xi_0, 2t_1) = N_{1,opt}H = \left(\sqrt{H}/\xi_0\right)^{2/t_2} Z^{(0)}_{2t_1}. \quad (2.3-53)$$

Remark 10

Recently, Iranian [15] and Indian [16] researchers have obtained more precise analytical expressions of $N_{1,opt}$. The functional dependencies $N_{1,opt}$ on the noise dispersion were similar to Eq. (2.3-52). It was shown that calculating the optimal window length requires knowledge of the signal [15]. The authors proposed an algorithm to solve this problem. Also, they briefly discussed numerous theoretical SGF studies.

To validate Eq. (2.3-52), we used the models (Eqs. (2.3-1) and (2.3-2)), distorted by normal noise. The numerical experiment was repeated 10^5 times. The mean value ($\bar{F}_0(0)$) and the standard deviation (ε_r) of the peak maximum were calculated. The total error (Eq. (2.3-51)) was $\varepsilon_{num} = 100\sqrt{\varepsilon_s{}^2 + \varepsilon_r^2}$, where ε_s =1 − $\bar{F}_0(0)$, was estimated.

Table 2.3-2. Characteristic SGF constants

n	2t	$C_{2t}^{(n)}$	Gaussian		Lorentzian		$l_{2t}^{(n)}$
			$P_{2t}^{(n)}$	$Z_{2t_1}^{(n)}$	$P_{2t}^{(n)}$	$Z_{2t_1}^{(n)}$	
0	2	2.232x10^{-4}	0.021	2.049	0.085	1.502	1.500
	4	4.732x10^{-7}	1.21x10^{-3}	2.557	0.0218	1.639	1.875
	6	5.357x10^{-10}	5.318x10^{-5}	2.965	5.529x10^{-3}	1.717	2.188
1	2	0.025	0.28	1.453	0.45	1.309	3.464
	4	1.24x10^{-4} **9.39x10^{-7} (Lorentz)**	0.023	2.2158	**0.128**	1.654	8.660
	6	2.53x10^{-7}	8.64x10^{-4}	2.9113	0.035	1.803	15.156
	8	2.79x10^{-10}	7.65x10^{-5}	3.084	0.018	1.756	22.736
2	2	1.786x10^{-2}	0.2971	1.757	0.857	1.321	26.833
	4	7.891x10^{-5}	0.0364	2.3455	0.455	1.537	93.921
	6	1.517x10^{-7}	2.716x10^{-3}	2.7977	0.196	1.648	211.34
	8	1.611x10^{-10}	1.439x10^{-4}	3.1793	0.075	1.716	387.49
4	4	5.463x10^{-5}	0.0588	2.1251	1.468	1.250	5040.4
	6	1.964x10^{-7}	0.0106	2.5015	1.901	1.349	27,726
	8	3.000x10^{-10}	9.828x10^{-4}	4.1012	1.533	2.054	90x10^5

Figure 2.3-7a demonstrates an example of the dependencies of the systematic, random, and total errors on the SGF width (N_1). According to the well-known facts, the dependencies $\varepsilon_s(N_1)$ and $\varepsilon_r(N_1)$ are increasing and decreasing functions, respectively; the plot $\varepsilon_{num}(N_1)$ has an explicit minimum.

Table 2.3-3 shows that the values $N_{1,min}$, which were obtained by the numerical experiments, and estimated by the two-term approximation (Eq. (2.3-52)), differ significantly, if the signal-to-noise ratio ξ_0 is small, and the FSG has a large polynomial order $2t_1$.

To overcome this drawback, we used a more precise approximation (Eq. (2.3-29)) of the frequency characteristic. Then the output signal was the inverse FT ($\mathcal{F}$) of the time domain data:

$$F_{sm}^{(0)}(\nu) = \mathcal{F}\{F_0(\omega)/(1+\Psi(x, N_1, 2t_{1,}\Delta 2t_1)\}, \quad (2.3-54)$$

where $\Psi(x, N_1, 2t_{1,}\Delta 2t_1) = C_{2t_1}(N_1 x)^{2t_1+\Delta 2t_1}$.

We supposed that N_1 is very large.

The systematic error in the peak maximum was

$$\varepsilon_s = F_0(0) - F_{sm}^{(0)}(0) =$$

$$\mathcal{F}\{F_0(\omega)\Psi(x, N_1, 2t_{1,}\Delta 2t_1)/(1+\Psi(x, N_1, 2t_{1,}\Delta 2t_1)\}. \quad (2.3-55)$$

Figure 2.3-7b demonstrates the high precision of this method. Unfortunately, the estimation of the systematic error ε_s (Eq. (2.3-55)) requires increment $\Delta 2t_1$ to be determined the and the FT to be performed numerically. Generally, an analytical solution is impossible. Using the frequency characteristics of the FSG, we only eliminated the first drawback. The advantage of this method is that the multiple repetitions of the smoothing process are not needed as in the numerical estimations (Fig. 2.3-7a).

Table 2.3-3. Optimal width ($N_{1,opt}$) of the smoothing SGF

<table>
<tr><th rowspan="3">t_1</th><th rowspan="3">H
X
100</th><th colspan="8">ξ_0</th></tr>
<tr><th rowspan="2">10</th><th>$\frac{\sqrt{H}}{\xi_0}$</th><th rowspan="2">50</th><th>$\frac{\sqrt{H}}{\xi_0}$</th><th rowspan="2">100</th><th>$\frac{\sqrt{H}}{\xi_0}$</th><th rowspan="2">200</th><th>$\frac{\sqrt{H}}{\xi_0}$</th></tr>
<tr><th>ε_{min}^{sm}
>2%</th><th>ε_{min}^{sm}
<2%</th><th>ε_{min}^{sm}
>2%</th><th>ε_{min}^{sm}
<2%</th></tr>
<tr><td rowspan="2">1</td><td rowspan="2">2</td><td>40/39</td><td rowspan="6">141x
10^{-4}</td><td>28/27</td><td rowspan="6">28x
10^{-4}</td><td>24/23</td><td rowspan="6">141x
10^{-5}</td><td>21/21</td><td rowspan="6">71x
10^{-5}</td></tr>
<tr><td>29/31</td><td>20/21</td><td>17/17</td><td>15/15</td></tr>
<tr><td rowspan="2">2</td><td rowspan="2">2</td><td>66/69</td><td>52/51</td><td>47/45</td><td>42/39</td></tr>
<tr><td>43/53</td><td>32/35</td><td>30/31</td><td>27/25</td></tr>
<tr><td rowspan="4">3</td><td rowspan="2">2</td><td>90/103</td><td>74/79</td><td>69/71</td><td>63/63</td></tr>
<tr><td>52/73</td><td>43/51</td><td>40/45</td><td>37/39</td></tr>
<tr><td rowspan="2">5</td><td>38/43</td><td rowspan="2">224x
10^{-4}</td><td>31/31</td><td rowspan="2">45x
10^{-4}</td><td>29/27</td><td rowspan="2">22x
10^{-4}</td><td>27/25</td><td rowspan="2">11x
10^{-4}</td></tr>
<tr><td>22/33</td><td>18/21</td><td>17/19</td><td>15/17</td></tr>
</table>

$N_{1,opt}$ must be odd. Data for the Gaussian and Lorentzian peaks are given in the upper and the lower rows, respectively. The optimal widths, obtained numerically, are given after the slash.

Further, the minimum smoothing error will be estimated using the precise approximation (Eq. (2.3-28)) by the necessary selection of the coefficient C_{2t_1} and the power $2t_1 + \Delta 2t_1$.
By substituting Eq. (2.3-52) of the two-terms approximation into Eq. (2.3 −51), we have

$$\varepsilon_{min}^{sm} = E_{2t1}^{(0)} \left(\sqrt{H}/\xi_0\right)^{\psi}, \qquad (2.3-56)$$

where $E_{2t1}^{(0)} = \left(P_{2t1}^{(0)}\right)^{1/(4t_1+5)} \sqrt{4t_1+5} \left[\left(l_{2t}^{(0)}\right)^2/(4t_1+4)\right]^{\psi/2}$ (Table 2.3-4); $\psi = (4t_1+4)/(4t_1+5)$.

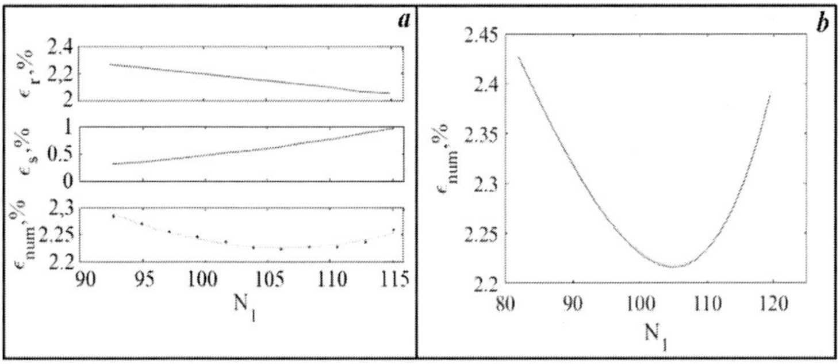

Figure 2.3-7. Estimation of the smoothing errors: (a) numerically ($2t_1 = 6$, $H = 0.02, \sqrt{H}/\xi_0 = 10$). $N_{1,min} = 105$. $min\{\varepsilon_{num}\} = 2.225\%$. (b) using precise and approximated the frequency characteristic (Eq. (2.3-54), $\Delta 2t_1 = 1.69$), $min\{\varepsilon_{num}\} = 2.219\%$ and 2.217%, respectively.

Table 2.3-4 includes the maximum values of H/ξ_0 for the smoothing polynomial of some degree, which gives $\varepsilon_{min}^{sm} \leq 2\%$ (Eq. (2.3-56)).

Inequalities ε_{min}^{sm}>2% or ε_{min}^{sm}<2% (Table 2.3-3) were obtained by comparing the values $\sqrt{H}/\xi_0$, given in Table 2.3-3, with the corresponding values in Table 2.3-4. Table data show that the dependence of ε_{min}^{sm} on the polynomial degree $2t_1$is weak. $\varepsilon_{min}^{sm} < 2-3\%$, if $2t_1 = 6$ and 4 for the Gaussians and Lorentzians, respectively.

Multiple smoothing

Using Eq. (2.3-48), we have

$$\varepsilon(r) = \sqrt{\left(rP_{2t1}^{(0)}(N_1H)^{2t_1+2}\right)^2 + \left(l_{t_1}^{(0)}\right)^2/(N_1\xi_0^2 r^{2/(4t_1+3)})}. \qquad (2.3-57)$$

$$N_{1,opt}(r) = N_{1,opt} r^{-2\psi/(4t_1+3)}. \qquad (2.3-58)$$

$$\varepsilon_{min}^{sm}(r) = \varepsilon_{min}^{sm} r^{-\chi}, \qquad (2.3-59)$$

where $\chi = 1/[(4t_1+3)(4t_1+5)]$. So, $\varepsilon_{min}^{sm}(r)$ is a very slowly decreasing function. This fact is clear because the multiple smoothing performed by the broad SGFs, not only strongly suppresses noise but also strongly distorts the input signal. In practice, the multiple smoothing carried out by the short SGFs has advantages over single filtering.

Table 2.3-4. Parameters of Eqs. (2.3-56) and (2.3-63)

n	2t	$E_{2t}^{(n)}$		max $\sqrt{H}/\xi_0$ for smoothing: ε_{min} ≤0.02 and for differentiation ε_{min} ≤0.05	
		Gaussian	Lorentzian	Gaussian	Lorentzian
0	2	1.111	1.298	0.0109	0.0091
	4	1.222	1.529	0.116	0.0091
	6	1.309	1.721	0.0118	0.0088
1	2	0.9025	1.1773	0.0063	0.0040
	4	1.0616	1.8356	0.0150	0.0071
	6	1.1763	2.6913	0.0193	0.0069
	8	1.5771	4.0952	0.0166	0.0053
2	2	1.2301	2.0070	7.1 x10^{-4}	2.5 x10^{-4}
	4	1.7763	4.0887	0.0030	7.8 x10^{-4}
	6	2.4016	7.2188	0.0041	8.7 x10^{-4}
	8	3.0787	11.5135	0.0045	7.9 x10^{-4}
4	4	1.6477	4.9285	5.9 x10^{-4}	5.8 x10^{-5}
	6	3.9484	17.4779	4.8 x10^{-4}	3.5 x10^{-5}
	8	131.0907	809.0107	4.6 x10^{-6}	2.7 x10^{-7}

Differentiation SGF

Similar to Eq. (2.3-51), the relative total error for $m >> 1$ is

$$\varepsilon = \sqrt{\varepsilon_s^2 + \varepsilon_r^2} =$$

$$\sqrt{\left(P_{2t}^{(n)}\right)^2 (NH)^{4t} + \left(l_{2t}^{(n)}/D^{(n)}\right)^2 \left(\sqrt{H}/\xi_0\right)^2/(NH)^{2n+1}}, \qquad (2.3-60)$$

where $P_{2t}^{(n)} = C_{2t}^{(n)}\left|R^{(2t+n)}/R^{(n)}\right|$; $D^{(n)}$=$\left|\Delta F_{extr}^{(n)} w^n/F_0\right|$ is the distance between the extrema of the derivatives of the models (Eqs. (2.3-1) and (2.3-2)). $D^{(1)} \approx [2.86, 2.6]$; $D^{(2)} \approx [8.02, 10]$; $D^{(4)} \approx [149.2, 546]$ for the Gaussians and Lorentzians, respectively.

Similar to Eq. (2.3-52), the optimal filter width is

$$N_{opt} = \left\{[(2n+1)/(4tH^{b-1})]\left[l_{2t_1}^{(n)}/\left(\xi_0 P_{2t_1}^{(n)} D^{(n)}\right)\right]^2\right\}^{1/b}, \qquad (2.3-61)$$

where $b = 4t + 2n + 1$.

It follows from Eq. (2.3-61), that for $m >> 1$, the relative optimal width of the differentiation SGF, expressed as a part of the peak FWHM:

$$\Xi_{opt}^{(n)}(H,\xi_0,2t) = N_{opt}H = N_{opt}h/w = \left(\sqrt{H}/\xi_0\right)^{2/b} Z_{2t}^{(n)}, \qquad (2.3-62)$$

where $Z_{2t}^{(n)} = \left\{[(2n+1)/(4t)]\left[l_{2t_1}^{(n)}/\left(P_{2t}^{(n)}D^{(n)}\right)\right]^2\right\}^{1/b}$.

Substituting Eq. (2.3-62) into Eq. (2.3-60), we obtain the minimum error:

$$\varepsilon_{min}^{(n)} = E_{2t}^{(n)}\left(\sqrt{H}/\xi_0\right)^{4t/b}, \qquad (2.3-63)$$

where $E_{2t}^{(n)} = \sqrt{\left(P_{2t}^{(n)}\right)^2\left(Z_{2t}^{(n)}\right)^{4t} + \left(l_{2t}^{(n)}/D^{(n)}\right)^2/\left(Z_{2t}^{(n)}\right)^{2n+1}}$.

Table 2.3-5 contains the theoretical values of the optimal widths of some differentiation SGFs, obtained according to Eq. (2.3-61). The optimal widths were evaluated by numerical experiments, similar to those involved in studying of the smoothing filters. Table 2.3-5 represents some N_{opt} values for the 2nd-order differentiation filters. These optimal values are larger than the theoretically-obtained corresponding data.

Maximum values $\sqrt{H}/\xi_0$ were estimated for the smoothing and differentiation SGFs, provided ε_{min} ≤0.02 and ε_{min} ≤0.05, respectively (Table 2.3- 4). They were compared with corresponding values given in Table 2.3-5. The comparison shows that the widths, designated in Table 2.3-5 by stars, correspond to the total relative smoothing errors in the peak maximum: ε_{min} ≤2%, and the total relative errors of the differentiation in the extrema, ε_{min} ≤5%.

Table 2.3-5 demonstrates that, for the 2nd-order derivative, the total error decreases if the polynomial power increases from 2 to 4. However, the filter is significantly broadened.

For the rest of the data, the total minimum errors of the 1st- and the 2nd-order derivatives may be more than 10%. However, for the 4th-order derivatives, the errors are substantial. The use of high-order derivatives requires a tiny sampling interval (very wide filters) and/or a high signal-to-noise ratio of the raw data. The drawback of broad filters is the non-accurate processing of the extreme left and right points of the data vectors.

Multiple differentiation

Suppose that the 1st-order differentiation SGF is repeated n times. Then, using Eqs. (2.3-44), (2.3-47) and (2.3-49), for $N \gg 1$, we have

$$\varepsilon(n) = \sqrt{\left(nP_{2t}^{(1)}(NH)^{2t}\right)^2 + \left(l_{2t}^{(1)}\right)^{2a_{2t}^{(n)}}\Big/\left[\xi_0(D^{(n)}H^n)^2N^{3a_{2t}^{(n)}}\right]}. \qquad (2.3-64)$$

The width of the optimal filter:

$$N_{opt}^{(n)} = \left\{\left[3a_{2t}^{(n)}\left(l_{2t}^{(1)}\right)^{2a_{2t}^{(n)}}\right] / \left[4t\left(\xi_0 D^{(n)} n P_{2t}^{(1)}\right)^2 H^{4t+2n}\right]\right\}^c, \qquad (2.3-65)$$

where $c = 1/(4t + 3a_{2t}^{(n)})$.

The optimal relative width of the differentiation SGF, expressed as a part of the peak FWHM, is

$$\Xi_{opt}^{(n)}(H, \xi_0, 2t) = N_{opt}^{(n)} H = \left(H^{1.5a_{2t}^{(n)}-n}/\xi_0\right)^{2c} Y_{2t}^{(n)}, \qquad (2.3-66)$$

where $Y_{2t}^{(n)} = \left\{\left[3a_{2t}^{(n)}\left(l_{2t}^{(1)}\right)^{2a_{2t}^{(n)}}\right] / \left[4t\left(D^{(n)} n P_{2t}^{(1)}\right)^2\right]\right\}^c$.

By substituting Eq. (2.3 −66) into Eq. (2.3 −64), we have

$$\varepsilon_{min}(n) = E_{2t}(n)\left(H^{1.5a_{2t}^{(n)}-n}/\xi_0\right)^{4t/\left(4t+3a_{2t}^{(n)}\right)}, \qquad (2.3-67)$$

where $E_{2t}(n) = \sqrt{\left[nP_{2t}^{(1)}\left(Y_{2t}^{(n)}\right)^{4t}\right]^2 + \left[l_{2t}^{(1)} / \left(D^{(n)}\left(Y_{2t}^{(n)}\right)^{1.5a_{2t}^{(n)}}\right)\right]^2}$.

For $n = 1$, Eq. (2.3-67) is similar to Eq. (2.3-63).

As an example, Table 2.3-6 contains the optimal width and minimum total relative error of the 2nd-order differentiation obtained by the self-convolution of the 1st-order differentiation SGF. These values were compared with those estimated by the single filter ($2t = 2$) (Table 2.3-7). The optimal width of the combined filter is two times the 1st-order differentiation filter width. The convolution decreases the total error due to the significant broadening of the combined filter. However, for $2t = 4$ and 6, the convolution increases the errors.

Remark 11

We recommend the tutorial [17] to well-prepared readers. The book [18] details the error analysis in some applications of derivative spectra. The SGF have near-optimal smoothing properties for white and pink noises [19] (Chapter 2.5).

Numerical differentiation of unequally spaced experimental chemical data was firstly performed using the movable strip technique [20], which was similar to the Lanczos method [7]. Later, Antonov et al. developed polynomial smoothing and the differential filters of the irregularly distant points [21, 22]. Sampling with varying intervals occurs when the linear wavelength scale is transformed into the wavenumbers' non-linear range (Chapter 1.2).

These step-by-step filters (SBSF) are built on the moving polynomial, which approximates data segments consisting of consecutive points. The coefficients of the SBSFs are not constants since the sampling interval of the x-axis is a non-linear. An essential disadvantage of the SBSFs is considerable matrix calculations. It is impossible to tabulate the filter coefficients and obtain their analytical expressions as they have done for the SG method.

Table 2.3-5. Optimal width (N_{opt}) of the differentiation SGF

n	t	H, %		ξ_0 = 50: $\frac{\sqrt{H}}{\xi_0}$ x 10^4	50: G	50: L	ξ_0 = 100: $\frac{\sqrt{H}}{\xi_0}$ x 10^4	100: G	100: L	ξ_0 = 200: $\frac{\sqrt{H}}{\xi_0}$ x 10^4	200: G	200: L
1	1	0.5		14	45*	40*	7.1	37*	33*	3.5	30*	27*
		1		20	25*	22*	10.0	20*	18	5.0	17*	15*
	2	0.5		14	134*	100*	7.1	119*	88*	3.5	104*	78*
		1		20	72*	53*	10.0	63*	47*	5.0	55*	41*
	3	0.5		14	243*	150*	7.1	221*	137*	3.5	202*	125*
		1		20	127*	79*	10.0	116*	72*	5.0	106*	65*
2	1	0.5	theor	14	82	61	7.1	70*	53	3.5	60*	45
					6.7%	10.9%		4.9%	8.0%		3.6%	5.9%
			exper		**89**	**69**		**77**	**59**		**67**	**51**
					7.6%	**12%**		**5.6%**	**8.5%**		**4.1%**	**6.3%**
		1	theor	20	44	33	10.0	38	28	5.0	32*	24
					7.8%	12.7%		5.7%	9.3%		4.2%	6.8%
			exper		**49**	**37**		**41**	**33**		**35**	**2**
					8.9%	**13%**		**6.5%**	**9.9%**		**4.8%**	**7.3%**
	2	0.5	theor	14	171*	112	7.1	154*	101*	3.5	138*	91*
					3.1%	7.2%		2%	4.7%		1.3%	3.1%
			exper		**193**	**139**		**173**	**119**		**153**	**107**
					3.3%	**6.4%**		**2.2%**	**4.4%**		**1.5%**	**2.9%**
		1	theor	20	90*	59	10.0	81*	53	5.0	73*	48*
					3.9%	8.9%		2.5%	5.8%		1.7%	3.8%
			exper		**103**	**73**		**91**	**65**		**81**	**57**
					4.1%	**7.8%**		**2.7%**	**5.3%**		**1.8%**	**3.6%**
	3	0.5		14	259	152	7.1	238	140	3.5	220*	129*
		1		20	135	79	10.0	124	73	5.0	114*	67*
4	2	0.5		14	196	116	7.1	181	106	3.5	167*	98*
		1		20	102	60	10.0	94	55	5.0	87	51
	3	0.5		14	268	144	7.1	251	135	3.5	235*	127*
		1		20	138	75	10.0	130	70	5.0	122	65
	8	0.5		14	485	243	7.1	459	230	3.5	434*	218
		1		20	249	125	10.0	236	118	5.0	223	112

In practice N_{opt} must be odd. The rows after N_{opt} include ε_{min}. * $\varepsilon_{min} \leq 5\%$

Table 2.3-6. The optimal width and minimum total relative error of the 2nd-order differentiation obtained by the self-convolution of the 1st-order differentiation SGF

2t	H, %	Parameters	ξ_0					
			50		100		200	
			G	L	G	L	G	L
2	0.5	$N_{opt}^{(2)}$	52	44	44	38	38	33
		$\varepsilon_{min}(2), \%$	5.4	4.8	4.0	3.5	2.9	2.6
	1	$N_{opt}^{(2)}$	28	24	24	21	21	18
		$\varepsilon_{min}(2), \%$	6.4	5.7	4.7	4.2	3.5	3.1
4	0.5	$N_{opt}^{(2)}$	143	106	129	96	116	86
		$\varepsilon_{min}(2), \%$	17.7	9.0	11.6	5.9	7.5	3.8
	1	$N_{opt}^{(2)}$	76	56	68	50	61	45
		$\varepsilon_{min}(2), \%$	21.8	11.1	14.2	7.3	9.3	4.7
6	0.5	$N_{opt}^{(2)}$	260	163	239	150	220	138
		$\varepsilon_{min}(2), \%$	>100	41	>100	25	>100	15.3
	1	$N_{opt}^{(2)}$	114	84	123	77	114	71
		$\varepsilon_{min}(2), \%$	>100	50.0	>100	30.5	>100	18.6

G and L stand for the Gaussian and Lorentzian peaks, respectively.

Table 2.3-7. The optimal widths and minimum total relative errors of the 2nd-order differentiation obtained by the self-convolution (SConv) of the 1st-order differentiation SGFs and the single filter (Sng)

H, %	SGF	Parameters	ξ_0					
			50		100		200	
			G	L	G	L	G	L
0.5	SConv	N_{opt}	105	89	89	77	77	67
		$\varepsilon_{min}, \%$	5.4	4.8	4.0	3.5	2.9	2.6
	Sng	N_{opt}	82	61	70	53	60	45
		$\varepsilon_{min}, \%$	6.7	10.9	4.9	8.0	3.6	5.9
1	SConv	N_{opt}	115	99	99	87	87	75
		$\varepsilon_{min}, \%$	6.4	5.7	4.7	4.2	3.5	3.1
	Sng	N_{opt}	44	33	38	28	32	24
		$\varepsilon_{min}, \%$	7.8	12.7	5.7	9.3	4.2	6.8

G and L stand for the Gaussian and Lorentzian peaks, respectively.

CHAPTER FOUR

SMOOTHING AND DIFFERENTIATION USING THE FAST FOURIER TRANSFORM

This chapter intends to demonstrate simple methods of smoothing and differentiation of noisy spectra using the Fast Fourier Transform (FFT). FFT reduces the complexity of computing the discrete FT of N- point vector from $O(N^2)$ to $O(NlogN)$. Curious readers, we refer to the entertaining story of this remarkable algorithm [1], indicating that "nil sub sole novum".

The Fourier differentiation methods are discussed and illustrated by numerous examples in the textbook: therefore, we have only demonstrated the FFT regularized differentiation and suggest that our readers complete the exercise.

Smoothing of noisy data using Fourier Transform (FT) has a long history: from Lanczos' empirical algorithms [2] to Tikhonov's mathematically rigorous regularization methods [3], which are continuously developing (e.g., [4]). In spectroscopy, there are pioneering studies of FT processing, which involve smoothing and differentiation [5, 6]. However, the FT smoothing is not widespread in analytical spectroscopy since it was supplanted by the Savitzky-Golay (SG) polynomial filters [7].

In recent decades, interest in the problem of the optimal smoothing of analytical signals and, particularly, spectra, has disappeared. Obtaining smoothed spectra has become a routine for modern spectral instruments. However, the reliability of the peak parameters, which are measured in smoothed spectra, as a rule, is not questioned. This fact contradicts the importance of the error analysis in analytical spectrometry [8].

Theory

Let $F\ (\lambda)$ be a noisy spectrum, where $\lambda = \{\lambda_1, \lambda_2, ..., \lambda_N\}$ is a grid of equally-spaced points. Our goal is to find a smooth function $F_{sm}(\lambda_i)$ that approximates $F\ (\lambda_i)$ appropriately.

According to the penalized linear regression approach [9]:

$$\hat{F}_{sm} = argmin_{F_{sm}}\{||F - F_{sm}||_2^2 + \gamma||\mathcal{D}F_{sm}||_2^2\}, \qquad (2.4-1)$$

where $||\ ||_2$ stands for the Euclidean norm; γ is the smoothing (regularization) parameter; $\mathcal{D}$ is the penalty matrix. By varying γ, one can control the trade - off between smoothness of the signal $\hat{F}_{sm}$ and goodness of fit. The solution of Eq. (2.4-1) in the time domain is obtained using the least-squares method [9].

Another possible regularized solution of the smoothing procedures in the Fourier domain (Eq. (2.4-2)) is based on the discrete Inverse Fourier transform (IFT) (MATLAB function FourieSmooth in Appendix SW2):

$$F_{sm}(\lambda_i) = real\left(IFT\{\widetilde{U}(\omega_k)/[1 + M(\omega_k)]^2\}\right), \qquad (2.4-2)$$

where real denotes the real FT part; $\widetilde{U}(\omega)$ is the FT of the $F\ (\lambda)$; ω is the angular (Fourier frequency); $M(\omega) = \alpha\omega^{2p}$ is the stabilizer; α is the regularization parameter. This simplest analytical expression of the stabilizer was taken from a broad class of functions [3]. The regularization parameter α plays the same role as γ in Eq. (2.4-1).

It is well-known that there are numerous sets of regularized solutions similar to Eq. (2.4-2) [3]. To select the "best" variant, one must establish its criterion, select the "best" form of the stabilizer, and estimate the optimal regularization parameter. In the lack of *a priori* information, this task is impossible.

Suppose that the deterministic, true spectrum ($F_T(\lambda)$) and the random noise ($\eta\ (\lambda)$) are known. Also, $\eta\ (\lambda)$ and $F_T(\lambda)$ are uncorrelated statistically. Then, the optimal solution, which minimizes the Euclidian distance between the true and the smoothed spectrum, is obtained using the FTs of F_T and η [3].

Another method uses the spectroscopic criteria, such as the minimum relative errors of the peak intensities or their areas. In this case, the estimation of the optimal regularization parameter has no technical problems.

If accurate data for $F_T(\lambda)$) and $\eta\ (\lambda)$ are unknown, then one can only use some *a priori* assumption about these functions [3].

An empirical method includes:

a. Modelling possible spectral elements, e.g., doublets of known shapes (Gaussian, Lorentzian), distorted by known noise (normal, pink).
b. Estimation of the "best" solution using a spectroscopic criterion.
c. Estimation of the approximate solution using some empirical criterion.
d. If the difference between solutions obtained by "b" and "c" is acceptable, then the empirical criterion may be used in practice.

Various numerical methods for estimation of the quasi-optimal regularization parameters were suggested and compared during the last decades. Among them, the most popular is the L-curve method [10].

In the L-curve method, the dependence of the Euclidian norms $S = \left\{\|\hat{F}_{sm}\|_2\right\}(\alpha)$ versus $Q = \left\{\|\hat{F}_{sm} - F\|_2\right\}(\alpha)$ is plotted in the log-log scale. S and Q are similar to the members of Eq. (2.4-1) if $\mathcal{D}$=1. The optimal value of the regularization parameter α corresponds to the "corner" of the L-curve. The corner is the point with an extremum curvature. If the range of the α values is small, then the log-scale is unnecessary. According to the differential calculus, the curvature is

$$C(\alpha) = [S'Q'' - S''Q']/[(S')^2 + (Q')^2]^{3/2}, \qquad (2.4-3)$$

where all derivatives are calculated relative to α.

Unfortunately, a numerical estimation of this point may be disturbed by noisy data.

Here, we introduce a new empirical method developed by numerical experiments.

It seems reasonable to assume that the analytical signal (AS) (a spectrum) is an *ergodic* process. Its statistical properties can be deduced from a single, sufficiently long, random sample of this signal. In this case, the following functional (Eq. (2.4-4)) (which resembles the Tikhonov functional $F(\alpha)$ for the first-order kernel (Appendix G) [3]), has a minimum over α.

$$\psi(\alpha, p) = \textstyle\sum_k \{M(\omega_k)\widetilde{U}(\omega_k)/[1 + M(\omega_k)]\}^2, \qquad (2.4-4)$$

where $\omega_k = [1{:}\,len1, -len1 + 1{:}\,1{:}\,0];$, $len = 2len1$ is the signal length (see MATLAB function FourieSmooth in Appendix SW2).

We suppose that the quasi-optimal value $\alpha_{opt} = argmin\{\psi(\alpha, p)\}$.

The Fourier spectrum (Fig. 2.4-1) of the noise-free symmetrical doublet (Fig. 2.4-2a) shows that the low-frequency Fourier harmonics concentrate on the most useful information [2]. Therefore, the cut-off for the high-frequency components by the FT filter may significantly reduce the noise, while the useful signal is only slightly disturbed. However, this conclusion is right if the noise power is uniformly distributed over all angular frequencies similar to the white normal noise (Fig. 2.4-3). The low-frequency harmonics of the pink noise distort the signal more strongly than the high-frequency components (Fig. 2.4-3).

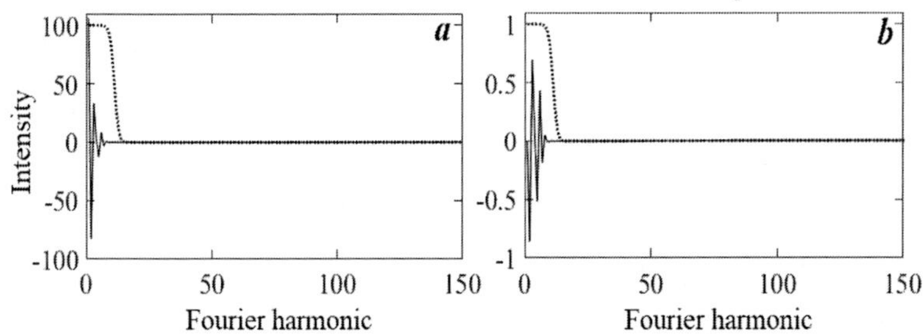

Figure 2.4-1. The real (a) and imaginary (b) FT components of the symmetrical Gaussian doublet (Fig.2.4-2a).The FT filter (dotted line) is multiplied by 100 (a).

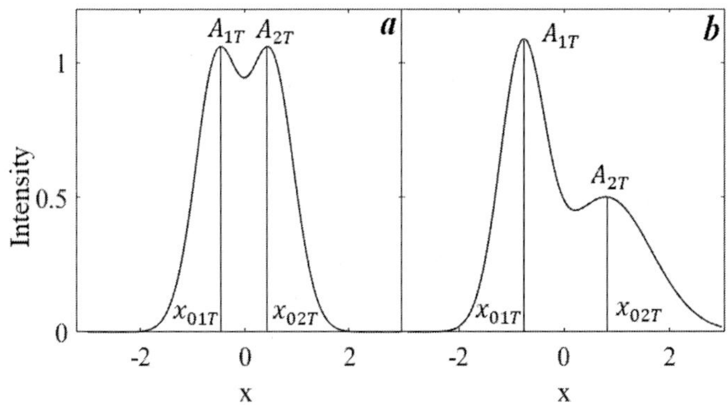

Figure 2.4-2. Gaussian doublets.

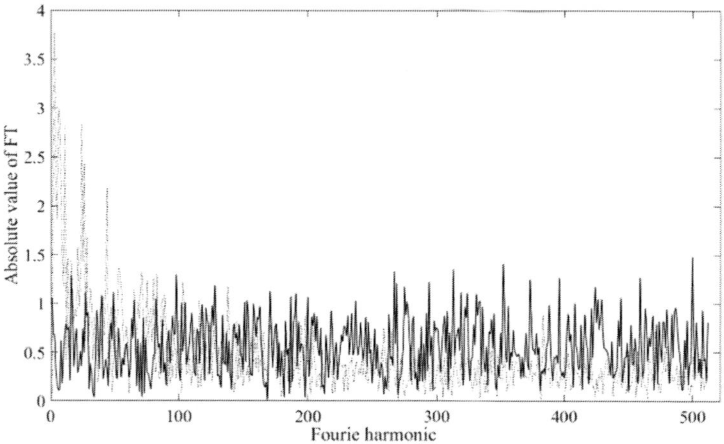

Figure 2.4-3. Absolute value of the FT spectrum of the normal and pink noise (solid and dotted curves, respectively). $\sigma = 0.02$.

Numerical experiment

Models

1. Gaussian doublets with dimensionless parameters (Fig. 2.4-1):

$$F_d = A_1\, exp\left[-k\left(\frac{x+x_0}{w_1}\right)^2\right] + A_2\, exp\left[-k\left(\frac{x-x_0}{w_2}\right)^2\right] + \eta\,, (2.4-5)$$

where $k = 4ln2; \eta$ is the normal or pink noise with mean zero and standard deviation $\sigma = 0.02$, $x = [-3{:}\,0.02{:}\,3]$. The pink noise was generated by the program in [11].

a. Symmetrical doublet: $A_1 = A_2 = 1$; $w_1 = w_2 = 1$; $x_0 = 0.52$. If the noise is absent, then the peak amplitudes are $A_{1T} = A_{2T} = 1.0599$; and the peak positions are $x_{01T} = -0.44, x_{02T} = 0.44$.

b. Asymmetrical doublet: $A_1 = 1, A_2 = 0.5$; $w_1 = 1, w_2 = 2$; $x_0 = 0.80$. If the noise is absent, then the peak amplitudes $A_{1T} = 1.0881, A_{2T} = 0.5008$; and the peak positions $x_{01T} = -0.76, x_{02T} = 0.78$.

2. Absorption UV spectrum of anthracene [12, 13] (see below).

Criteria

1. The minimum total relative error of the peak intensities is

$$Err(p,\alpha) = \sqrt{RelErr_1^2 + RelErr_2^2}, \qquad (2.4-6)$$

where $RelErr_i^2 = [\Delta_i^2 + \xi_i^2]/A_{iT}^2$ is the squared relative error; $i = 1,2$; $\Delta_i = A_{iT} - \hat{A}_{im}$ is the systematic error of the peak amplitude (the

difference between the true value and the average of the mean value); $\xi_i = std(A_{im})$ is the random error (the standard deviation).
2. The minimum of the L-curve curvature (Eq. (2.4-3), Fig. 2.4-4). Since the ranges of the x and y arguments are small, then the log-log scale is unnecessary.
3. The minimum of the functional $\psi(\alpha,p)$ (Eq. (2.4-4), Fig. 2.4-5). In an abnormal case (Fig. 2.4-5c), the minimum of $\psi(\alpha,p)$ was not observed. Therefore, the maximum position of the 1st-order derivative over α was estimated.

All experiments were repeated 10^4 times. The noise was regenerated in each repetition, and the program provided the condition for the resolved doublet. This condition is needed for estimating the total error (Eq. (2.4-6)).

In the first group of the experiments, for a given p (Eq. (2.4-2)), the functions $Err(\alpha)$, L-curves and their curvatures ($C(\alpha)$), and the functionals ($\psi(\alpha)$), which were obtained in each repetition, were averaged (e.g., Figs. 2.4-4 and 2.4-5). The above criteria were applied to the mean data. So, α_{opt} was estimated according to each criterion, and then the $Err(\alpha_{opt})$ was calculated (Tables 2.4-1 and 2.4-2).

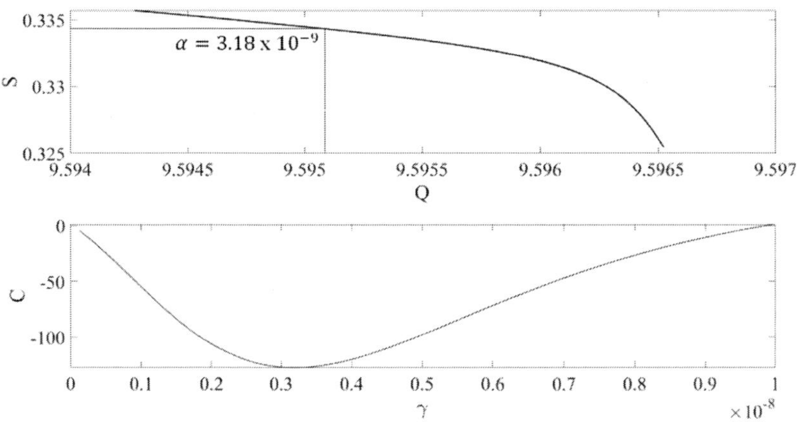

Figure 2.4-4. The L-curve (a) and its curvature (b) $\alpha = 3.18 \times 10^{-9}$ corresponds to the minimum curvature. Mean data of 10^4 repetitions; the doublet was resolved in each repetition. $p = 4$. $\sigma = 0.02$.

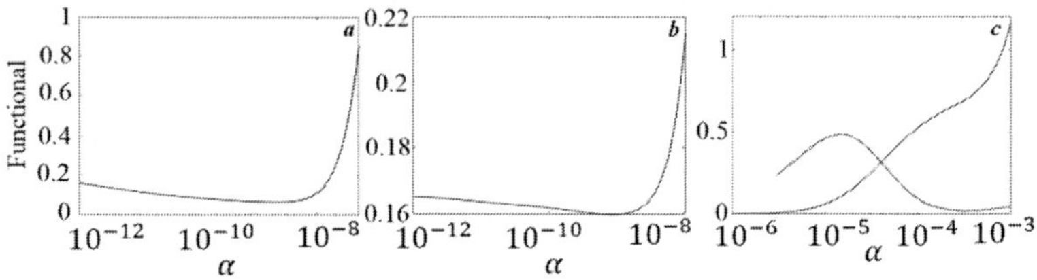

Figure 2.4-5. Functional (Eq. (2.4-4)) for the symmetrical Gaussian doublet (Fig. 2.4-2a). (c) Abnormal case. The derivative of the functional over α (dotted curve). Mean data of 10^4 repetitions in each of them, the doublet was resolved. Normal (a) and pink (b, c) noise. $\sigma = 0.02$. $p = 4$ (a, b) and 1(c).

In the second group, we studied the uncertainties in determining peak positions and amplitudes in the FFT- smoothed doublet spectra using the L-curve method. The numerical experiment included the following steps:
1. Calculation of the L-curves and their curvature $C(\alpha)$, for a given p (Eq. (2.4-2)), in each repetition.
2. Estimating the optimal regularization parameter (α_{opt}) at the point corresponding to the L-curve curvature minimum.
3. Calculation of the peak positions (x_{01} and x_{02}) and the total relative error (Err) for the optimal smoothed doublet.
4. Estimating the average values and standard deviations of the peak positions and the errors over all of the repetitions (Tables 2.4-3 and 2.4-4).

In both groups, the uncertainties in determining peak positions and intensities are due only to the smoothing. Peak overlapping also causes significant uncertainties that have been carefully studied [8].

Results and Discussion

Analysis of the data, obtained for the models 1a and 1b, shows the following:
1. If the stabilizer order p increases, then the optimal value of the regularization parameter α_{opt} decreases, and the minimum total error reduces. However, a large $p \geq 4$ causes saturation. That is, there was no gain in accuracy.
2. The errors, due to the pink noise, are more than two times larger than those of the normal noise. This result is following the above conclusions.
3. The errors obtained using the functional and L-curve methods do not differ significantly from those obtained by the spectroscopic criterion (Eq. (2.4-6)). That is, these methods may be successfully used in practice.

Table 2.4-3 includes the errors when determining the relative peak positions in the FFT smoothed doublets using the L-curve method. The table data includes the mean peak shifts from their correct positions (x_{0T}) when noise is absent: $\overline{\Delta\hat{x}_0/h} = \left(\overline{\hat{x}_0} \pm x_{0T}\right)/h$, and the standard deviations $\sigma(\hat{x}_0)/h = std(\hat{x}_0)/h$. $x_{0T} \neq x_0$ due to the peak overlapping.

From these data, we conclude that, for the symmetrical doublet (model 1a), the impact of the FFT smoothing on the measurement errors of the peak positions is small. For $p \geq 3$, the 95% significant interval is less than $2h$, which corresponds to 4% of the peak width. This conclusion is also valid for the first, narrow peak of the asymmetrical doublet (model 1b). However, the 95% relative uncertainties in determining the position of the broad, weak second peak may be up to 30-50% of its width.

The relative uncertainties (due to the smoothing) in determining the peak intensities of the asymmetrical doublet, corrupted by the pink noise, are no more than 3-5% with a significance level of 95% (Table 2.4-4).

Table 2.4-1. Data for the model 1a

p	Criterion					
	Spectroscopic (Eq. (2.4-6))		Functional (Eq. (2.4-3))		*L*-curve (Eq. (2.4-4))	
	α_{opt1}	$Err_{\alpha_{opt1}}$, %	α_{opt2}	$Err_{\alpha_{opt2}}$, %	α_{opt3}	$Err_{\alpha_{opt3}}$, %
1	$4.50\text{x}10^{-4}$	0.9318	$6.10\text{x}10^{-4}$	1.0071	$4.65\text{x}10^{-4}$	0.9374
	$5.20\text{x}10^{-4}$	1.8755	$\underline{3.27\text{x}10^{-4}}$*	1.4399	$3.76\text{x}10^{-4}$	1.8954
2	$1.13\text{x}10^{-5}$	0.7421	$9.10\text{x}10^{-6}$	0.7666	$8.11\text{x}10^{-6}$	0.7545
	$1.80\text{x}10^{-5}$	1.7179	$1.88\text{x}10^{-6}$	1.900	$1.08\text{x}10^{-5}$	1.7510
3	$2.40\text{x}10^{-7}$	0.6783	$1.45\text{x}10^{-7}$	0.6895	$1.62\text{x}10^{-7}$	0.6796
	$4.05\text{x}10^{-7}$	1.6530	$5.00\text{x}10^{-8}$	1.7874	$2.55\text{x}10^{-7}$	1.6895
4	$4.90\text{x}10^{-9}$	0.6384	$2.40\text{x}10^{-9}$	0.6617	$3.18\text{x}10^{-9}$	0.6517
	$8.40\text{x}10^{-9}$	1.6409	$1.03\text{x}10^{-9}$	1.7305	$5.47\text{x}10^{-9}$	1.6333
5	$9.60\text{x}10^{-11}$	0.6191	$3.90\text{x}10^{-11}$	0.6362	$6.10\text{x}10^{-11}$	0.6330
	$1.62\text{x}10^{-10}$	1.6188	$2.02\text{x}10^{-11}$	1.7112	$1.09\text{x}10^{-10}$	1.6124
6	$1.80\text{x}10^{-12}$	0.6149	$5.87\text{x}10^{-13}$	0.6379	$1.08\text{x}10^{-12}$	0.6213
	$3.10\text{x}10^{-12}$	1.6063	$2.40\text{x}10^{-12}$	1.6812	$2.14\text{x}10^{-12}$	1.6136

Upper and lower rows represent data for normal and pink noises, respectively.
* obtained using the minimum of the 1st-order derivative.
$\alpha_{opt1} = argmin\{Err(p,\alpha)\}$, $\alpha_{opt2} = argmin\{\psi(\alpha,p)\}$, $\alpha_{opt3} = argmin\{C(\alpha,p)\}$.

Table2.4- 2. Data for the model 1b

p	Criterion					
	Spectroscopic (Eq. (2.4-6))		Functional (Eq. (2.4-3))		*L*-curve (Eq. (2.4-4))	
	α_{opt1}	$Err_{\alpha_{opt1}}$, %	α_{opt2}	$Err_{\alpha_{opt2}}$, %	α_{opt3}	$Err_{\alpha_{opt3}}$, %
1	7.70x10^{-4}	1.8381	6.57x10^{-4}	1.8809	4.97x10^{-4}	2.006
	1.00x10^{-3}	3.4318	3.44x10^{-4}*	3.9007	4.06x10^{-4}	3.8111
2	1.76x10^{-5}	1.2138	1.20x10^{-5}	1.2108	9.48x10^{-6}	1.2762
	2.98x10^{-5}	2.8316	2.80x10^{-6}	3.3640	1.42x10^{-5}	2.9539
3	4.90x10^{-7}	1.0702	2.29x10^{-7}	1.0551	2.36x10^{-7}	1.0999
	8.48x10^{-7}	2.6488	8.60x10^{-8}	2.9314	4.31x10^{-7}	2.6752
4	1.14x10^{-8}	1.0075	3.99x10^{-9}	1.0023	5.52x10^{-9}	1.0200
	2.08x10^{-8}	2.6044	1.45x10^{-9}	2.8040	1.10x10^{-8}	2.6011
5	2.40x10^{-10}	0.9776	6.53x10^{-11}	0.9633	1.13x10^{-10}	0.9968
	4.40x10^{-10}	2.5221	3.42x10^{-11}	2.7168	2.45x10^{-10}	2.5625
6	4.80x10^{-12}	0.9630	9.49x10^{-13}	0.9720	2.13x10^{-12}	0.9732
	9.20x10^{-12}	2.5121	5.50x10^{-13}	2.6713	5.45x10^{-12}	2.5124

Upper and lower rows represent data for normal and pink noises, respectively.
* obtained using the minimum of the 1st-order derivative.
$\alpha_{opt1} = argmin\{Err(p, \alpha)\}$, $\alpha_{opt2} = argmin\{\psi(\alpha, p)\}$, $\alpha_{opt3} = argmin\{C(\alpha, p)\}$.

Comparison of FFT and SG smoothing

The model 1a, which was disturbed by the normal noise ($\sigma = 0.02$), was smoothed by FFT ($p = 4$) and SG filters (the polynomial power is two). The number of repetitions was 10^4. As we supposed [14], the comparison of the measurement uncertainties, which appeared because of the FFT and SG-smoothing, does not show a significant difference between these methods (Table 2.4-5).

Table 2.4-3. Estimation errors of the peak positions using the L-curve method

p	Model 1a				Model 1b			
	Normal noise		Pink noise		Normal noise		Pink noise	
	$\frac{\overline{\Delta\hat{x}_0}}{h}$	$\frac{\sigma(\Delta\hat{x}_0)}{h}$	$\frac{\overline{\Delta\hat{x}_0}}{h}$	$\frac{\sigma(\Delta\hat{x}_0)}{h}$	$\frac{\overline{\Delta\hat{x}_0}}{h}$	$\frac{\sigma(\Delta\hat{x}_0)}{h}$	$\frac{\overline{\Delta\hat{x}_0}}{h}$	$\frac{\sigma(\Delta\hat{x}_0)}{h}$
1	-0.08	1.79	0	2.24	-0.17	1.33	-0.15	1.64
	0.06	1.80	0.01	2.23	-10.7	18.1	-11.0	18.5
2	-0.21	0.87	-0.14	1.22	-0.19	0.66	-0.16	0.82
	0.21	0.87	0.13	1.22	-13.7	19.4	-14.6	19.8
3	-0.26	0.70	-0.19	0.99	-0.21	0.59	-0.16	0.70
	0.26	0.70	0.19	0.99	-12.8	19.0	-14.3	19.5
4	-0.28	0.65	-0.25	0.93	-0.23	0.57	-0.17	0.65
	0.28	0.66	0.26	0.91	-12.0	18.6	-13.9	19.4
5	-0.31	0.63	-0.25	0.87	-0.23	0.56	-0.16	0.64
	0.30	0.63	0.24	0.87	-11.5	18.4	-14.3	19.4
6	-0.32	0.62	-0.27	0.85	-0.25	0.57	-0.16	0.62
	0.32	0.62	0.27	0.85	-11.8	18.6	-13.8	19.1

Upper and lower rows represent $\overline{\Delta\hat{x}_0/h} = \left(\overline{\hat{x}_{01}} \pm x_{01T}\right)/h$, and $\sigma(\hat{x}_0)/h = std(\hat{x}_0)/h$ for the first and the second peak, respectively. 10^4 repetitions.

Smoothing of the anthracene absorption spectrum

The anthracene absorption spectrum, which was taken from the open database PhotochemCAD 3 [12, 13], was smoothed using the functional (Eq. (2.4-4), $p = 2$, $\alpha = 6 \times 10^{-9}$) (Fig. 2.4-6). The relative absolute difference between the smoothed and non-smoothed spectra was less than 5% (Fig. 2.4-6). The mean and the standard deviation of the difference were -4.43×10^{-9} and 0.0020, respectively. Therefore, we consider this difference to be a quasi-noise in the anthracene spectrum. The Fourier spectrum of this noise shows that the noise is not white (Fig. 2.4-7). The harmonic amplitudes are an increasing function of the frequency.

Table 2.4-4. Estimation errors of the peak amplitudes using the L-curve method

p	Model 1a				Model 1b			
	Normal noise		Pink noise		Normal noise		Pink noise	
	$\frac{\overline{\Delta\hat{A}}}{A_T}$, %	$\sigma(\Delta\hat{A})$, %	$\frac{\overline{\Delta\hat{A}}}{A_T}$, %	$\sigma(\Delta\hat{A})$, %	$\frac{\overline{\Delta\hat{A}}}{A_T}$, %	$\sigma(\Delta\hat{A})$, %	$\frac{\overline{\Delta\hat{A}}}{A_T}$, %	$\sigma(\Delta\hat{A})$, %
1	-0.11	0.65	0.11	1.35	-0.41	0.65	0.26	1.33
	-0.11	0.65	0.11	1.33	1.50	1.08	2.76	2.23
2	-0.070	0.54	-0.12	1.22	-0.17	0.51	-0.28	1.21
	-0.076	0.54	-0.12	1.24	0.76	0.91	1.76	2.06
3	-0.039	0.49	-0.088	1.19	-0.11	0.47	-0.23	1.17
	-0.043	0.49	-0.072	1.19	0.56	0.83	1.44	1.97
4	-0.023	0.47	-0.046	1.19	-0.07	0.45	-0.22	1.13
	-0.029	0.47	-0.060	1.17	0.50	0.81	1.38	1.99
5	-0.020	0.45	-0.097	1.18	-0.05	0.43	-0.22	1.13
	-0.012	0.46	-0.072	1.17	0.46	0.80	1.28	1.95
6	-0.007	0.45	-0.080	1.15	0.08	1.17	-0.20	1.12
	-0.017	0.45	-0.091	1.15	1.48	2.06	1.30	1.95

Upper and lower rows represent $\overline{\Delta\hat{A}}/A_T = \left(\bar{\hat{A}} - A_T\right)/A_T$ and $\sigma(\hat{A})/A_T = std(\hat{A})/A_T$ for the first and the second peak, respectively. 10^4 repetitions.

The smoothed (the quasi-noise-free (QNF)) anthracene spectrum was disturbed by normal noise (($\sigma = 0.05$), and further smoothed (Fig. 2.4-8). The normalization parameters' values, which were obtained by the functional and L-curve methods, are close to each other. Therefore, the smoothed spectra are similar. Therefore, the smoothed spectra are similar.

The smoothed spectra demonstrated small processing distortions of the spectral curve.

We estimated the confidence intervals of the peak position and the intensities of the FFT-smoothed QNF anthracene spectrum disturbed by normal noise (Table 2.4-6). This numerical experiment was similar to that of the doublet models. The FWHMs of the peaks are roughly$(35 - 40)\,h$. The maximum relative uncertainties in determining peak positions and peak intensities, using the smoothed spectrum, were approximately 15-16% of the peak width, and about 8% of the weak peak intensity.

The original anthracene spectrum contains low pseudo-random noise with $\sigma = 0.002$. Therefore, the table data, obtained for $\sigma = 0.01$, are useful for estimating the upper boundaries of the QNF anthracene spectrum's measurement uncertainties. So, we suppose that the absolute values of the peak positions' estimation errors are less than $2h$, and those of the intensities are smaller than 2%.

Table 2.4-5. FFT and SG smoothing of the model 1a

Peak	FTT			SG		
	$\Delta\hat{x}_0/h$	$\sigma(\Delta\hat{x}_0)/h$	Err, %	$\Delta\hat{x}_0/h$	$\sigma(\Delta\hat{x}_0)/h$	Err, %
Left	-0.28	0.32	0.66	0.32	0.87	1.41
Right	0.66	0.87		-0.32	0.88	

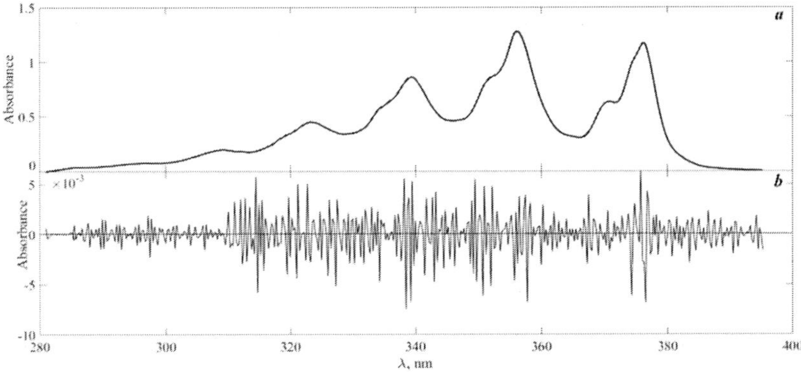

Figure 2.4-6. The quasi noise-free Anthracene spectrum (a) and the "noise" (b).

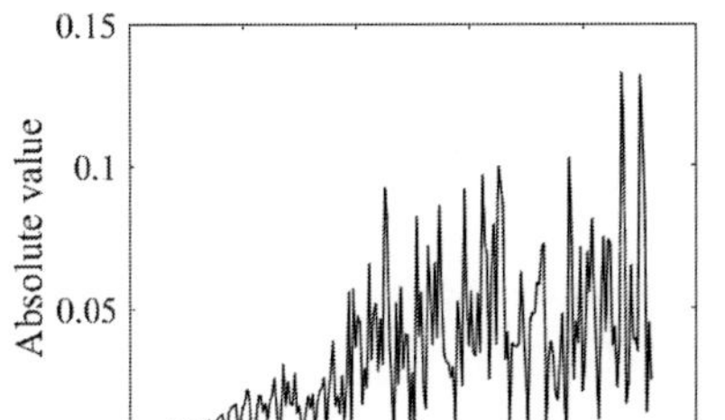

Figure 2.4-7. FFT of the "noise" (Fig. 2.4-6b).

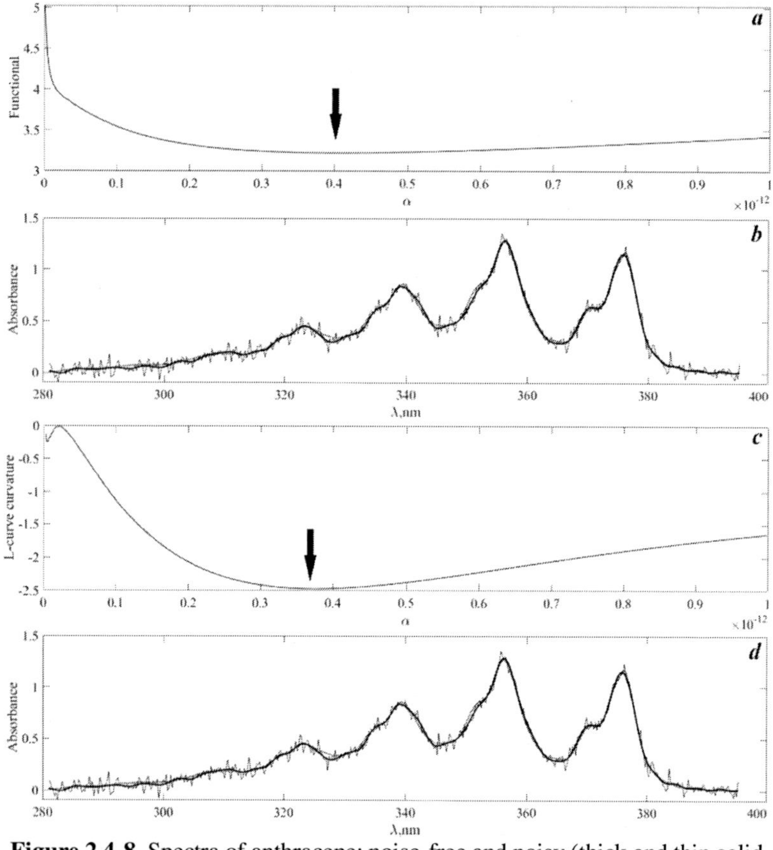

Figure 2.4-8. Spectra of anthracene: noise-free and noisy (thick and thin solid curves, respectively) ($\sigma = 0.05$) and smoothed (dotted curve) using the functional and L-curve methods (a, b and c, d panels, respectively). The arrows point to the optimal values of the regularization parameters.

Exercise 2.4-1

Readers are invited to study the FTT smoothing of the Gaussian peak corrupted by normal noise with mean zero and standard deviation σ. (Appendix SW2).

In some cases, the minimum of the functional $\psi(\alpha, p)$ is not observed.

1. How do the filter's parameters ($2p$ and α) depend on σ ?
2. Does the width of the peak and the length of its wings affect the smoothing results?

Table 2.4-6. The relative 95% confidence intervals (CI) of the estimated peak positions and intensities of the FFT-smoothed Anthracene spectrum using the L-curve method

Peak, nm	σ					
	0.01		0.02		0.05	
	$\frac{CI_{pos}}{h}$	CI_{ampl}, %	$\frac{CI_{pos}}{h}$	CI_{ampl}, %	$\frac{CI_{pos}}{h}$	CI_{ampl}, %
376	[-1.60, 0.025]	[-2.02, -0.47]	[-1.51, -0.39]	[-2.59 -0.07]	[-2.30, -0.06]	[-4.31, 1.65]
356	[-0.028, 0.028]	[-1.19, 0.40]	[-0.32, 0.37]	[-1.96, 0.68]	[-0.87, 1.25]	[-3.89, 1.99]
339	[-1.23, 0.62]	[-1.14, 0.78]	[-1.36, 0.68]	[-2.01, 1.51]	[-2.09, 1.43]	[-4.39, 3.93]
323	[-0.69, 2.30]	[-1.88, 1.55]	[-1.36, 3.08]	[-3.24, 3.20]	[-5.75, 3.69]	[-6.49, 8.19]

p=4. 10^4 repetitions.

FFT regularized differentiation

The FFT regularized n^{th}-order derivative is estimated by Eq.(2.4-2), which was modified by introducing the Fourier operator of the differentiation (Appendix G):

$$F^{(n)}(\lambda_i) = real\left(IFT\{(j\omega_k)^n \widetilde{U}(\omega_k)/[1 + M(\omega_k)]^2\}\right), \qquad (2.4-7)$$

where $j = \sqrt{-1}$.

We simplified the accurate Tikhonov regularized solution (Appendix G, Eq. (G-3)); the stabilizer $M(\omega_k)$ only included the maximum power of the frequency.

The quasi-optimal value of the regularization parameters was estimated by the functional

$$\psi(n, \alpha, p) = \sum_k \{(j\omega_k)^n M(\omega_k) \widetilde{U}(\omega_k)/[1 + M(\omega_k)]\}^2, \qquad (2.4-8)$$

similar to Eq. (2.4-4). However, the power was increased.

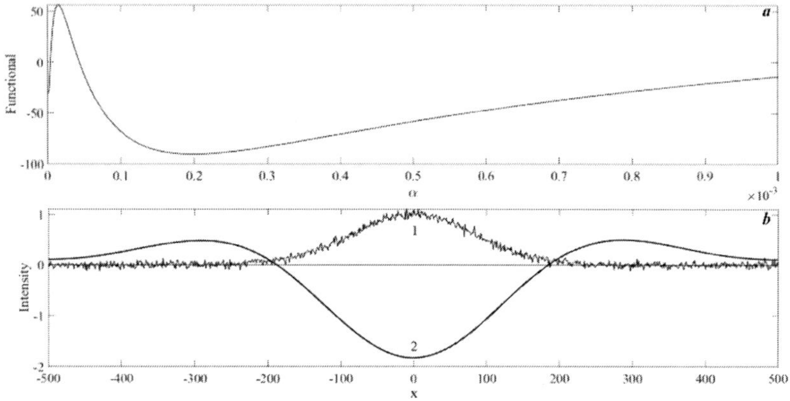

Figure 2.4-9. The functional (a) used for the quasi-optimal differentiation (b2) of the Gaussian peak (b1). $w = 200, \sigma = 0.05, 2p = 8$.

Exercise 2.4-2

Readers are invited to study the 1st- and 2nd-order FTT differentiation of the Gaussian peak corrupted by the normal noise with mean zero and standard deviation σ. (Appendix SW2).

In some cases, the minimum of the functional $\psi\,(n, \alpha, p)$ is not observed.

1. How do the filters parameters ($2p$ and α) depend on σ ?
2. How do the filters parameters depend on the differentiation order ?
3. Does the width of the peak and the length of its wings affect the differentiation results?

Concluding remarks

Suppose one has a noisy spectrum. This FFT- smoothed spectrum can be taken as a model for modelling experiments similar to those performed for Anthracene studies. This approach allows the estimation of the confidence intervals of the uncertainties when determining peak positions and intensities using smoothed spectra. The standard deviation of the noise may be calculated empirically from the noisy spectrum.

However, we prefer to decompose a noisy spectrum to its elementary components and estimate all of the peak parameters [8]. Despite some technical problems, such as the initial peak parameters estimation, this method eliminates all disturbances caused by smoothing and peak overlapping to a large extent. The confidence intervals of the peak parameters, which were obtained by decomposition, establish the ranges of possible accurate parameters.

CHAPTER FIVE

A HYPOTHETICALLY OPTIMAL NON-RECURSIVE SMOOTHING FILTER

This chapter is a slightly modified open access article [1]. We developed a hypothetically optimal non-recursive digital smoothing filter, which minimizes the sum of the squared norms of the residuals and the smoothed noise. It has been proven that (1) the optimal smoothing filter of constant and linear signals is the moving average filter; (2) Savitzky-Goley and Fourier filters have near-optimal smoothing properties for white and pink noises.

Introduction

The most popular smoothing filters used in analytical applications are the Savitzky-Golay polynomial filters (SG) [2]. These least-squares filters, which are well-known in mathematics, have a very long history [3, 4]. The computation of the SG filters was optimized [5-8], and their filtering properties were studied [5, 9-11]. The polynomial smoothing filters were also improved by Tikhonov (ridge) regularization [12-14] (the old version was named as "A method of graduation based on probability" [3] or "penalized least squares"). Chebyshev filters based on the minimax criterion were used in spectroscopy [15, 16]. The classical Finite Impulse Response (FIR) filters [17, 18] and adaptive techniques [19, 20] were taken from Digital Signal Processing [21].

In parallel, spline [22], Fourier [23-25], and wavelet smoothing [26, 27] methods were intensively applied to the analytical measurements. The matrix method, which is based on singular value decomposition, was suggested by [28, 29]. The "exotic" Kahrunen-Loève Transform was used by [30]. In a more sophisticated method [31], the noise data structure was taken into account for the implementation of the smoothing procedure.

Smoothing is an essential part of the pre-processing in the multivariate calibration [32]. The possible pitfalls of applying this procedure [33] are theoretically related to the smoothing in statistics [34].

There is no perfect smoothing method applicable to all numerical data since the quality of the processing depends on the form of the noisy signal, the type of noise, and the signal-to-noise ratio. Efficient smoothing is an ill-posed inverse problem [35], which requires additional information about the signal and the noise for choosing the quasi-optimal parameters of the smoothing procedure, e.g., the polynomial power and the length of polynomial filters [6], the regularization parameter for the penalized least squares [14] and Fourier smoothing [24], and the cut-off frequency of the FIR filters [20], etc. However, the parameters that define the smoothing quality are based on some criterion (a global minimum of the objective function [36]). Generally, the objective function is defined as the weighted sum of the goodness-of-fit measure and the "roughness" of the smoothed signals [13-15]. The least-squares goodness-of-fit measure is equal to the squared Euclidean norm of the discrepancies between the measured and the smoothed signal. The weighting factor is the regularization parameter. In the partial case of the SG filter, the norm is defined on the piece of data which is equal in length to the filter length, and the regularization parameter is zero [2].

Choosing the quasi-optimal parameters of smoothing algorithms was comprehensively discussed and illustrated by numerous examples. However, the analysis of the published materials clearly shows that, usually, the visual quality of the smoothed noisy signals has been considered as "the best proof" of the smoothing quality of the numerical algorithm (e.g., [29]).

Suppose that the quasi-optimal parameters of a given digital smoothing method, applied to a given noisy signal, have been well established. Then the following question arises: "How close is the smoothing quality to the theoretically best quality obtained by a hypothetically perfect method of smoothing?" This chapter will now provide an answer to the question.

Theory

Let the measured analytical signal be

$$\boldsymbol{F}_m = \boldsymbol{F} + \boldsymbol{\eta}, \qquad (2.5-1)$$

where the components of the vector $\boldsymbol{F} : f_i;\ i = 1,2, \dots, L$ are obtained by sampling the uniform grid where step h of the x-axis; $\boldsymbol{\eta} = \{\eta_i\}$ ise normal noise (independent on $\boldsymbol{F}$) with a mean of zero; the vector $\boldsymbol{\eta}$ components (η_i) are uncorrelated and have the same standard deviation σ.

Suppose that $\boldsymbol{F}_m$ is smoothed by the linear symmetrical non-recursive digital filter. The vector of the filter coefficients $\boldsymbol{w} = \{w_j\}, j = 1,2,\ldots;$ $N = 2m + 1$, and $w_k = w_{N-k+1}, k = 1,2,\ldots,m$.

Our goal is to find the optimal set of N coefficients w_j, which minimize the object function:

$$\theta = \|\boldsymbol{F}_{red} - \boldsymbol{F}_m \boldsymbol{W}\|^2, \qquad (2.5-2)$$

where $\|\ \|$ is the Euclidean norm; $\boldsymbol{F}_{red}$ is defined in the range: $(m + 1, m + 2, \ldots, L - m)$;

$$\boldsymbol{W} = \begin{vmatrix} w_1 & 0 & \cdots & 0 \\ w_2 & w_1 & \cdots & \vdots \\ \vdots & w_2 & \cdots & 0 \\ w_N & \vdots & \cdots & w_1 \\ 0 & w_N & \cdots & w_2 \\ \vdots & \vdots & \cdots & \vdots \\ 0 & 0 & \cdots & w_N \end{vmatrix}.$$

By substituting Eq. (2.5-1) into Eq. (2.5-2) and ignoring the term $2(\boldsymbol{F}_{red} - \boldsymbol{FW})^T \boldsymbol{\eta W}$, whose average value is negligible compared to other terms, we have

$$\theta = \|\boldsymbol{F}_{red} - \boldsymbol{FW}\|^2 + \|\boldsymbol{\eta W}\|^2. \qquad (2.5-3)$$

Since the average valuc $\overline{\eta_\iota \eta_J} = \begin{cases} \sigma^2, i = j \\ 0, i \neq j \end{cases}$,

$$\|\boldsymbol{\eta W}\|^2 = \boldsymbol{\Phi}_\eta \boldsymbol{w}^T \boldsymbol{w} \boldsymbol{\Phi}_\eta^T = (L - N + 1)\sigma^2 \|\boldsymbol{w}\|^2, \qquad (2.5-4)$$

where $\boldsymbol{\Phi}_\eta = \begin{vmatrix} \eta_1 & \eta_2 & \cdots & \eta_N \\ \eta_2 & \eta_3 & \cdots & \eta_{N+1} \\ \vdots & \vdots & \vdots & \vdots \\ \eta_{L-N+1} & \eta_{L-N+2} & \cdots & \eta_L \end{vmatrix}$.

In what follows, we suppose that in the last and the first m points $f_i << \sigma$.

The column vector of the derivatives of θ over the coefficients w_j:

$$\boldsymbol{\theta}'_w = -2\boldsymbol{\Phi}_F^T(\boldsymbol{F}_{red} - \boldsymbol{FW})^T + 2(L - N + 1)\sigma^2 \boldsymbol{w}^T \boldsymbol{I}, \qquad (2.5-5)$$

where matrix $\boldsymbol{\Phi}_F$ has the structure of the matrix $\boldsymbol{\Phi}_\eta$; $\boldsymbol{I}$ is the identity matrix (N x N).

Since all elements of the 2nd-order derivative vector $\boldsymbol{\theta}''_w$ are positive, the zeroes of the $\boldsymbol{\theta}'_w$ correspond to the minimum of the objective function over the filter coefficients.

By substituting $\boldsymbol{FW} = \boldsymbol{\Phi}_F \boldsymbol{w}^T$ into Eq. (2.5-5), which represents the zero vectors in the minimum of the objective function, we have

$$(\boldsymbol{\Phi}_F^T \boldsymbol{\Phi}_F + (L - N + 1)\sigma^2 \boldsymbol{I})\boldsymbol{w}^T = \boldsymbol{\Phi}_F^T \boldsymbol{F}_{red}^T. \qquad (2.5-6)$$

If $\sigma = 0$, then the trivial solution of Eq. (2.5-6):

$w_i = \begin{cases} 1, i = m+1 \\ 0, i \neq m+1 \end{cases}$ has no practical interest.

Generally, the solution to Eq. (2.5-6) is

$$\boldsymbol{w}^T = (\boldsymbol{\Phi}_F^T \boldsymbol{\Phi}_F + (L - N + 1)\sigma^2 \boldsymbol{I})^{-1} \boldsymbol{\Phi}_F^T \boldsymbol{F}_{red}^T. \qquad (2.5-7)$$

Matrix $\boldsymbol{\Phi}_F^T \boldsymbol{\Phi}_F$ is badly conditioned. Matrix $(L - N + 1)\sigma^2 \boldsymbol{I}$ is the regularization term in the matrix inverse.

Eq. (2.5-7) shows that the coefficients of the optimal filter are the convolution of the true analytical signal with the matrix kernel $(\boldsymbol{\Phi}_F^T \boldsymbol{\Phi}_F + (L - N + 1)\sigma^2 \boldsymbol{I})^{-1} \boldsymbol{\Phi}_F^T$. The kernel depends on the analytical signal and on the dispersion of the noise. The convolution is performed in the boundaries of the filter width.

Eq. (2.5-7) was analytically solved only in some particular cases.

For $f_i = 1$ $(i = 1,..,L)$, we used the elegant inversion procedure suggested by J. Falta [37]:

$$(\boldsymbol{\Phi}_F^T \boldsymbol{\Phi}_F/(L - N + 1) + \sigma^2 \boldsymbol{I})^{-1} =$$

$$1/[\sigma^2(N + \sigma^2)] \begin{vmatrix} N - 1 + \sigma^2 & -1 & \cdots & -1 \\ -1 & N - 1 + \sigma^2 & \cdots & -1 \\ \vdots & \vdots & \vdots & \vdots \\ -1 & -1 & \cdots & N - 1 + \sigma^2 \end{vmatrix}.$$

Substituting the last equation into Eq. (2.5-7), we have

$$\boldsymbol{w} = 1/(N + \sigma^2)|1, \quad 1, \quad \cdots \quad 1|. \qquad (2.5-8)$$

Usually σ^2 is negligible compared to N; therefore, the filter (Eq. (2.5-8)) is the moving average.

J. Domsta obtained the analytical expression of the inverse matrix (Eq. (2.5-7)) in the linear case $f_i = f_0 + hi$ [37] (see Appendix H).

Computer modelling

The model was the symmetrical Gaussian doublet (Fig. 2.5-1) corrupted by noise:

$$\boldsymbol{F}_G = exp[-4ln2(\boldsymbol{x} + d)^2] + exp[-4ln2(\boldsymbol{x} - d)^2], \qquad (2.5-9)$$

where the components of vector $\boldsymbol{x}$ were taken uniformly in the range [-6, 6] with step 0.01; $d = 0.47$ is half of the separation of the peaks. To completely eliminate the effect of the loss of the start and the endpoints while digital smoothing, the object function θ (Eq. (2.5-3)) was estimated in the range [-1.5, 1.5]. The length of the reduced range was L_{red} =301 points. At the edges of this interval, the doublet's intensity was less than 0.2% of its maximum value. So, the doublet wings plunged into the noise with the standard deviation σ =0.005, 0.01, 0.02 and 0.05.

The signal was smoothed in the Fourier domain:

$$\boldsymbol{F}_{G_{sm}} = real\{IFT[FT(\boldsymbol{F}_G) * Filter(\omega)]\}, \qquad (2.5-10)$$

where $real$ is a real part of Fourier transform (FT); IFT is the inverse FT; $Filter(\omega) = 1/(1 + \alpha\omega^6)$; α is the regularization parameter; ω is the angular frequency.

Pink noise was generated by the Matlab program [38]. The same noisy signal was separately smoothed by the SG, optimal, and Fourier filters. Processing with new regenerated noise was repeated 20 times. For each smoothing method, the average error $\varepsilon = \sqrt{\theta/L_{red}}$ and its standard deviation (σ_ε) were estimated. Due to the same noise component, the errors obtained by these methods were strongly correlated. Therefore, the three σ_ε values were very close to each other. The radius of the 95% confidence interval of ε ($\Delta\varepsilon$) was calculated using the maximum σ_ε. The error dependencies ($\varepsilon(m)$) for the SG and optimal filters, and $\varepsilon(\alpha)$ for the Fourier method were estimated.

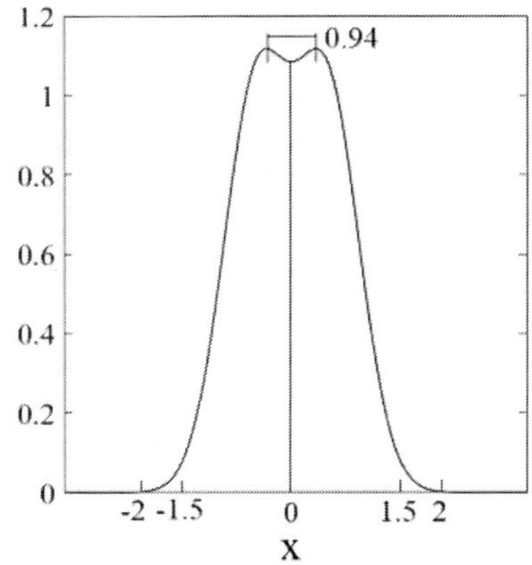

Figure 2.5-1. The Gaussian doublet

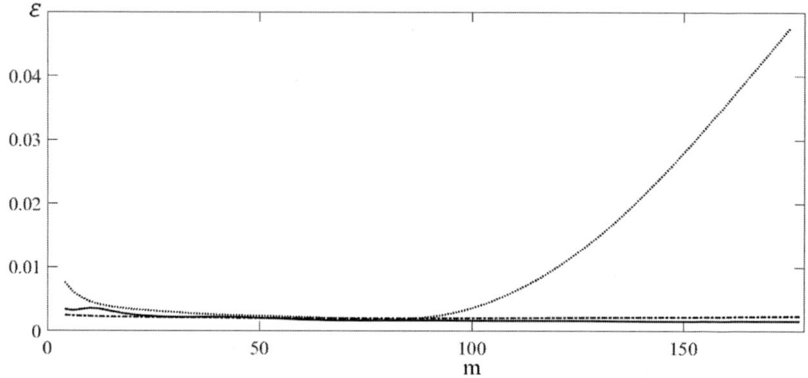

Figure 2.5-2. The error dependencies for the optimal, SG (power=6) and FT smoothing filters (solid, dotted, and dash-dotted curves, respectively. $\sigma = 0.01$.

Results and Discussion

The data, given in Figure 2.5-2 and Table 2.5-1 allows us to draw the following conclusion:

1. If the SG filter's power increases, then the error ε_{min} reduces, but the filter width (m_{opt}) increases [6].

2. For the optimal filter, $\varepsilon_{min_{opt}} \leq \varepsilon_{min}$. This inequality is valid for other filters, including the non-recursive ones [13, 15, 16]. We found that this inequality is valid for a large set of simulated and real noisy spectra.

3. For the pink noise, $\varepsilon_{min_{pink}}$ is significantly larger than that for the white noise $\varepsilon_{min_{white}}$.

4. The error $\varepsilon_{min_{opt}}$ is approximately proportional to the standard deviation of the noise σ.

5. If the optimal filter's width increases, its frequency characteristic becomes closer to the rectangle (Fig. 2.5-3), whose Fourier transform (the impulse characteristic of the filter) is $sinc(x) = sin(x)/x$.

Table 2.5-1. Optimal parameters of the smoothing filters using for denoising the Gaussian doublet

Noise	σ x 10^3	Filter				
		SG $2t$			Optimal	Fourier
		2	4	6		
white		ε_{min}, x 10^4				
	5	12	10	9.0	***8.0***	10.0
	10	22	19	18	***16***	19.0
	20	42	39	37	***34***	38
	50	85	80	77	***70***	79
		m_{opt}				α_{opt} x 10^9
	5	22	43	68	***292***	2.6
	10	24	49	74	***302***	4.8
	20	30	55	82	***290***	9.5
	50	38	67	96	***287***	28
pink		ε_{min}, x 10^4				
	5	33	31	31	***31***	31
	10	62	61	60	***57***	61
	20	133	130	129	***129***	130
	50	306	292	291	***291***	292
		m_{opt}				α_{opt} x 10^9
	5	26	49	76	***325***	5.8
	10	30	55	82	***262***	11
	20	38	65	94	***160***	25
	50	46	79	110	***141***	40

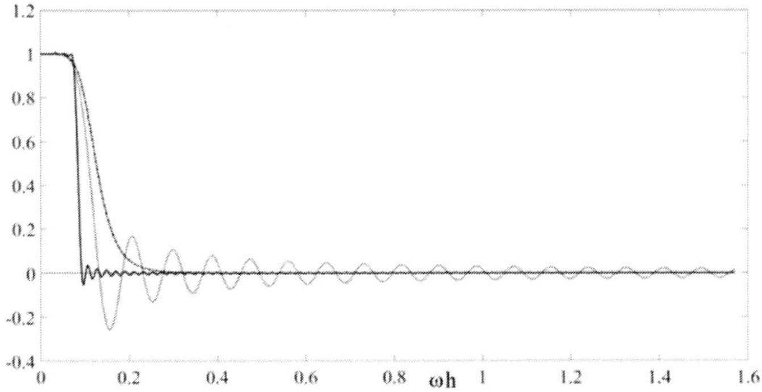

Figure 2.5-3. Frequency characteristics of the "best" smoothing filters: optimal ($m = 302$), SG (power=6, $m = 74$), and FT ($\alpha = 5\text{x}10^{-9}$): (solid dotted, and dash-dotted curves, respectively). $\sigma = 0.01$.

Conclusion

Numerical data, obtained by computer modelling, showed that the SG and Fourier filters have near-optimal smoothing properties. However, the optimal parameters of these filters (in the chapter's context) depend on the "pure" unknown measurement signal (free of noise) and the noise properties.

Generally speaking, since the solution to the optimal smoothing problem contains so many unknowns, it requires the correct initial conditions for further discussion: the optimal smoothing criteria, the signal which must be smoothed, and the noise characteristics.

Acknowledgments

The author is grateful to Prof. J. Domsta (University of Applied Sciences, Elblag, Poland) and to J. Falta (Czech Technical University, Prague) for solving some mathematical problems discussed in this chapter.

CHAPTER SIX

SMOOTHING AND DIFFERENTIATION USING A REGULARIZED CONTINUOUS FOURIER TRANSFORM

In this chapter, we continue the theoretical estimating of the pseudo-optimal parameters the differentiation and smoothing. However, the calculations are performed using a regularized continuous Fourier transform (CFT). The error data, obtained by CFT, are compared with the estimations performed using the Savitzky-Goley (SG) filters (Chapter 2.3) and the discrete FT (DFT) (Chapter 1.4).

Let us consider the Gaussian and the Lorentzian model peaks with unit maximum intensities:

$$F_{G0}(\nu) = \exp[-4ln2(\nu/w)^2], \qquad (2.6-1)$$

$$F_{L0}(\nu) = 1/[1+4(\nu/w)^2]. \qquad (2.6-2)$$

The integral FTs of these peaks in the frequency domain are

$$\tilde{F}_{G0}(x) = wC_G f_G(x), \qquad (2.6-3)$$

$$\tilde{F}_{L0}(x) = wC_L f_L(x), \qquad (2.6-4)$$

where $C_G = \sqrt{\pi/ln2}/2$; $f_G = exp(-x^2/16ln2)$; $x = \omega w$; ω is Fourier (angular) frequency; $f_L = exp(-0.5|x|)$; $C_L = \pi/2$.

Suppose that a noise corrupts the peaks. Then, the dispersion of the nth-order derivative (for smoothing $n = 0$) is [1]

$$\sigma_n^2 = (1/2\pi)\int_{-\infty}^{\infty} L(\omega)S(\omega)/[1+\alpha M(\omega)]^2\, d\omega, \qquad (2.6-5)$$

where, $L(\omega) = K(\omega)K(-\omega)$; $K(\omega) = (j\omega)^n$ is the operator of the precise differentiation; $j = \sqrt{-1}$; $S(\omega)$ is a spectral noise power; α is the regularization parameter; and $M(\omega) = \omega^{2p}$ is the stabilizer (Appendix G).

For the white noise with dispersion σ^2

$$S(\omega) = \sigma^2\tau, \qquad (2.6-6)$$

where $\tau = \pi/\omega_0$ is the time constant of the spectral instrument; ω_0 is the maximum Fourier frequency of the instrument's bandwidth. Therefore, the integration limits in Eq. (2.6-5) are $[-\omega_0, \omega_0]$. By substituting Eq.(2.6-6) into Eq.(2.6-5), we have

$$\sigma_n^2 = (\sigma^2/\omega_0)\int_0^{\omega_0} \omega^{2n}/[1+\alpha\omega^{2p}]^2\, d\omega. \qquad (2.6-7)$$

The systematic error of the nth-order derivative of the model peak, which Fourier transform is $\tilde{F}(\omega)$ is

$$\Delta_\nu = (1/2\pi)\int_{-\omega_0}^{\omega_0}\{\tilde{F}_K(\omega) - \tilde{F}_K(\omega)/[1+\alpha M(\omega)]\}\exp(-j\omega\nu)d\omega =$$
$$-(\alpha/2\pi)\int_{-\omega_0}^{\omega_0}\{\tilde{F}_K(\omega)\, M(\omega)/[1+\alpha M(\omega)]\}\exp(-j\omega\nu)\, d\omega, \quad (2.6-8)$$

where $\tilde{F}_K(\omega) = \tilde{F}(\omega)K(\omega)$.

The total error is

$$\Delta_{total}^{(n)}(\nu,\alpha,p) = \sqrt{\sigma_n^2 + {\Delta_\nu}^2}. \qquad (2.6-9)$$

Smoothing filters

We numerically estimated the $min\left\{\Delta_{total}^{(0)}(\nu,\alpha,p)\right\}$ over α and p in the peak maxima ($\nu = 0$). By substituting Eqs. (2.6-6)-(2.6-8) into Eq. (2.6-9), we have

$$\Delta_{total}^{(0)}(0,\alpha,p) = \sqrt{\frac{\sigma^2}{\omega_0}\int_0^{\omega_0}\frac{1}{(1+\alpha\omega^{2p})^2}d\omega + \left\{\frac{1}{\pi}\int_0^{\omega_0}\tilde{F}(\omega)\frac{\alpha\omega^{2p}}{1+\alpha\omega^{2p}}d\omega\right\}^2}. \quad (2.6-10)$$

Since the peak maximum is the unit, the relative total error is

$$\varepsilon_{total}^{(0)} = \Delta_{total}^{(0)}(0,\alpha,p). \qquad (2.6-11)$$

By substituting a dimensionless variable $x = w\omega$ into Eq. (19) and using Eqs. (2.6-10) and (2.6-11), we finally obtain

$$\varepsilon_{total}^{(0)} = \sqrt{\frac{\sigma^2}{x_0}\int_0^{x_0}\frac{1}{(1+\gamma x^{2p})^2}dx + \left\{\frac{C}{\pi}\int_0^{x_0}f(x)\frac{\gamma x^{2p}}{1+\gamma x^{2p}}dx\right\}^2}, \quad (2.6-12)$$

where $x_0 = w\omega_0$, $\gamma = \alpha w^{-2p}$.

To compare the total errors, estimated by Eq. (2.6-12), with the errors of the optimal SG smoothing filters, we took the upper limit of the integral (Eq. (2.6-12)) with x_0 equal to the maximum value corresponding to the discrete SG filters:

$$x_0 = \pi w/h = \pi/H. \qquad (2.6-13)$$

Results and discussion

Tables 2.6-1 and 2.6-2 include some optimal parameters of the Fourier smoothing of the Gaussian and Lorentzian peaks. These data, illustrated by Figures 2.6-1-2.6-6, allowed us to draw the following conclusions.

1. The errors estimated for the Lorentzian model are more significant than those for the Gaussian. This conclusion is due to the slowly damping tail of the Fourier spectrum of the Lorentzian curve compared to the quickly fading Fourier spectrum of the Gaussian (Eqs. (2.6-3) and (2.6-4)).

2. Figure 2.6-3 demonstrates very weak parabolic dependences $\varepsilon(\sigma)$ ($\sigma = 1/\xi_0$), which are very close to the linear plots.

3. $\varepsilon \sim 1/(1+p^2)$ (Fig. 2.6-4). If p increases, then the Fourier filter $1/(1+\gamma x^{2p})$ (Eq. (2.6-11)) approaches the rectangle window (Fig.2.6-2). Therefore, when p is significant, further increasing of p has a small impact on the error.

4. There is the following precise approximation:

$$\varepsilon_{min}^{(0)}(H)_{\substack{p=const,\\ \xi_0=const}} = E_s^{(0)}\left(\sqrt{H}/\xi_0\right) + E_i^{(0)}. \qquad (2.6-14)$$

Appendix I includes Table 2.6-3 and 2.6-4, which contain the slope $E_s^{(0)}$ and the intercept $E_i^{(0)}$.

Eq. (2.6-14) shows that the slope of the dependence $\varepsilon_{min}^{(0)}(H)$ similar to that of the SG method (Chapter 2. 3) [2]:

$$\varepsilon_{min}^{(0)} = E_{2t1}^{(0)}\left(\sqrt{H}/\xi_0\right)^{(4t_1+4)/(4t_1+5)}. \qquad (2.6-15)$$

However, the intercept appears.

5. By minimizing Eq. (2.6-12), we obtained the following dependences of the optimal regularization parameter $\gamma_{opt}^{(0)}$:

$$\gamma_{opt}^{(0)} = \left(-K_s^{(0)}\right)_{\xi_0,H} p + \left(-K_i^{(0)}\right)_{\xi_0,H}. \qquad (2.6-16)$$

$$log\left(\gamma_{opt}^{(0)}\right) = \left(L_s^{(0)}\right)_{p,H}\log(\sigma) + \left(-L_i^{(0)}\right)_{p,H}. \qquad (2.6-17)$$

$$log\left(\gamma_{opt}^{(0)}\right) = \left(-M_c^{(0)}\right)_p\left(\sqrt{H}/\xi_0\right)^2 + \left(M_s^{(0)}\right)_p\sqrt{H}/\xi_0 + \left(-M_i^{(0)}\right)_p. \qquad (2.6-18)$$

Tables 2.6-5- 2.6-10 (Appendix I) represent all constants.

The errors obtained using the discrete and integral FT are not very different from each other (e.g., Tables 2.6-1 and 2.6-2, $\xi_0 = 100$).

6. Only for the large power p of the Fourier stabilizer, the errors, obtained by Fourier smoothing, are less than those obtained by the SG filters.

Table 2.6-1. The optimal parameters of the Fourier smoothing of the Gaussian peak

p	Parameters	ξ_0				
		10	**20**	**50**	**100**	**200**
2	$\varepsilon_{min}^{(0)}$, %	2.4531	1.3305	0.5910	**0.3195** **(0.2800)*** **(0.3215)****	0.1726
	$\gamma_{opt}^{(0)}$	1.00x 10^{-4}	5.16x 10^{-5}	2.20x 10^{-5}	1.18x 10^{-5}	6.30x 10^{-6}
3	$\varepsilon_{min}^{(0)}$, %	2.2401	1.1892	0.5129	**0.2710** **(0.2624)*** **(0.2753)****	0.1431
	$\gamma_{opt}^{(0)}$	3.23x 10^{-6}	1.53x 10^{-6}	6.03x 10^{-7}	3.076x 10^{-7}	1.59x 10^{-7}
4	$\varepsilon_{min}^{(0)}$, %	2.1628	1.1369	0.4836	**0.2527** **(0.2533)*** **(0.2571)****	0.1319
	γ	9.30x 10^{-8}	4.00x 10^{-8}	1.43x 10^{-8}	6.90x 10^{-9}	3.45x 10^{-9}
5	$\varepsilon_{min}^{(0)}$, %	2.1284	1.1131	0.4700	**0.2441** **(0.2479)*** **(0.2476)****	0.1266
	$\gamma_{opt}^{(0)}$	2.50x 10^{-9}	9.60x 10^{-10}	3.07x 10^{-10}	1.39x 10^{-10}	6.55x 10^{-11}
6	$\varepsilon_{min}^{(0)}$, %	2.1112	1.009	0.4629	**0.2396** **(0.2445)*** **(0.2430)****	0.1238
	$\gamma_{opt}^{(0)}$	6.50x 10^{-11}	2.18x 10^{-11}	6.11x 10^{-12}	2.55x 10^{-12}	1.13x 10^{-12}

$H = 0.02$, * SG filter, ** Discrete FT

Table 2.6-2. The optimal parameters of the Fourier smoothing of the Lorentzian peak

p	Parameters	ξ_0				
		10	**20**	**50**	**100**	**200**
2	$\varepsilon_{min}^{(0)}$, %,	2.7932	1.5297	0.6853	**0.3720 (0.3406)* (0.3715)****	0.2015
	$\gamma_{opt}^{(0)}$	3.83x 10^{-5}	1.78x 10^{-5}	7.00x 10^{-6}	3.55x 10^{-6}	1.85x 10^{-6}
3	$\varepsilon_{min}^{(0)}$, %	2.6373	1.4217	0.6227	**0.3319 (0.3280)* (0.3310)****	0.1764
	$\gamma_{opt}^{(0)}$	5.37x 10^{-7}	2.04x 10^{-7}	6.45 10^{-8}	2.90x 10^{-8}	1.36x 10^{-8}
4	$\varepsilon_{min}^{(0)}$, %	2.5905	1.3879	0.6026	**0.3189 (0.3221)* (0.3187)****	0.1681
	$\gamma_{opt}^{(0)}$	6.80x 10^{-9}	2.04x 10^{-9}	5.03x 10^{-10}	1.93x 10^{-10}	7.94x 10^{-11}
5	$\varepsilon_{min}^{(0)}$, %	2.5738	1.3749	0.5945	**0.3135 (0.3190)* (0.3126)****	0.1647
	$\gamma_{opt}^{(0)}$	8.00x 10^{-11}	1.90x 10^{-11}	3.60x 10^{-12}	1.20x 10^{-12}	4.10x 10^{-13}
6	$\varepsilon_{min}^{(0)}$, %	2.5679	1.3695	0.5908	**0.3110 (0.3170)* (0.3139)****	0.1631
	$\gamma_{opt}^{(0)}$	9.30x 10^{-13}	1.71x 10^{-13}	2.42x 10^{-14}	6.50x 10^{-15}	1.95x 10^{-15}

$H = 0.02$, * SG filter, ** Discrete FT

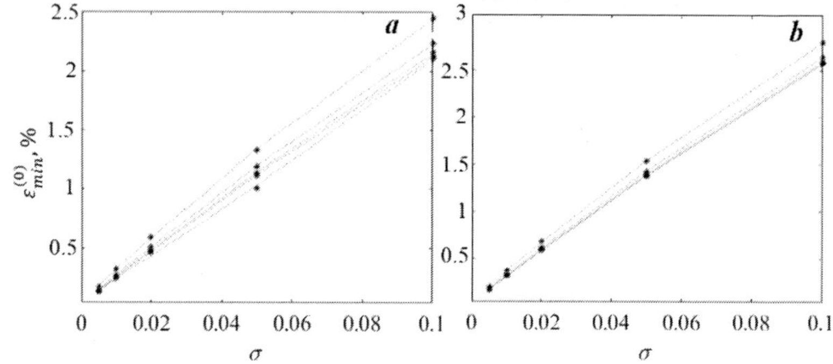

Figure 2.6-1. The error dependencies $\varepsilon_{min}^{(0)}$ (σ) for the Fourier smoothing of the Gaussian (a) and Lorentzian (b) peaks. $H = 0.02$.

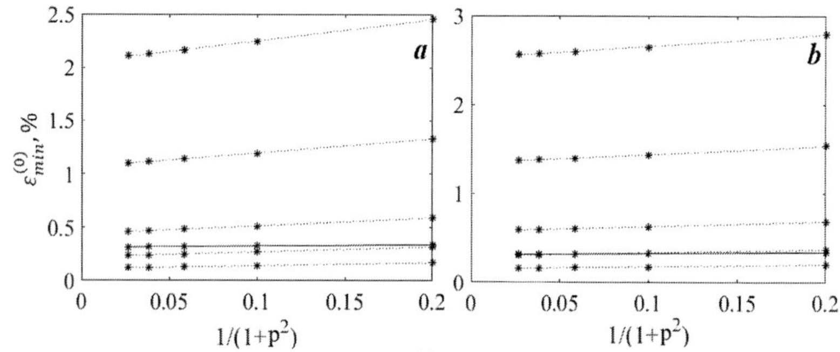

Figure 2.6-2. The error dependencies $\varepsilon_{min}^{(0)}$ (1/ (1+p^2)) for the Fourier smoothing of the Gaussian (a) and Lorentzian (b) peaks. σ =0.1, 0.05, 0.02, 0.01, 0.005 from the top to the bottom dotted curves. The SG data refer to the solid curve. (see Tables 2.6-1 and 2.6-2). $H = 0.02. \xi_0 = 100$.

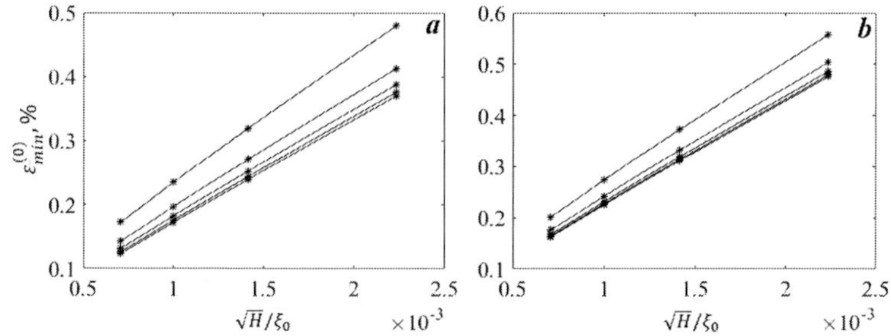

Figure 2.6-3. The error dependencies $\varepsilon_{min}^{(0)}$ $(\sqrt{H}/\xi_0)$ for the Fourier smoothing of the Gaussian (a) and Lorentzian (b) peaks. $p = 2 - 6$ from the top to the bottom curves. $\xi_0 = 100$.

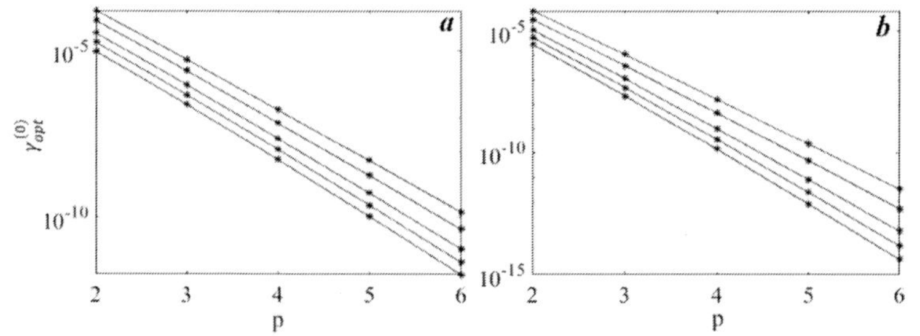

Figure 2.6-4. The dependencies $\gamma_{opt}^{(0)}$ (p) for the Fourier smoothing of the Gaussian (a) and Lorentzian (b) peaks. $\xi_0 = 10, 20, 50, 100,$ and 200 from the top to the bottom curves. $H = 0.05$.

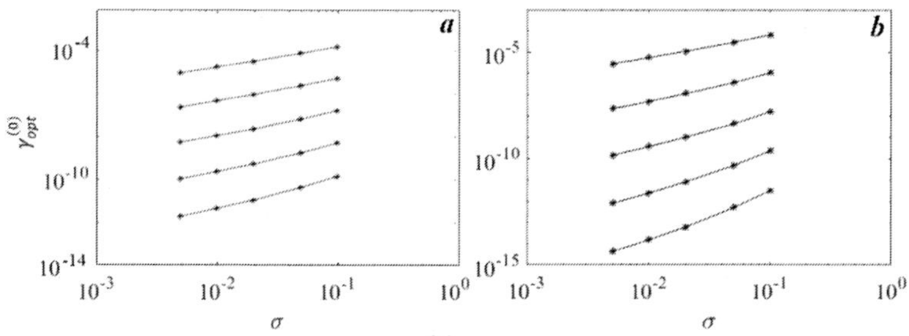

Figure 2.6-5. The dependencies $\gamma_{opt}^{(0)}$ (σ) for the Fourier smoothing of the Gaussian (a) and Lorentzian (b) peaks.$p = 2{:}\,6$ from the top to the bottom curves. $H = 0.05$.

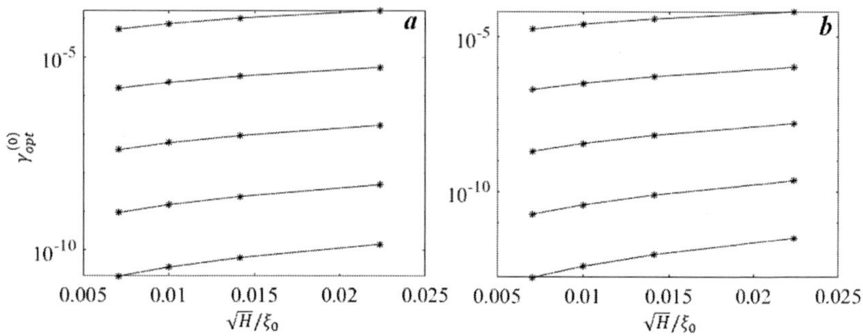

Figure 2.6-6. The dependencies $\gamma_{opt}^{(0)}$ $(\sqrt{H}/\xi_0)$ for the Fourier smoothing of the Gaussian (a) and Lorentzian (b) peaks. $p = 2{:}\,6$ from the top to the bottom curves.

Differentiation filters

We numerically estimated the $min\left\{\Delta_{total}^{(n)}(\nu,\alpha,p)\right\}$ over α and p in the extrema points $(\nu = 0)$ for the 2nd-order derivatives $(n = 2)$ of the Gaussian and Lorentzian peaks.

By substituting the operator of the 2nd-order differention $(-\omega^2)$ into Eq. (2.6-10), we have

$$\Delta_{total}^{(2)}(0,\alpha,p) = \sqrt{\frac{\sigma^2}{\omega_0}\int_0^{\omega_0}\frac{\omega^4}{[1+\alpha\omega^{2p}]^2}d\omega + \left(\frac{1}{\pi}\right)^2\left\{\int_0^{\omega_0}\left\{\tilde{F}(\omega)\frac{\omega^2\alpha\omega^{2p}}{[1+\alpha\omega^{2p}]}\right\}d\omega\right\}^2}\,. \quad (2.6-19)$$

By transforming the variable ω to the dimensionless variable ωw, similar to Eq. (2.6-12), we obtain

$$\Delta_{total}^{(2)} = \frac{1}{w^2}\sqrt{\frac{\sigma^2}{x_0}\int_0^{x_0}\frac{x^4}{(1+\gamma x^{2p})^2}dx + \left\{\frac{C}{\pi}\int_0^{x_0} f(x)\frac{\gamma x^{2p+2}}{1+\gamma x^{2p}}dx\right\}^2}\,, \quad (2.6-20)$$

where $x_0 = w\omega_0$, $\gamma = \alpha w^{-2p}$.

The total error relative to the absolute value of the derivative peak intensity in $\nu = 0$ measured from the zero-line is

$$\varepsilon_{total}^{(2)} = \Delta_{total}^{(2)}(0,\alpha,p)/\left(D^{(2)}/w^2\right), \quad (2.6-21)$$

where $D^{(2)} = 5.545w^2$ and $8w^2$ for the Gaussian and Lorentzian peaks, respectively.

Finally, we have

$$\varepsilon_{total}^{(2)} = \frac{1}{D^{(2)}}\sqrt{\frac{\sigma^2}{x_0}\int_0^{x_0}\frac{x^4}{(1+\gamma x^{2p})^2}dx + \left\{\frac{C}{\pi}\int_0^{x_0} f(x)\frac{\gamma x^{2p+2}}{1+\gamma x^{2p}}dx\right\}^2}. \quad (2.6-22)$$

Results and discussion

Tables 2.6-11 and 2.6-12 include some of the optimal parameters of the Fourier differentiation of the Gaussian and Lorentzian peaks, respectively. These data, illustrated by Figures 2.6-7-2.6.12, allowed us to conclude the following:

1. The errors estimated for the Lorentzian model are more significant than those for the Gaussian. This conclusion is due to the slowly damping tail of the Fourier spectrum of the Lorentzian peak compared to the quickly fading Fourier spectrum of the Gaussian peak (Eqs. (2.6 −3) and (2.6-4)).
2. Figure 2.6-7 demonstrates the linear dependences $\varepsilon(\sqrt{\sigma})$ $(\sigma = 1/\xi_0)$.
3. $\varepsilon \sim 1/(1+p^3)$ (Fig. 2.6-8). The power of p became larger than the one taken for the smoothing (Fig. 2.6-2).
4. If p increases, then the Fourier filter $1/(1+\gamma x^{2p})$ (Eq. (2.6-20)) approaches the rectangle window. Therefore, when p is sufficiently large, further increasing p has a small impact on the error.
5. There is the following precise approximation:

$$\varepsilon_{min}^{(2)}(H)_{\substack{p=const,\\ \xi_0=const}} = E_s^{(2)}\left(\sqrt{H}/\xi_0\right) + E_i^{(2)}, \quad (2.6-23)$$

where the slope $E_s^{(2)}$ and the intercept $E_i^{(2)}$ are given in Tables 2.6-13 and 2.6-14 (Appendix I).

Eq. (2.6-23) shows that the slope of the dependence $\varepsilon_{min}^{(2)}(H)$ is similar to Eq. (2.6-15), but the intercept appears.

6. The following dependences of the optimal regularization parameter $\gamma_{opt}^{(0)}$ are obtained by minimizing Eq. (2.6-22):

$$\gamma_{opt}^{(2)} = \left(-K_s^{(2)}\right)_{\xi_0,H} p + \left(-K_i^{(2)}\right)_{\xi_0,H}. \quad (2.6-24)$$

$$log\left(\gamma_{opt}^{(2)}\right) = \left(L_s^{(2)}\right)_{p,H}\log(\sigma) + \left(-L_i^{(2)}\right)_{p,H}. \quad (2.6-25)$$

$$log\left(\gamma_{opt}^{(2)}\right) = \left(-M_c^{(2)}\right)_p \left(\sqrt{H}/\xi_0\right)^2 + \left(M_s^{(2)}\right)_p \sqrt{H}/\xi_0 + \left(-M_i^{(2)}\right)_p. \quad (2.6-26)$$

All constants are given in Tables 2.6-(15-20) (Appendix I).

7. The errors obtained by Fourier differentiation, become less than those obtained by the SG filters but only for the large Fourier stabilizer's orders, (Tables 2.6-11 and 2.6-12).

Table 2.6-11. The optimal parameters of the Fourier differentiation of the Gaussian peak

p	Parameters	ξ_0				
		10	**20**	**50**	**100**	**200**
2	$\varepsilon^{(2)}_{min}$, %	12.7597	8.6380	5.0754	**3.3657 (2.945)***	2.2207
	$\gamma^{(2)}_{opt}$	2.530x 10^{-4}	1.518x 10^{-4}	8.020x 10^{-5}	5.046x 10^{-5}	3.212x 10^{-5}
3	$\varepsilon^{(2)}_{min}$, %	8.9024	5.6735	3.0749	**1.9171 (2.046)***	1.1887
	$\gamma^{(2)}_{opt}$	4.593x 10^{-6}	2.280x 10^{-6}	1.160x 10^{-6}	6.700x 10^{-7}	3.950x 10^{-7}
4	$\varepsilon^{(2)}_{min}$, %	7.6669	4.7370	2.4597	**1.4825 (1.676)***	0.8877
	$\gamma^{(2)}_{opt}$	9.5800x 10^{-8}	4.640x 10^{-8}	1.924x 10^{-8}	1.030x 10^{-8}	5.700x 10^{-9}
5	$\varepsilon^{(2)}_{min}$, %	7.1242	4.3247	2.1906	**1.2942 (1.484)***	0.7588
	$\gamma^{(2)}_{opt}$	2.040x 10^{-9}	8.850x 10^{-10}	3.200x 10^{-10}	1.610x 10^{-10}	8.300x 10^{-10}
6	$\varepsilon^{(2)}_{min}$, %	6.8468	4.1124	2.0513	**1.1969 (1.372)***	0.6926
	$\gamma^{(2)}_{opt}$	4.366x 10^{-11}	1.680x 10^{-11}	5.410x 10^{-12}	2.470x 10^{-12}	1.186x 10^{-12}

$H = 0.01$, * SG filter.

Table 2.6-12. The optimal parameters of the Fourier differentiation of the Lorentzian peak

p	Parameters	ξ_0				
		10	**20**	**50**	**100**	**200**
2	$\varepsilon_{min}^{(2)}$, %	21.0348	14.9831	9.3045	**6.3799 (5.643)***	4.3242
	$\gamma_{opt}^{(2)}$	7.156x 10^{-5}	3.870x 10^{-5}	1.825x 10^{-5}	1.072x 10^{-5}	6.450x 10^{-6}
3	$\varepsilon_{min}^{(2)}$, %	16.8271	11.4627	6.6952	**4.3752 (4.538)***	2.8220
	$\gamma_{opt}^{(2)}$	5.100x 10^{-7}	2.266x 10^{-7}	8.520x 10^{-8}	4.303x 10^{-8}	2.261x 10^{-8}
4	$\varepsilon_{min}^{(2)}$, %	15.5862	10.4155	5.9214	**3.7864 (4.098)***	2.3870
	$\gamma_{opt}^{(2)}$	4.160x 10^{-9}	1.500x 10^{-9}	4.400x 10^{-10}	1.910x 10^{-10}	8.700x 10^{-11}
5	$\varepsilon_{min}^{(2)}$, %	15.1006	9.9939	5.6041	**3.5436 (3.868)***	2.2078
	$\gamma_{opt}^{(2)}$	3.500x 10^{-11}	1.020x 10^{-11}	2.300x 10^{-12}	8.600x 10^{-13}	3.400x 10^{-13}
6	$\varepsilon_{min}^{(2)}$, %	14.8886	9.8004	5.4533	**3.4264 (3.741)***	2.1204
	$\gamma_{opt}^{(2)}$	3.000x 10^{-13}	7.000x 10^{-14}	1.200x 10^{-14}	3.800x 10^{-15}	1.280x 10^{-15}

$H = 0.01$, * SG filter.

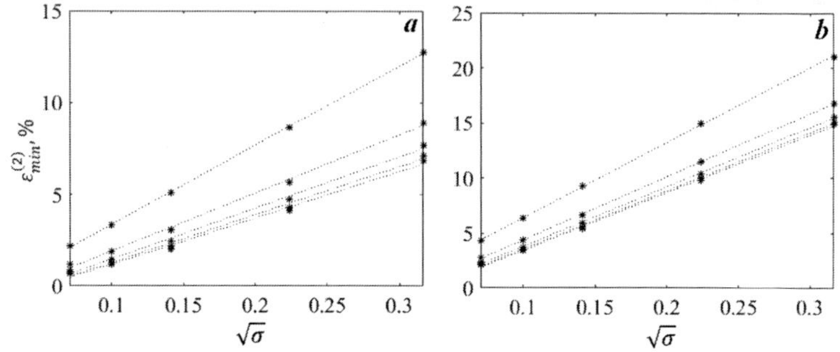

Figure 2.6-7. The errors of the 2nd-order Fourier differentiation of the Gaussian (a) and Lorentzian (b) peaks. $p = 2{:}\,6$ from the top to the bottom lines. $H = 0.01$.

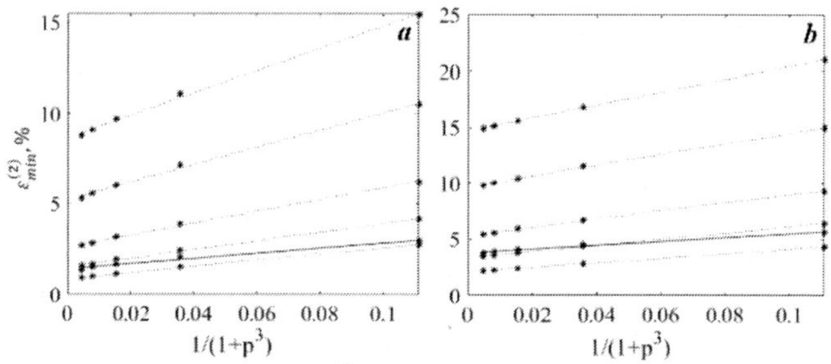

Figure 2.6-8. The errors of the 2nd-order Fourier differentiation of the Gaussian (a) and Lorentzian (b) peaks. $\sigma = 0.1, 0.05, 0.02, 0.01, 0.005$ from the top to the bottom lines. $H = 0.01$. $\xi_0 = 100$. The SG data are represented by the solid line.

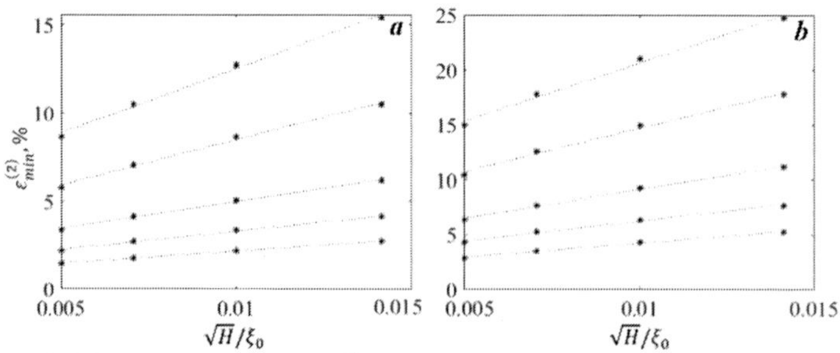

Figure 2.6-9. The errors of the 2nd-order Fourier differentiation of the Gaussian (a) and Lorentzian (b) peaks. $p = 2{:}\,6$ from the top to the bottom lines.

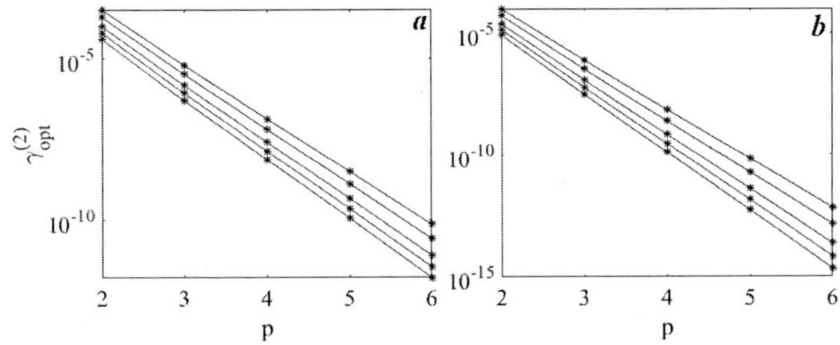

Figure 2.6-10. Dependencies of the optimal regularization parameter on p for the 2nd-order Fourier differentiation of the Gaussian (a) and Lorentzian (b) peaks. $\xi_0 = 10, 20, 50, 100$, and 200 from the top to the bottom lines. $H = 0.02$.

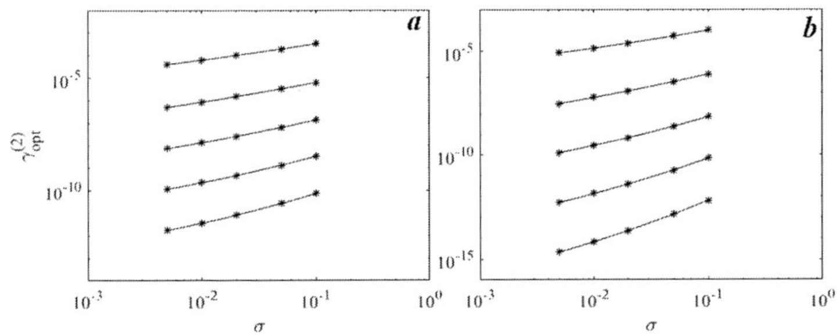

Figure 2.6-11. Dependencies of the optimal regularization parameter on σ for the 2nd-order Fourier differentiation of the Gaussian (a) and Lorentzian (b) peaks. $p = 2{:}\,6$ from the top to the bottom lines. $H = 0.02$.

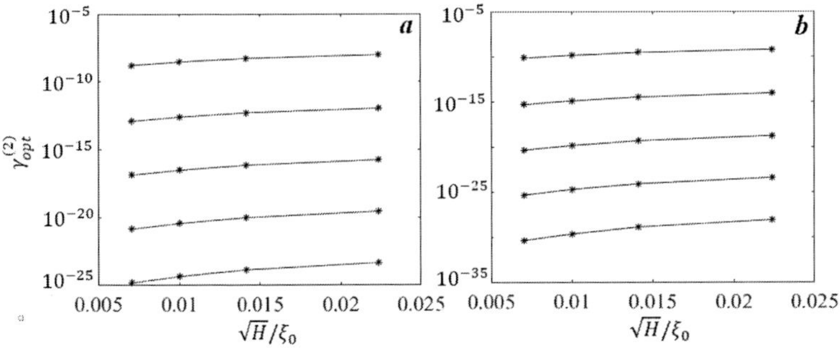

Figure 2.6-12. Dependencies of the optimal regularization parameter on $\sqrt{H}/\xi_0$ for the 2nd-order Fourier differentiation of the Gaussian (a) and Lorentzian (b) peaks. $p = 2{:}\,6$ from the top to the bottom lines.

Conclusion

In Chapter 2.5, we described a hypothetically optimal non-recursive digital smoothing filter, which minimizes the total error (the sum of the squared norms of the residuals, between accurate and smoothed signals, and of the smoothed noise). It has been proven theoretically that the optimal smoothing filter of constant and linear signals is the moving average filter. It was found that the SG and Fourier filters have near-optimal smoothing properties for white and pink noises.

The present chapter showed that the total errors of smoothing and the 2nd -order differentiation of the Gaussian and Lorentzian peaks, disturbed by the white noise, could slightly decrease using processing in the Fourier domain. However, in all cases, prior information about the peak's shape and width is needed. It is impossible to optimally process such data, as we do not have the exact information in advance! One can practically obtain a set of smoothed curves and their derivatives, as well as perform measurements of the parameters needed for analytical purposes. Only ranges of these parameters are reliable to some extent. This result is similar to the decomposition of the overlapping peaks [1].

CHAPTER SEVEN

SMOOTHING WITH SPLINES

Among the numerical procedures employed by analysts for smoothing and differentiation, the spline method is rarely used in practice [1-3]. The mathematical details and some applications of this method in spectroscopy are briefly discussed in Appendix A3. Here, we consider only some peculiarities of the method inherent to the processing of the analytical signals.

In the spline approximation, uniformly sampled data are "sewed" by piece-wised continuous polynomials. Some conditions of the continuity of the smoothed curve and its derivatives must be satisfied in the conjunctions (nodes).

The spline method's significant problems are the nodes selection, their distribution, and the polynomial spline's power. The quasi-optimal node selection criterion is the minimum mean square error of the spline fitting to the accurate data (or to the derivative). This selection is possible for the single peak of the known shape. However, for overlapped peaks that differ in intensities and widths, the optimization problem seems cumbersome. Usually, the nodes are condensed in the segment of the maximum peak curvatures (e.g., in the peak maxima, which are of the most interest). Therefore, the random errors increase.

Let us study the filtering properties of the cubic spline smoothing for the uniform nodes' distribution. Consider the frequency characteristic of this spline [3]:

$$G(\omega) = \frac{2}{h\omega^2}\left(\frac{sinz}{z}\right)^2\left[\frac{1 - cos2z}{\varphi + 2\mu cos2z + 2\beta cos4z}\right], \qquad (2.7-1)$$

where ω is the angular frequency; h is the sampling interval; $z = \omega h/2$; $\varphi = 2h/3 + 6\beta$; $\mu = h/6 + 4\beta$; $\beta = \alpha p/h^2$; α is the regularization parameter; p is the constant weighting factor (Appendix A, Eq. (A-3.6)). If $p = 1$, then Eq. (2.7-1) is readily transformed to

$$G(\omega) = 3\left(\frac{sinz}{z}\right)^4\left[\frac{1}{3 - 2sin^2z + \gamma sin^4z}\right], \qquad (2.7-2)$$

where $\gamma = 48\alpha/h^3$.

The low-frequency harmonics ($x = \omega h \ll 1$) contain most of the useful information (Appendix A2):

$$G_{Spl}(\omega) \cong 1/[1 + (\alpha/h^3)x^4]. \qquad (2.7-3)$$

The frequency characteristic of the smoothing Fourier filter corresponding to Eq. (2.7-3) (Chapter 2.4), is

$$G_{Four}(\omega) = 1/[1 + \alpha_F\omega^4] = 1/[1 + (\alpha_F/h^4)x^4]. \qquad (2.7-4)$$

If $\alpha = \alpha_F/h$, then Eqs. (2.7-3) and (2.7-4) become equal. So, the cubic smoothing spline's filtering properties at low frequencies are similar to those of the smoothing Fourier filter ($p = 2$). Figure 2.7-1 proves this conclusion.

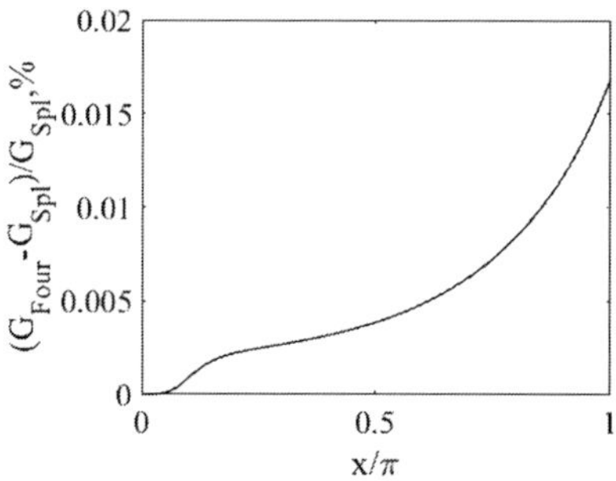

Figure 2.7-1. The relative difference of the frequency characteristics: $[G_{Four}(\omega) - G_{Spl}(\omega)]/G_{Spl}(\omega)$. $H = 0.02$; $\alpha_F = 10^{-5}$.

From the mathematical point of view, the nodes distribution with a constant interval is the simplest. However, this distribution produces false ripples similar to the SG differentiation [1].

Numerical experiment

Models

Three Gaussian doublets with dimensionless parameters (Fig. 2.7-2):

$$F_d = A_1\, exp\{-\beta[(x + x_0)/w_1]^2\} + A_2\, exp\{-\beta[(x - x_0)/w_2]^2\} + \eta, \quad (2.7-5)$$

where $\beta = 4ln2$; $x = [-4{:}\, 0.02{:}\, 4]$; η is the normal noise with mean zero and the standard deviation is $\sigma = 0.02$.

a. Symmetrical doublet: $A_1 = A_2 = 1$; $w_1 = w_2 = 1$; $x_0 = 0.52$. If the noise is absent, then the peak amplitudes $A_{1T} = A_{2T} = 1.0599$; and the peak positions $x_{01T} = -0.44, x_{02T} = 0.44$.

b. Asymmetrical doublet: $A_1 = 1, A_2 = 0.5$; $w_1 = 1, w_2 = 2$; $x_0 = 0.80$. If the noise is absent, then the peak amplitudes $A_{1T} = 1.0881, A_{2T} = 0.5008$; the peak positions $x_{01T} = -0.76, x_{02T} = 0.78$.

c.

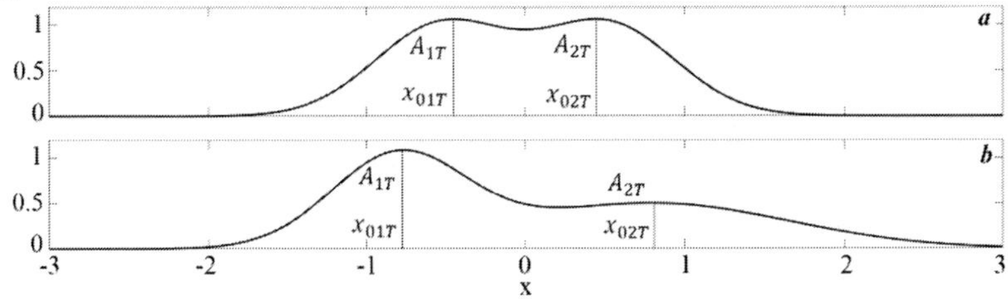

Figure 2.7-2. Gaussian model doublets.

Figure 2.7-3 represents the diagram of the experiment. For a given noisy doublet (Eq. (2.7-5)), the experiment involved the following steps in the program studySmoothSplines.m (Appendix SW-4):

1. Doublet generation (gaussDoublet.m).
2. Generation of the random breaks (knots) (randBreaksGen.m).
3. Fitting spline to the data (open source Matlab code: splinefit.m [4]).
4. Checking of the doublet resolution (existence of the minimum between the peak maxima).
5. Error estimation.

We estimated the mean total relative error of the 10^4 measurements of the peak intensities:

$$\delta = \sqrt{\delta_1^2 + \delta_1^2}, \qquad (2.7-6)$$

where the squared relative error $\delta_i^2 = [\Delta_i^2 + \sigma_i^2]/A_{iT}^2$; $i = 1,2$; $\Delta_i = A_{iT} - \hat{A}_{im}$ is the systematic error of the peak amplitude (a difference between the true value and the mean value); $\sigma_i = std(A_{im})$ is the random error (the standard deviation).

Table 2.7-1 includes the mean peak shifts from their accurate positions (x_{0T}) (when noise is absent): $\overline{\Delta\hat{x}_0/h} = (\overline{\hat{x}_0} - x_{0T})/h$ and the standard deviation $\sigma(\hat{x}_0)/h = std(\hat{x}_0)/h$. $x_{0T} \neq x_0$ due to the peak overlapping.

Table 2.7-2 represents the mean total relative errors (Eq. (2.7-6)) obtained for the asymmetrical doublet (model b).

Exercise 2.7-1

Readers are invited to calculate the following parameters: $\overline{\Delta\hat{x}_0/h}$ and $\sigma(\hat{x}_0)/h$ for model b, and compare their values with those given in Table 2.7-1.

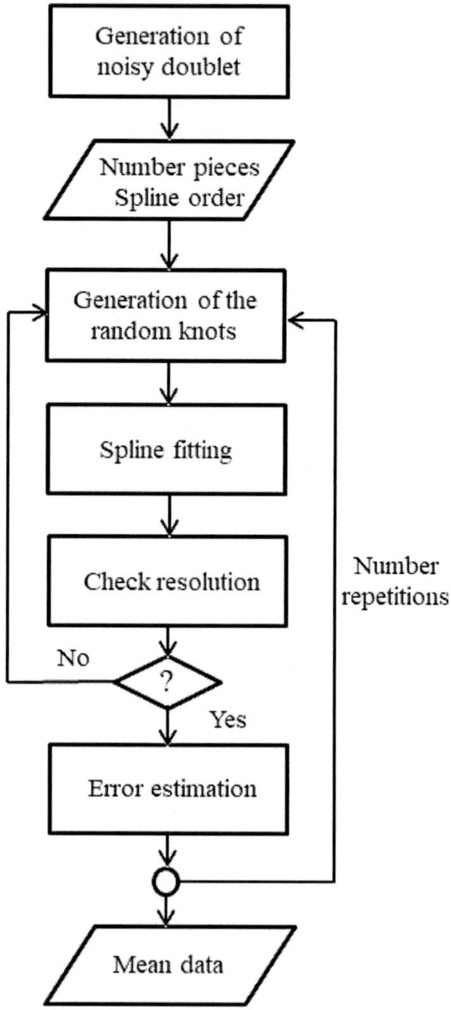

Figure 2.7-3. Diagram of the experiment.

Table 2.7-1. Symmetrical Gauss doublet (model a). Error data

Spline order	Parameter	Number pieces							
		20	30	40	50	60	70	80	90
4	$\delta,\%$	1.88	0.75	0.42	0.28	**0.28**	0.47	0.55	0.44
	$\Delta\hat{x}_{01}/h$	-1.5	-0.9	-1.5	-1.4	-1.3	-2.1	-2.9	-2.0
		1.1	0.6	0.5	0.5	0.4	0.4	0.3	0.1
	$\Delta\hat{x}_{02}/h$	0.89	0.0	-0.3	-	-1.4	-1.1	-1.7	-3.0
		1.0	0.6	0.5	0.4	0.7	0.3	0.5	0.2
5	$\delta,\%$	0.81	0.48	0.62	0.28	**0.19**	0.23	0.53	0.85
	$\Delta\hat{x}_{01}/h$	-0.1	-1.0	-1.5	-1.2	-1.7	-3.0	-2.0	-3.0
		0.8	0.2	0.5	0.4	0.5	0.1	0.1	0.1
	$\Delta\hat{x}_{02}/h$	-0.2	0.3	0.02	-0.9	-1.0	-1.3	-2.4	-2.7
		0.9	0.5	0.2	0.4	0.5	0.4	0.7	0.5
6	$\delta,\%$	1.00	0.71	0.51	0.22	**0.19**	0.37	0.33	1.02
	$\Delta\hat{x}_{01}/h$	-1.2	-1.0	-1.3	-1.0	-2.0	-2.0	-2.5	-
		0.5	0.1	0.5	0.1	0.2	0.2	0.5	0.03
	$\Delta\hat{x}_{02}/h$	0.5	0.01	-0.2	-0.5	-1.2	-1.0	-1.9	-1.7
		0.5	0.1	0.4	0.5	0.4	0.1	0.3	1.3
7	$\delta,\%$	0.48	0.48	0.55	0.24	0.16	**0.14**	0.73	0.65
	$\Delta\hat{x}_{01}/h$	-0.4	-1.0	-1.9	-	-1.7	-2.5	-2.1	-2.9
		0.5	0.0	0.3	0.04	0.4	0.5	0.3	0.3
	$\Delta\hat{x}_{02}/h$	0.2	0.1	0.0	-	-1.2	-1.1	-1.1	-3.0
		0.5	0.3	0.0	0.1	0.4	0.2	0.4	0.1
8	$\delta,\%$	0.68	0.72	0.65	0.21	**0.16**	0.37	0.32	1.02
	$\Delta\hat{x}_{01}/h$	-1.0	-1.0	-1.0	-1.0	-1.0	-2.1	-2.0	-2.0
		0.2	0.02	0.1	0.02	0.1	0.3	0.2	0.06
	$\Delta\hat{x}_{02}/h$	0.2	0.0	-0.0	-0.9	-2.0	-1.0	-2.0	-1.6
		0.4	0.0	0.1	0.3	0.1	0.1	0.1	0.7

The mean and the standard deviations of the $\Delta\hat{x}_{0i}/h$ are in the upper and lower rows, respectively.

Tabular data allow us to draw the following conclusions:

1. The mean total relative error (δ) (Eq. (2.7-6)) of the peak intensities measurements in a given doublet smoothed by the spline method depends on the spline order (t) and the number of the pieces (N_p). The functions $\delta_{N_p=const}(t)$ and $\delta_{t=const}(N_p)$ have minima (see bold, underlined numbers).
2. The errors for the asymmetrical doublet are more significant than those for the symmetrical one.
3. The uncertainties in determining the peak intensities increase, when the doublet is corrupted by the pink noise, compared to the distortions,

that caused by the white noise. This result is in full concordance with that obtained in the previous chapters.
4. The absolute errors when determining the peak positions are usually less than two-three sampling intervals.

Table 2.7-2. Asymmetrical Gauss doublet (model b). The mean total relative errors

Spline order	Number pieces							
	20	**30**	**40**	**50**	**60**	**70**	**80**	**90**
4	1.06	**0.62**	0.66	0.89	1.22	1.67	1.91	1.90
	1.77	**1.46**	1.80	1.80	2.19	2.68	3.13	3.18
5	0.94	**0.55**	0.80	1.56	1.28	2.22	2.36	2.53
	1.02	1.63	2.37	2.51	3.15	3.54	3.41	2.78
6	0.84	**0.59**	0.65	0.91	1.05	1.56	2.26	1.73
	1.09	1.58	1.82	1.80	2.66	2.94	3.46	3.68
7	1.02	**0.54**	0.78	1.43	1.06	1.76	2.26	2.46
	0.92	1.41	2.28	2.11	3.13	3.94	3.30	2.69
8	0.94	**0.60**	0.71	1.19	1.28	1.90	2.38	1.88
	0.96	1.68	2.00	1.95	2.22	2.91	3.54	3.64
9	1.06	**0.52**	0.77	1.66	0.91	1.57	2.32	2.24
	0.89	1.20	1.90	2.34	3.21	4.03	3.31	3.08

The data, obtained for the white and pink noises, are in the upper and lower rows, respectively.

Exercise 2.7-2

Readers are invited to plot the noisy doublets smoothed by the spline and Fourier methods (program plotSplineSmoothedDoublet.m, Appendix SW-4). For a given doublet, you can change: (a) the standard deviation of the noise, the spline order, and the breaks; (b) the power and the regularization parameter of the Fourier smoothing.

1. Pay attention to the fact that the fitting errors $ErrFit = \left\|F_d - F_d^{(sm)}\right\|$ of the spline and Fourier smoothing are different in each program run. Why? What is your conclusion about the reliability of a single numerical experiment?
2. Try to choose the Fourier filter parameters so, that the fitting errors of the spline and Fourier smoothing are close to each other.
3. How do the spline parameters affect the smoothed doublet?

CHAPTER EIGHT

DIFFERENTIATION USING WAVELETS

Signals produced by analytical instruments are, in essence, random. Therefore, their description by deterministic mathematical expressions is only an appropriate approximate mathematical model. Random signals contain stationary and nonstationary parts. The first part's statistical properties do not change with time, while those of the second one do. For example, the mathematical expectation (mean) of the stationary process does not depend on time. The covariation function of this process only depends on the time intervals.

The wavelet transform (WT) was invented to analyze time series (processes) which have a nonstationary Fourier spectrum [1]. The power of Fourier harmonics is time-dependent. The main problem of the Fourier Transform (FT) analysis of such processes was that, in the time domain, a signal does not provide any spectral information. However, Fourier harmonics include no data about their time localization. Also, the Fourier decomposition of the discontinuous functions that have singular points is poor due to the Gibbs phenomenon (Appendix A). In this case, the reason is the same: Fourier coefficients do not indicate the location of these points.

The short-time windowed FT, which was based on a moving segment, was an inaccurate and inefficient time-frequency localization method. J. Morlet made the revolutionary step of introducing a different window basis function, a wavelet, for analyzing various frequency bands [1]. Further, the wavelet transform has been dramatically improved and has been successfully applied in signal processing.

From 1989, WT was mainly employed in analytical chemistry for denoising and data compression in flow injection analysis; high-performance liquid chromatography; UF-VIS, IR, mass and NMR spectrometry; and voltammetry [2-10]. Chinese scientists have made a significant contribution to the development and application of this method [2-6].

Theory

The WT principle is the decomposition of a signal $F(t)$ into a set of wavelet basis functions ($\psi(t)$)). It is assumed that the signal is square-integrable. That is, the integral of its squared module is finite.

The continuous WT (CWT) [3] is

$$W_\psi\{F\} = C\int_{-\infty}^{\infty} F(t)\psi_{a,b}^*(t)dt, \qquad (2.8-1)$$

where $\psi_{a,b}(t) = (1/\sqrt{s})\psi[(t-d)/s]$; s and d are the scale and the dilation parameters, respectively; the asterisk designates complex conjugation. The normalizing constant C must satisfy the admissibility condition:$\int_{-\infty}^{\infty}|\psi(\omega)|^2/|\omega|d\omega < \infty$), where ω is the Fourier frequency.

A demonstrative example of a family of continuous wavelets is the set of the 1st-to 5th-order derivatives of the Gaussian function (Chapter 1.2) [5]:

$$\mathcal{F}_g^{(n)}(x) = (-1)^{n+1}\frac{d^n}{dx^n}\mathcal{F}_g^{(0)}(x) = K^{(n)}(x)\mathcal{F}_g^{(0)}(x), \qquad (2.8-2)$$

where $K^{(n)}(x) = -x, (1-x^2), (3x-x^3)\,, (6x^2-x^4-3)$, and $(10x^3-15x-x^5)$ for $n = 1-5$, respectively; $\mathcal{F}_g^{(0)}(x) = \exp(-x^2/2)$.

The normalizing coefficient of $\mathcal{F}_g^{(n)}(x)$ is $1/\sqrt{2\pi(n-1)!}$. The 2nd-order derivative is named the Mexican hat wavelet [11].

Figure 2.8-1 shows the real part of the Morlet wavelet [11]:

$$\psi(\eta) = C\exp(j\omega_0\eta)\exp(-\eta^2/2) \qquad (2.8-3)$$

and its FT [11],

$$\psi(\omega_1) = CH(\omega)exp[-(s\omega-\omega_0)^2/2], \qquad (2.8-4)$$

where $\eta = t/s$; $C = 1/\pi^4$; $\omega_1 = s\omega$; ω_0 is the dimensionless frequency (ω_0=6 satisfies the admissibility condition);

$$H(\omega) = \begin{cases} 1, \omega > 0 \\ 0, \ \omega \le 0 \end{cases}.$$

Similar to the discrete FT (Appendix A), there is a discrete variant of the CWT, which is called the discrete WT (DWT), in which the wavelets (Eq. (2.8-1)) are discretely sampled [12].

The mathematical tools used for the DWT are challenging to understand for non-professionals. For example, Mallat constructed wavelet functions using a multiresolution signal decomposition based on orthogonal projections [13]. Wavelets are often generated in the Fourier domain. Mathematically, their Fourier representation is more straightforward than the one in the time domain.

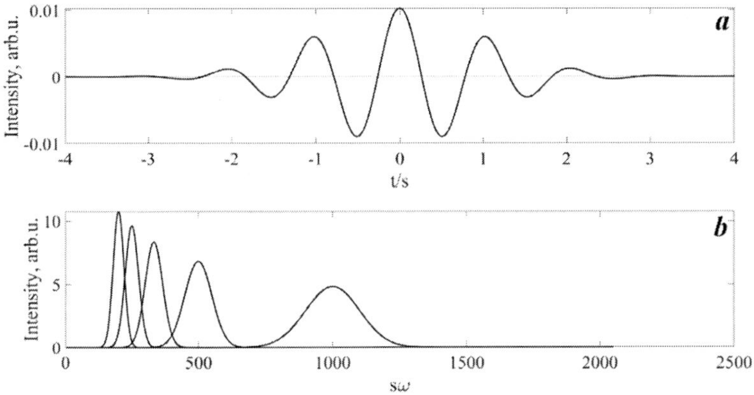

Figure 2.8-1. The Morlet wavelet (Eq. (2.8-3)) (a) and its FT components (b). *s*=0.01, 0.02,…, 0.05 from the left to the right curves. FT was plotted using software [11].

The convergence of the WT decomposition of a function $f(t)$ depends on the number (p) of the vanishing (zero) moments of the wavelet function $\psi(t)$. That is, for $0 \leq k < p$:

$$\int_{-\infty}^{\infty} t^k \, \psi(t) dt = 0. \qquad (2.8-5)$$

Eq. (2.8-5) shows that $\psi(t)$ is orthogonal to the $(p-1)^{th}$-order polynomial. In other words, the convolution of the polynomial of the maximum degree $p-1$ with $\psi(t)$ is zero. The Gaussian wavelets $\mathcal{F}_g^{(p)}(x)$ (Eq. 2.8-2) satisfy the condition defined by Eq. (2.8-5).

This property of $\mathcal{F}_g^{(p)}$ is similar to that of the coefficients of the Savitzky-Golay $(p-1)^{th}$-order differentiation filter (SGF) (Chapter 2.3). The SGF coefficients are the derivatives of the orthogonal polynomial in the middle point of the segment used for approximation. Analogously, the wavelet function with n vanishing moments is the nth-order derivative of a smoothing function $\theta(t)$ [5]:

$$\psi(t) = (-1)^n \, d^n\theta(t)/dt^n. \qquad (2.8-6)$$

The Gaussian derivatives (Eq. (2.8-2)) are examples of such functions.

According to Eq. (2.8-1), the WT of the function $F(t)$ is its convolution with the wavelet function $\psi(t)$. The convolution (designated by the asterisk) has the following properties [5]:

$d^n\{x(t) * y(t)\}/dt^n = d^n x(t)/dt^n * y(t) = y(t) * d^n y(t)/dt^n$.
Therefore, from Eq. (2.8-6), we have

$$W_\psi\{F\} = F * \psi = F * a^n \frac{d^n\theta(t)}{dt^n} = a^n \frac{d^n\{F * \theta(t)\}}{dt^n}. \qquad (2.8-7)$$

So, the WT of a function F, performed by a wavelet with n vanishing moments, is the estimation of the nth-order derivative of this function smoothed by $\theta(t)$.

The more apparent integral form of Eq. (2.8-7) is the folding of the function F and the derivative of the wavelet $\psi_{a,b}(t)$ [9]:

$$W_\psi\left\{\frac{d^n F(t)}{dt^n}\right\} = (-1)^n \int_{-\infty}^{\infty} F(t) \frac{d^n}{dt^n}\{\psi^*_{a,b}(t)\}dt. \qquad (2.8-8)$$

Eq. (2.8-8) justifies the fine structure analysis method of the signal $F(t)$ using the wavelet derivatives [9].

Following the classic approach [9], the matrix notation of the nth-order derivative of the function F is

$$\boldsymbol{F}^{(n)} = \boldsymbol{D}_n \boldsymbol{F}, \qquad (2.8-9)$$

where $\boldsymbol{D}_n$ is the nth-order derivative operator.

By multiplying both sides of Eq. (2.8-9) on the pth-order derivative wavelet matrix $\boldsymbol{W}_{\psi_p}$, we obtain

$$\boldsymbol{W}_{\psi_p} \boldsymbol{F}^{(n)} = \boldsymbol{W}_{\psi_p} \boldsymbol{D}_n \boldsymbol{F} = \boldsymbol{W}_{\psi_{p+n}} \boldsymbol{F}. \qquad (2.8-10)$$

The least-square approximation of $\boldsymbol{F}^{(n)}$:

$$\widehat{\boldsymbol{F}}^{(n)} = \left(\boldsymbol{W}^T_{\psi_p} \boldsymbol{W}_{\psi_p}\right)^{-1} \boldsymbol{W}^T_{\psi_p} \boldsymbol{W}_{\psi_{p+n}} \boldsymbol{F}. \qquad (2.8-11)$$

The term $\boldsymbol{W}_{\psi_{p+n}} \boldsymbol{F}$, which coincides with Eq. (2.8-6), was used to study the resolution enhancement in WT spectrometry [9].

The CWT Gaussian wavelet derivative of the Gaussian peak in the time domain is readily estimated by the inverse FT of the CWT derivative obtained in the frequency domain:

$$\hat{F}_G^{(1)}(x) = (1/2\pi) \int_{-\infty}^{\infty} F_G(\omega) \psi_G^{(1)}(\omega)\, exp(j\omega x)\, d\omega, \qquad (2.8-12)$$

where $F_G(\omega) \sim w exp(-p^2/16ln2)$ is the FT of the Gaussian peak; $p = \omega w$; $\psi_G^{(1)}(\omega) \sim -p_s exp(-p_s^2/2)$; $p_s = \omega s$.

By substituting $F_G(\omega)$ and $\psi_G^{(1)}$ into Eq. (2.8-12), we have

$$\hat{F}_G^{(1)}(x) \sim (s/\gamma) \int_{-\infty}^{\infty} p_\gamma\, exp(-p_\gamma^2/16ln2) exp(j\omega x)\, d\omega, \qquad (2.8-13)$$

where $\gamma = \sqrt{1 + 8ln2(s/w)^2}$; $p_\gamma = \gamma w \omega = w_\gamma \omega$.

So, $\hat{F}_G^{(1)}(x)$ is the CWT derivative of the broaden Gaussian peak with width of

$$w_\gamma = w\sqrt{1 + 8ln2(s/w)^2} \cong w[1 + 4ln2(s/w)^2], \qquad (2.8-14)$$

if $s/w \ll 1$.

Figure 2.8-2 shows that if the derivative curve broadens, its intensity decreases, while the scaling factor (s) increases. If $s \leq 0.2$, then the plot is close to the precise derivative $F_{PR}^{(1)}$.

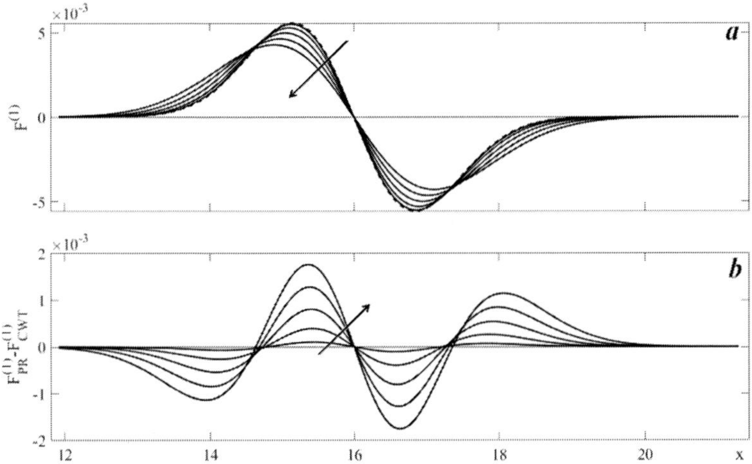

Figure 2.8-2. (a) The precise and CWT Gauss derivatives of the Gaussian peak ($w = 2$) (dashed and solid curves, respectively). (b) The deviations from the precise derivative. Scale $s = 0.2, 0{,}4, \ldots, 1.0$ in the arrow's direction. All plots are normalized to the absolute value of the area under curves.

The choice of a wavelet, which is appropriated to analyze a given process, is of the most significant importance [9]. The shape of the user-defined regularized wavelet must be adaptive to the structure of the data.

Interested readers can find in-depth descriptions of different wavelet basis functions (families) and WT methods in [3, 12]. The regularization procedures improved the approximation using wavelets [14, 15]. However, this issue is beyond the scope of this textbook, which is mainly dedicated to a chemical audience.

Numerical experiment

The models (Table 2.8-1) were disturbed by the white, blue, pink, and red noise with a zero mean and standard deviation of 0.02. The noise was generated using MATLAB software [16]. We performed the 1st-order differentiation using the SG method, the DWT (according to the algorithm and MATLAB program [16], based on the Haar wavelet), and the CWT based on the Gaussian 1st-order derivative wavelet. Obtained data included a large set of derivatives calculated 10^4 times, each time with a new regenerated noise.

Table 2.8-1. Models.

Gaussian	F_G	$exp\{-arg_1^2\}$
	$F_G^{(1)}$	$-2\sqrt{ln2}arg_1F_G$
Lorentzian	F_L	$\{1+arg_2^2\}^{-1}$
	$F_L^{(1)}$	$-2arg_2\{1+arg_2^2\}^{-2}$
Triangle	F_T	$\begin{cases} 0, & x=[1,14],[18,32] \\ 0.5x-7, & x=[14,16] \\ -0.5x+9, & x=[16,18] \end{cases}$
	$F_T^{(1)}$	$\begin{cases} 0, & x=[1,14],[18,32] \\ 0.5, & x=[14,16] \\ -0.5, & x=[16,18] \end{cases}$
Sum of Gaussian, Lorentzian, and sigmoid shapes	F_S	$\exp\{-arg_3^2\}+\{1+arg_4^2\}^{-1}+\{1+exp[-arg_5]\}^{-1}$
	$F_S^{(1)}$	$-(4\sqrt{ln2}/1.3)arg_3\,exp\{-arg_3^2\}$ $-(4/1.3)arg_4\{1+arg_4^2\}^{-2}-$ $3exp(-arg_5)\{1+exp(-arg_5)\}^{-2}$

$arg_1=2\sqrt{ln2}(x-16)/2$; $arg_2=2(x-16)/2$; $arg_3=2\sqrt{ln2}(x-5)/1.3$: $arg_4=2(x-15)/1.3$; $arg_5=3(x-25)$; $x=[1,32]$; $h=32/N$. $N=2048$; $N=5120$ for F_s [16], which is similar to the model [5].

Two criteria were adopted to compare these differentiation procedures (CWT, DWT, and SG). The first one was the minimum total error:

$$\xi_{dif}=\sqrt{\sigma_{dif}^2+\Delta_{dif}^2}, \qquad (2.8-15)$$

where σ_{dif}^2 is the variance of $\hat{F}_G^{(1)}$; $\Delta_{dif}^2=\int_{x_1}^{x_2}\left[F_G^{(1)}(x)-\overline{\hat{F}_G^{(1)}}(x)\right]^2dx$ is the squared systematic error; $F_G^{(1)}$ and $\hat{F}_G^{(1)}$ were normalized to the absolute value of the area under the curve. The simple integration of the precise derivative showed that this value equals two. The 1st-order derivative must be negligible out of the integration interval $[x_1, x_2]$. The bar stands for the mean data obtained in the repetitions.

As an example, Figure 2.8-3 represents the plots $\xi_{CWT}(s)$, $\sigma^2_{CWT}(s)$ and $\Delta_{CWT}(s)$, which were obtained for the noisy Gaussian peak (F_G). These plots confirm the well-known fact that the minimum of the total error is achieved by the trade-off between systematic and random errors. However, surprisingly, there are minima of the systematic errors observed in the ranges of the significant random distortions. These minima are useful for the optimization of the differentiation procedure if the random noise is very small.

Figures 2.8-4-2.8-6 and Table 2.8-2 illustrate some of the results obtained by the numerical experiments. We conclude that

1. The total errors of the SG and the CWT differentiation are close to each other. Let us explain this fact theoretically. It follows from Eqs. (2.8-1) and (2.8-2) that the CWT derivative, which was estimated using the Gauss wavelet, is

$$\hat{F}(x_1) \sim \int_{-\infty}^{\infty} F(x)[(x-x_1)/s]exp\{-[(x-x_1)/s]^2\}dx. \qquad (2.8-16)$$

Eq. (2.8-16) is approximately equals to the sum calculated in the range $[-x_{lim}{:}\, h{:}\, x_{lim}]$, where h is the sampling interval:

$$\hat{F}(x_1) \sim h\sum_{x=-x_{lim}}^{x_{lim}} F(x)[(x-x_1)/s]exp\{-[(x-x_1)/s]^2\}. \quad (2.8-17)$$

Suppose that in the narrow interval $[-x_0+x_1, x_0+x_1]$, symmetrical around a point x_1, $(x-x_1)/s \ll 1$. Then

$$\hat{F}(x_1) \sim h\sum_{x=-x_0+x_1}^{x_0+x_1} F(x)\,(x-x_1)/s. \qquad (2.8-18)$$

Using corresponding the x-scale indexes, we have

$$\hat{F}(i_1) \sim h\sum_{i=-i_0+i_1}^{i_0+i_1} F(i)W_i, \qquad (2.8-19)$$

where the constant coefficient $W_i = (i-i_1)h/s$ varies in the range $[-x_0/s, x_0/s]$.

The weights W_i of the convolution (Eq. (2.8-27)) resemble the coefficients of the 1st-order differentiation SGF built on the 2nd-degree polynomial (Chapter 2.3):

$$W_i^{SG} = [-m/M, m/M], \qquad (2.8-20)$$

where $= (2m+1)(m+1)m/3$.

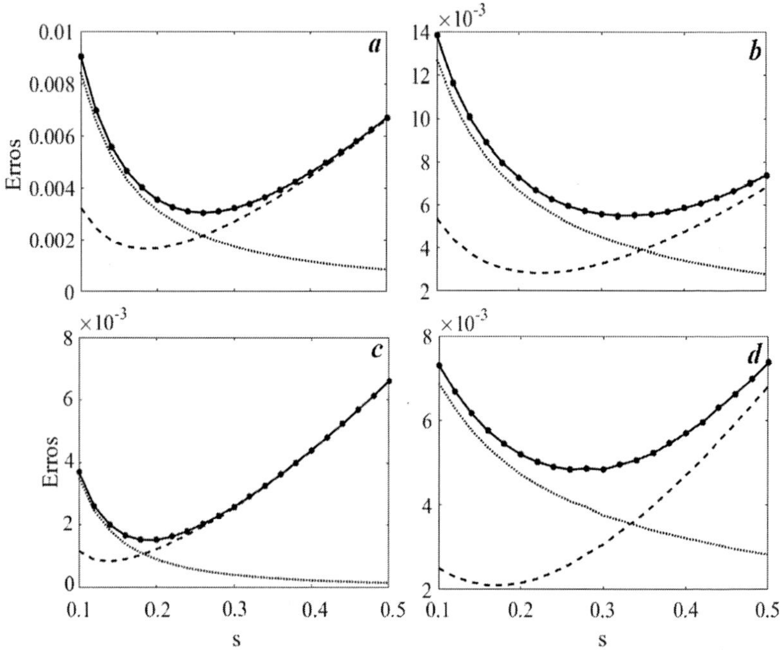

Figure 2.8-3. Systematic, random, and total errors (dashed, dotted, and solid lines, respectively) of the CWT Gauss differentiation of the Gaussian peak disturbed by the white, pink, blue, and red noises ($\sigma = 0.02$). (a-d, respectively). For these panels,

$$\{10^4 * min(\xi_{dif}), s_{opt}\} = \{30, 0.26; 55, 0.32;\ 15, 0.20;\ 48, 0.26\}.$$

Data obtained for the Gaussian peak disturbed by the normal noise (Table 2.8-2) show that the optimal value of the scale factor (s_{opt}) weakly depends on the sampling interval h. It is clear, since the range $[-x_0/s, x_0/s]$ is only determined by the precision of the linear approximation and does not depend on h. Insignificant changes of s_{opt} are due to the influence of the sampling interval on the accuracy of the numerical procedures.

2. The difference between the precise and the CWT derivatives of the Gaussian peak is less than that of the Lorentzian peak (Fig. 2.8-4, Table 2.8-2)). The SG method demonstrates a similar behavior.
3. As a rule, the DWT differentiation has the most significant total errors.

Table 2.8-2. The minimum total errors (x10⁴) of the 1st-order differentiation

Model	Noise	Processing		
		CWT	SG	DWT
Gaussian peak	white	30 (0.26)	30 (26)	32
		22 (0.25)*	22 (36)*	25*
		18 (0.24)**	17 (47)**	27**
	pink	55 (0.32)	53 (32)	55
	blue	15 (0.20)	17 (20)	26
	red	48 (0.28)	47 (26)	48
Lorentzian peak	white	41 (0.22)	40 (22)	47
	pink	70 (0.28)	68 (28)	69
	blue	21 (0.16)	24 (18)	43
	red	59 (0.24)	58 (22)	62
Gaussian, Lorentzian and Sigmoid Functions***	white	20 (0.18)	20 (42)	27
	pink	41 (0.22)	40 (56)	61
	blue	8.8 (0.12)	10 (30)	11
	red	35 (0.18)	34 (46)	38
Triangle peak	white	128 (0.14)	132 (12)	168
	pink	152 (0.16)	155 (14)	173
	blue	104 (0.090)	113 (9)	168
	red	122 (0.080)	124 (8)	171

The number in parentheses are the optimal parameters s and m for the CWT and SG ($2t = 2$) methods, respectively. The DWT level is 6. $h = 0.0156$, (*) 0.0104, (**) 0.0078 and (***) 0.0064. The maximum signal-to-noise ratio is 50.

4. Both pink and red noise has a power spectrum, which dominates at low Fourier frequencies and causes the most significant error. However, blue noise, whose the power spectrum increases at high frequencies, is suppressed by the smoothing multipliers of the differentiation filters. Therefore, in this case, the total errors are minimal.

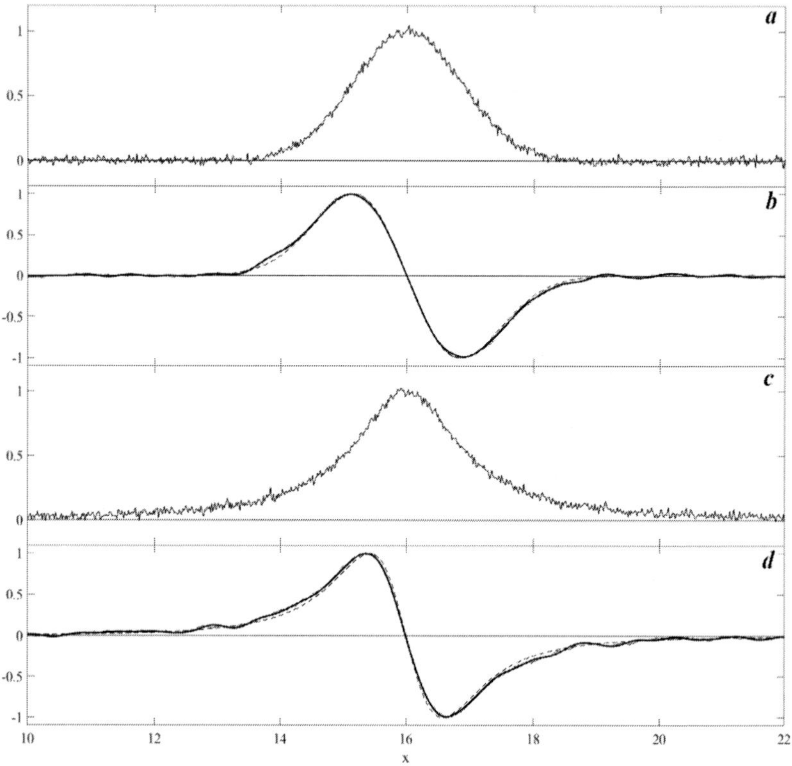

Figure 2.8-4. The Gaussian (a) and Lorentzian (c) peaks disturbed by the normal noise and their 1st-order derivatives (panels b and d, respectively): precise and estimated by the SG filter and the CWT (dashed, dotted, and solid curves, respectively). Processing parameters are given in Table 2.8-2.

5. The derivative of the triangle function (Fig. 2.8-5, panels b and d) significantly differs from the precise curve. The smoothed derivative is obtained by the DWT method (panel d) but at the cost of increasing systematic errors.
6. All figures demonstrate weak ripples in the WT derivatives. The regularized wavelet derivatives also showed such artifacts [15].

Our numerical experiments showed that, in the framework of the standard models, which are used in analytical spectrometry, wavelet differentiation has no advantage over the SG method.

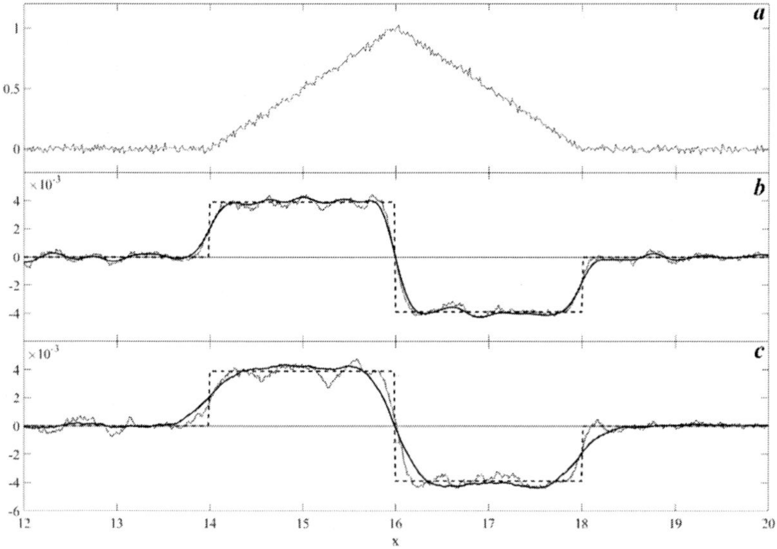

Figure 2.8-5. The triangle peak disturbed by the normal noise (a), and its 1st-order CWT and DWT derivatives (solid curves in panels b and b, respectively). The precise and SG derivatives are shown by dashed and dotted curves, respectively. Processing parameters are given in Table 2.8-2

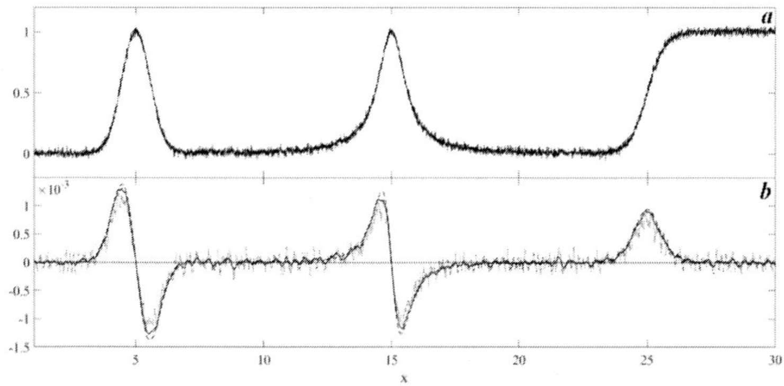

Figure 2.8-6. (a) The Gaussian, Lorentzian, and sigmoid peaks disturbed by the normal noise (from the left to the right) (Eq. (2.8-15)); (b) Their 1st-order derivatives: precise and estimated by the SG filter and CWT (dashed, dotted, and solid curves, respectively). Processing parameters are given in Table 2.8-2.

Exercise 2.8-1.

Readers are invited to study the 1st-order derivatives of the models (Table 2.8-1) obtained by the CWT, DWT, SG, and FT methods (plotD1WTSGFT.m, Appendix SW-5). By varying the parameters of the differentiation procedures, you can trace how the shape of the curves and the noise change, as well as how they correspond to the exact derivative. Which method do you prefer?

CHAPTER NINE

OPTICAL DIFFERENTIATION: SLIT MONOCHROMATORS

Optical differentiation is directly performed in the optical part of the spectral instrument, usually in monochromators. If the light source is a laser, then differentiation may be carried out by the wavelength or frequency modulation of the laser beam. In the multiplex spectrometers based on coding (e.g., [1]), a special mask may, in principle, be used for differentiation.

Dual-wavelength techniques and the wavelength modulation in the slit monochromators were the most popular instrumental methods of the optical differentiation in 1960-80s [2]. In the dual-wavelength optical differentiation (DWOD) device, the difference between the two signals, having close wavelengths λ_i and λ_{i+1}, is measured:

$$\Delta A_{i+1} = A(\lambda_{i+1}) - A(\lambda_i), \qquad (2.9-1)$$

where $i = 1, 2, \ldots$

So, DWOD is similar to the finite-difference numerical procedure (Chapter 2.2). The DWOD monochromator includes two coaxial grids shifted by a small angle. The beams, which are diffracted by these grids, are alternately directed on the detector. However, the optical densities $A(\lambda_i)$ must be generated before subtraction (Eq. (2.9-1)).

DWOD should not be confused with the routine dual-wavelength quantitative analysis, which is widely used in analytical spectrometry (e.g., [3]). In the last method, only two, not close wavelengths, serve as analytical points.

The disadvantage of DWOD is that it produces only the 1st-order derivative.

The theory of the optical modulation

Optical modulation (wavelength (WM) and frequency (FM)) is a kind of modulation spectroscopy (MS). MS allows researchers to increase the sensitivity of the analysis and reduce (wholly or partly) the impact of the interference signals. Generally speaking, the light emitted or absorbed by a sample is modulated by periodical distortions of the sample state or the light beam. For example, in the first case, the pressure, the electric, or magnetic fields are changed. Since these changes are often not possible in practice, the light distortions (WM or FM) are the standard methods of MS. Figure 2.9-1 illustrates the modulation principle in a simple way. The signal (e.g., the light intensity) is modulated by changing the spectral abscissa argument:

$$\nu_1 = \nu - \nu_c = \Delta\nu cos\omega_m t, \qquad (2.9-2)$$

where $\nu_c = \nu$ in the lack of modulation; $\Delta\nu$ and ω_m are the amplitude and the angular frequency of the modulation, respectively.

Suppose that the amplitude $\Delta\nu$ is so small that the spectral curve may be considered to be linear in the interval $[a, b]$. In this case, the amplitude of the spectrum, which is modulated by frequency ω_m, is proportional to the slope of the curve $A(\nu)$ at the point ν_c; that is, to the 1st-order derivative $A^{(1)}(\nu_c)$.

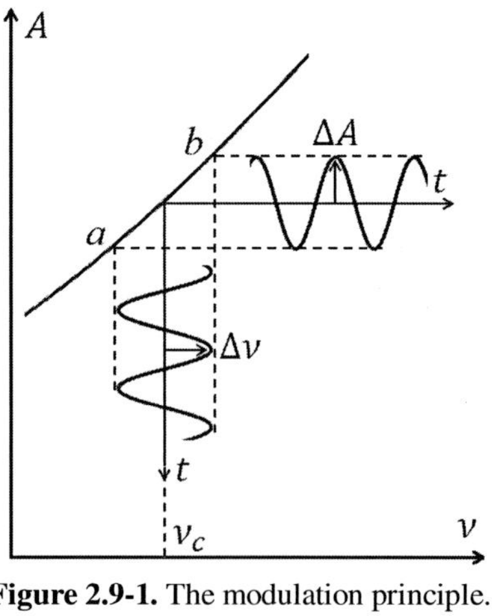

Figure 2.9-1. The modulation principle.

Let us consider a general approach to generating the function's derivatives via the arbitrary modulation of argument ν (Fig. 2.9-2) [4].

Taking Eq. (2.9-2) into account, we have the Taylor series near $\nu_1 = 0$:

$$A(\nu) = \sum_{n=0}^{\infty} A^{(n)}(\nu_1)\,((\Delta\nu)^n/n!)cos^n\omega_m t =$$

$$\sum_{n=0}^{\infty} A^{(n)}(\nu_1)\,((\Delta\nu)^n/n!)[(exp(\omega_m t)+exp(-j\omega_m t))/2]^n =$$
$$\sum_{n=0}^{\infty} Z^{(n)} \sum_{k=0}^{n} C_n^k exp[j(n-2k)\omega_m t], \qquad (2.9-3)$$

where $j=\sqrt{-1}$; $Z^{(n)} = A^{(n)}(\nu_1)\,(\Delta\nu/2)^n/n!$; $C_n^k = n!/[(n-k)!\,k!]$. The system multiplies $A(\nu_1)$ by the reference signals and filters the result in the range $[-\pi/\omega_m, \pi/\omega_m]$. Therefore, the output is

$$a_i = (\omega_m/2\pi)\int_{-\pi/\omega_m}^{\pi/\omega_m} A(\nu_1) r_i(t)\,dt = \sum_{n=0}^{\infty} Z^{(n)} \sum_{k=0}^{n} C_n^k r_{i,k}^{(n)}, \quad (2.9-4)$$

where $r_{i,k}^{(n)} = (\omega_m/2\pi)\int_{-\pi/\omega_m}^{\pi/\omega_m} r_i(t)\,exp[j(n-2k)\omega_m t]dt.$

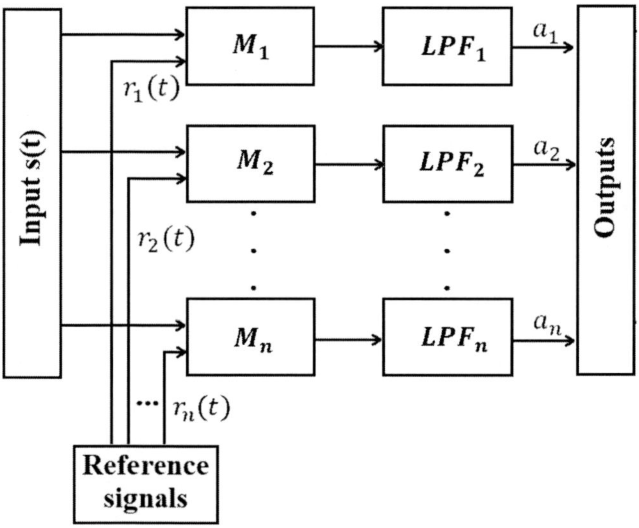

Figure 2.9-2. Diagram of the modulation device [4]. ***M*** is the multiplier. ***LPF*** is the low-pass filter.

The reference functions $r_i(t)$, which are represented by the infinite Fourier series, were selected so that $a_i = Z^{(i)}$. To simplify the solution, $A(\nu_1)$ was approximated by the polynomial with power $2t$. For example, Table 2.9-1 represents $r_i(t)$ for $2t = 8$ [4].

According to Eq. (2.9-4) and Table 2.9-1, the terms $r_{i>8}(t)$ compensate the impact of the derivatives with an order greater than eight. This conclusion is only valid if the modulation amplitude is small enough to guarantee a statistical validity of the curve approximation by the 8th-order polynomial.

For the white noise of the input signal, the mean squared error of the nth-order derivative [4] is

$$\overline{\sigma_n^2(t)} \sim \sum_{k=n}^{2t} r_{nk}^2. \qquad (2.9-5)$$

Therefore, if the polynomial order increases, then the noise in the derivatives also increases.

Table 2.9-1. Calculation of $r_i(t)$.

$r_1(t)$		1	0	-3	0	5	0	-7	0		$\cos\omega_m t$
$r_2(t)$		0	1	0	-4	0	9	0	16		$\cos 2\omega_m t$
$r_3(t)$		0	0	1	0	-5	0	14	0		$\cos 3\omega_m t$
$r_4(t)$	=	0	0	0	1	0	-6	0	20	x	$\cos 4\omega_m t$
$r_5(t)$		0	0	0	0	1	0	-7	0		$\cos 5\omega_m t$
$r_6(t)$		0	0	0	0	0	1	0	-8		$\cos 6\omega_m t$
$r_7(t)$		0	0	0	0	0	0	1	0		$\cos 7\omega_m t$
$r_8(t)$		0	0	0	0	0	0	0	1		$\cos 8\omega_m t$

Consider Eq. (2.9-3) in a more detailed form:

$$A(\nu) = A(\nu_1) + A^{(1)}(\nu_1)\Delta\nu cos\omega_m t +$$
$$A^{(2)}(\nu_1)[(\Delta\nu)^2/4](1 + cos2\omega_m t) +$$
$$A^{(3)}(\nu_1)[(\Delta\nu)^3/24](3cos\omega_m t + cos3\omega_m t) + \cdots \qquad (2.9-6)$$

In the conventional method, the nth-order optical derivative is obtained using the lock-in amplifier (Appendix J). The harmonic $cosn\omega_m t$ and the reference signal $r_n(t) = cosn\omega_m t$ are multiplied and low-pass filtered.

In the general approach [4], a combination of different harmonics was used. However, in the conventional method, the modulation amplitude $\Delta\nu$ must be small enough to suppress the impact of the derivatives of the order more than n. For example, in Eq. (2.9-6) the factor of the 1st-order harmonic $cos\omega_m t\{A^{(1)}(\nu_1)\Delta\nu + A^{(3)}(\nu_1)\,(\Delta\nu)^3/8 + \cdots\}$ contains the 3rd - and higher-order derivatives. Therefore, the signal-to-noise ratio of the conventional method is less than that of the general approach [4].

An advanced theoretical description of FM and WM spectroscopy [5] oriented toward physic, is cumbersome for chemical analysts. Therefore, we tried to simplify the explanation using our old textbook [2].

In a more general case, than that described by Eq. (2.9-2), the sinusoidal modulated optical field intensity [5] is

$$E(t) = E_0 exp(j\omega t) = E_0 exp(j\omega_0 t) \sum_{n=-\infty}^{\infty} J_n(\Delta v) \, exp(jn\omega_m t), \qquad (2.9-7)$$

where $\omega = \omega_0 t + \Delta v sin\omega_m t$; ω_0 is the carrier frequency; J_n is the nth-order Bessel function, which characterizes the Fourier components of the modulated signal.

In typical WM spectroscopy, the amplitude Δv is large. Therefore, many sidebands at $\omega_0 \pm n\omega_m$, which are stronger than the carrier, are observed. In FM spectroscopy, the amplitude Δv is so small that there is only one set of sidebands at $\omega_0 \pm \omega_m$. The higher-order components are negligible.

The authors of the theoretical analysis [5] assumed that the field (Eq. (2.9-7)) was absorbed by a sample characterized by a complex transmission function $T(\omega_n)$. The transmitted optical field was

$$E(t) = E_0 exp(j\omega_0 t) \sum_{n=-\infty}^{\infty} T(\omega_n) J_n(\Delta v) \, exp(jn\omega_m t). \qquad (2.9-8)$$

The algebraic expressions of intensity $|E_{\omega_m}(t)|^2$ were obtained for the WM and FM spectroscopy.

In the partial case of the derivative WM spectrometer [6], the authors found the Fourier transform (FT) of the modulated signal:

$$\tilde{A}(\omega') = \tilde{A}^{(0)}(\omega')\tilde{G}_M(\varphi_M)\Psi(\varphi_J, \omega_m), \qquad (2.9-9)$$

where $\Psi(\varphi_J, \omega_m) = J_0(\varphi_J) + 2\sum_{k=1}^{\infty} \Phi_k$,

$\Phi_k = J_{2k}(\varphi_J) cos(2k\omega_m t) - jJ_{2k-1}(\varphi_J) sin((2k-1)\omega_m t)$;

$\varphi_M = \omega' s$; $\varphi_J = \omega' \Delta v$; ω' is the Fourier frequency, whose units are reversible to those of v; $\tilde{A}^{(0)}(\omega')$ and $\tilde{G}_M(\varphi_M)$ are the FT of the source signal and the instrumental function of the monochromator (which width is s), respectively.

The term Φ_k (Eq. (2.9-9)) is the detailed form of the sum in Eq. (2.9-7). For the triangle instrumental function,

$$\tilde{G}_M(\varphi_M) = [sin(\varphi_M)/(\varphi_M)]^2. \qquad (2.9-10)$$

Eq. (2.9-9) shows that the amplitude of the nth- harmonic depends on the term $J_n(\varphi_J)$, which defines the filtering properties of the optical differentiator. If $\varphi_J \ll 1$, then [7]

$$2J_n(\varphi_J) = 2\sum_{k=0}^{\infty} \frac{(-1)^k}{k!(k+n)!} \left(\frac{\varphi_J}{2}\right)^{k+n} \approx \vartheta^{(n)} \varphi_J^n \left\{1 - \frac{\varphi_J^2}{4(n+1)}\right\}, \qquad (2.9-11)$$

where $\vartheta^{(n)} = 1/(n! \, 2^{n-1})$.

Suppose that the relative systematic error of the nth-order differentiation is not higher than 10%. In other words, the second term of Eq. (2.9-11) is less than 0.1. Then

$$\varphi_J = \omega' \Delta\nu \leq 0.63\sqrt{n+1}. \quad (2.9-12)$$

Let the range of φ_M be limited by the first zero of the instrumental function (Eq. (2.9-10)):

$$\varphi_M = \omega' s \leq \pi. \quad (2.9-13)$$

Then from Eqs. (2.9-12) and (2.9-13), we have the maximum modulation amplitude:

$$\Delta\nu \leq 0.2s\sqrt{n+1}. \quad (2.9-14)$$

If the line width $w >> s$, then approximately

$$\Delta\nu \leq 0.2C_l w\sqrt{n+1}, \quad (2.9-15)$$

where the coefficient C_l depends on the line shape.

For the relative low light power variation, the nth- harmonic output (coefficient) of the lock-in amplifier is [8]

$$H_n(\nu_c) \sim \int_{-\pi/\omega_m}^{\pi/\omega_m} T(\nu_c + \Delta\nu cos\omega_m t)cos(n\omega_m t)dt. \quad (2.9-16)$$

Dyroff [9] used the ideal laser-based model, which did not change its output intensity during tuning. The FT of Eq. (2.9-16) (the frequency response) for the model is

$$\widetilde{H}_n(k) \sim \vartheta^{(n)}(\Delta\nu)^n \widetilde{T}^{(n)}(k) J_n(2\pi\Delta\nu k)/(\pi\Delta\nu k)^n. \quad (2.9-17)$$

The properties of the harmonic spectra (Eq. (2.9-17)) were studied. Two properties were similar to those of the nth-order derivative: the zero mean value and orthogonality to polynomials of degree lower than n (for the Savitzky-Goley algorithm). The author showed that the transmission spectrum equals the zeroth-harmonic corrected with higher harmonic spectra:

$$T(\nu) = H_0(\nu) - H_2(\nu) + H_4(\nu) + \ldots + (-1)^n H_{2n}(\nu) + \ldots \quad (2.9-18)$$

Eq. (2.9-18) is a partial case of the general expression, which describes all the instrumental distortions of spectral devices [2].

Since real lasers demonstrate nonlinear behavior with respect to the laser current, the light intensity modulation overtones appear. Therefore, the nonlinear model was developed [8].

Dissertation [8] cited known approaches to the solution of Eq. (2.9-16).

A new method [8] based on the approximation of the transmission by a piecewise polynomial was suggested. In principle, this method is similar to a general approach to generating derivatives (Fig. 2.9-2) [4]. However, the domains (time and Fourier), and mathematics were different.

In order to perform a detailed study of the filtering properties, we considered a general diagram of the two-beam optical differentiator (Fig. 2.9-3). This device includes a common monochromator with WM equipment.

In the single-beam mode, the input slit of the monochromator is illuminated by the radiation

$$I_1 = I_{01}(1+\alpha) + \eta_1, \qquad (2.9-19)$$

where I_{01} is a deterministic term; η_1 is a noise produced by background emission or absorption; and α is a fluctuating multiplier, which has one-sided power spectral density,

$$SP_\alpha(\omega) = a_\alpha + b_\alpha/\omega, \qquad (2.9-20)$$

where a_α and b_α are the white and the flicker noise constants, respectively.

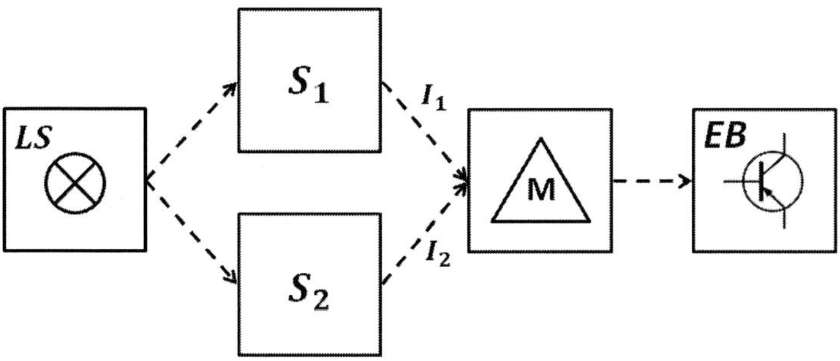

Figure 2.9-3. Diagram of the optical differentiator. LS is the light source, S-sample, M-monochromator, and EB-electronic block.

We suppose that the range of ν is small, therefore, the noise does not depend on ν.

The spectral density of $\eta_1$is

$$SP_\eta(\omega) = a_\eta + b_\eta/\omega, \qquad (2.9-21)$$

where b_η is due to the drifts of the sample and transmittance of the atmosphere, and other factors independent of I_{01}.

The combined effect of α, η_1, and the electronic block (EB), adjusted to the EB input, is expressed by the spectral density,

$$SP_{EB}(\omega) = A + B/\omega, \qquad (2.9-22)$$

where $A = \theta^2\left(a_{\alpha M}\overline{I_{01}^2} + a_{\eta M}\right) + a_d\theta\overline{I_{01}} + a_{EB}$ and $B = \theta^2\left(\eta_{\alpha M}\overline{I_{01}^2} + \eta_{\eta M}\right) + \eta_d\theta\overline{I_{01}} + \eta_{EB}$ are the white and flicker noise constants, respectively; θ is the transmittance of the monochromator;

subscripts M and d designate the impacts of the monochromator and the light detector, respectively; $\overline{I_{01}}$ is the mean light intensity at the detector in the range $\nu_c \pm \Delta\nu$.

The central frequency of the lock-in amplifier is $n\omega_m$, and the time constant of EB is $\tau \gg \tau_d \leq \pi/n\omega_m$, where τ_d is the time constant of the detector.

Let β be the transfer factor of EB. Then the noise power of the nth-order derivative at the EB output

$$\overline{U_0^2} = (\beta^2/\pi)\int_0^{\pi/\tau} SP_{EB}(\omega)d\omega =$$
$$(\beta^2/\pi)\int_0^{\pi/\tau}\{A + B/(\omega + n\omega_m)\}d\omega =$$
$$\beta^2\{A/\tau + (B/\pi)ln(1 + \pi/n\omega_m\tau)\} \approx$$
$$(\beta^2/\tau)(A + B/n\omega_m), \quad (2.9-23)$$

since $n\omega_m \gg \pi/\tau$, and $ln(1 + \pi/n\omega_m\tau) \approx \pi/n\omega_m\tau$.

The signal of the nth-order derivative is equal to the FT of Eq. (2.9-9) [6].

If the condition expressed by Eq. (2.9-14) is satisfied, then [2]:

$$A_{01}^{(n)}(\nu_c) \approx \vartheta^{(n)}(\Delta\nu)^n I_{01}^{(n)}(\nu_c), \quad (2.9-24)$$

where $\vartheta^{(n)}$ is defined in Eq. (2.9-11); $I_{01}^{(n)}(\nu_c)$ is the nth-order derivative of the source signal convoluted by the instrumental function.

Taking the transfer factor of EB (β) into account, we obtain the signal-to-noise ratio from Eqs. (2.9-23) and (2.9-24):

$$\xi_0^{(n)} = A_{01}^{(n)}(\nu_c)/\sqrt{\overline{U_0^2}} = \vartheta^{(n)}(\Delta\nu)^n I_{01}^{(n)}(\nu_c)\psi_{noise}, \quad (2.9-25)$$

where $\psi_{noise} = \sqrt{\dfrac{\tau}{A + B/n\omega_m}} = \begin{cases} \sqrt{\tau/A}, for\ the\ white\ noise. \\ > \sqrt{\pi/B}, for\ the\ flicker\ noise. \end{cases}$

Eq. (2.9-25) shows that the signal-to-noise ratio may be improved if the modulation amplitude increases, but the conditions, expressed by Eqs. (2.9-14) and (2.9-15), are satisfied. In the single-point measurement (non-scanning mode), the increase of the EB time constant may significantly suppress the white noise. However, the measurement time also increases. This mode is impossible for RC differentiation. However, the non-scanning numerical differentiation of data segments, obtained by the diode-array spectrometer, can compete with the optical differentiation in the presence of the white noise.

Rough estimates [2] showed that in the scanning mode, the relationship between the signal-to-noise ratios is

$$\xi_0^{(n)}/\xi_{RC}^{(n)} \sim \sqrt{(A + B/\omega_{sd})/(A + B/\omega_m)}, \quad (2.9-26)$$

where ω_{sd} is the frequency of the synchronous detection in a scanning spectrometer. It follows from Eq. (2.9-23), that for the white noise, the signal-to-noise ratios of the RC and optical differentiation are approximately equal. For the flicker noise, the optical method is preferred over the RC differentiator if the modulation frequency $\omega_m \gg \omega_{sd}$.

C. Dyroff [9] discussed the shapes of the 2nd-order optical derivatives of the Gaussian and Lorentzian peaks. Interested readers can find a lot of the literature sources concerning this issue in this study.

Consider the double-beam optical differentiator (Fig. 2.9-3) [2]. In this device, the radiation, transmitted by the samples S_1 and S_2, alternatively with interval Δt, illuminates the input slit of the monochromator. The monochromator's output noise is the mean squared difference of the noise of the light intensities in the following two channels:

$$\overline{\Delta I^2} = \overline{[\delta I_1(t) - \delta I_2(t+\Delta t)]^2} = \overline{\delta I_1^2(t)} + \overline{\delta I_2^2(t+\Delta t)} - 2\overline{\delta I_1(t)\delta I_2(t+\Delta t)}. \qquad (2.9-27)$$

If the noise is a stationary random process then, according to the Wiener-Khinchin theorem:

$$\delta \overline{I_1^2} = (1/\pi)\int_0^{\pi/\tau} SP_1(\omega)d\omega, \overline{I_2^2} = (1/\pi)\int_0^{\pi/\tau} SP_2(\omega)d\omega,$$

$$\overline{\delta I_1(t)\delta I_2(t+\Delta t)} = (1/\pi)\int_0^{\pi/\tau} SP_{12}(\omega)cos(\omega\Delta t)d\omega, \qquad (2.9-28)$$

where SP_1, SP_2, and SP_{12} are the spectral power densities of the noises δI_1, δI_2, and their joint density, respectively. If $SP_{12} = a_{12} + b_{12}/\omega$, then the covariance term in Eq. (2.9-28) has the same structure as Eq. (2.9-22), since $|cos(\omega\Delta t)| \leq 1$. Therefore, taking the linearity of the spectrometer into account, the impacts of each SPs are independent. So, similar to Eq. (2.9-26), we have

$$\xi_0^{(n)}/\xi_{RC}^{(n)} \sim \sqrt{(A_1 + B_1/\omega_{sd})/(A_1 + B_1/\omega_m)}, \qquad (2.9-29)$$

where each term A_1 and B_1 is related to the sum of SP_1, SP_2, and SP_{12}.

According to Eqs. (2.9-26) and (2.9-28), the conclusions that were drawn for the single and double-beam modes, are identical. Experimental data [10] showed that the signal-to-noise ratios of the RC and optical differentiation were equal for the white noise.

Optical modulation technique in slit monochromators

The central unit of the optical differentiator is the modulation device. Direct modulation was performed in the radiation source [11] (Chapter 2.10, Laser modulation spectroscopy). In pioneering studies, the wavelength modulation was carried out by low-frequency (10-15Hz) vibrations of the monochromator's dispersing elements (prism or diffraction grating) [12, 10]. This technical solution has many drawbacks, such as a low modulation frequency due to the large inertia of the vibration element, and instability caused by possible distortions of optical calibration and adjusting. Modulation based on the vibrating monochromator slit suffers from the same problems.

To eliminate the above problems, an oscillating plane-parallel plate (refractor plate) (Fig. 2.9-4) was mounted behind the entrance slit of the monochromator [13-15]. The modulation frequency was 145 Hz.

The plate, whose width is L and the refractive index is n, displaces the beam by

$$\Delta = Lsin(\alpha - \beta)/cos\beta = Lsin\alpha(1 - 1/n) \cong \alpha L(n-1)/n \quad (2.9-30)$$

for a small α.

Unfortunately, the modulation amplitude which was produced by the plate is not constant since the dispersion of the monochromator and the refractive index depend on the wavelength. Also, the transmittance of the plate depends on the spectral range. Therefore, a set of plates must be used.

Usually, the plate modulator was driven by a sinusoidal waveform. However, the rectangle (stepped) modulation doubled the signal-to-noise ratio of the 2nd-order harmonic for the furnace background shot-noise [16]. Stepped WM was successfully used for background suppression in atomic emission analysis by the independent selection analytical points on each side of the spectral line [17]. Technically, the stepped modulator is a simple device for low frequencies up to 17 Hz [17].

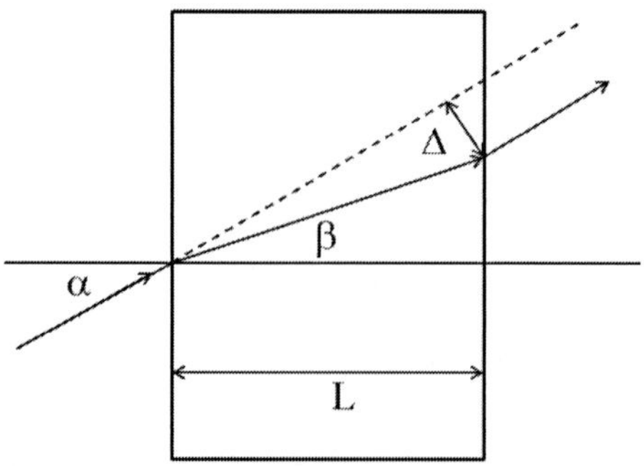

Figure 2.9-4. Beam displacement in the plane-parallel plate.

A rotating quartz mechanical chopper was used to simplify the stepped modulator and to increase its frequency (up to 160 Hz) (Fig. 2.9-5, panel a) [18]. The disk consisted of four quadrants of different thicknesses: $L = 2.5, 2.5, 1,$ and 4mm for 1-4 parts, respectively. According to Eq. (2.9-30), the beam displacement $\Delta \sim \Delta L$ (Fig. 2.9-5, panel b) produces the rectangle modulation. However, the rectangle shape is rounded off when the finite width beam passes boundaries between sectors. To narrow the beam, the modulator was placed near the entrance slit. Also, the reflections from the boundaries must be eliminated [18].

The rotating chopper allowed researchers to increase the modulation amplitude compared to the oscillating plate. This property was most notable for the background correction of the continuum excitation source in atomic fluorescence spectrometry when the measurements were performed at the left and the right side from the analyte wavelength [19].

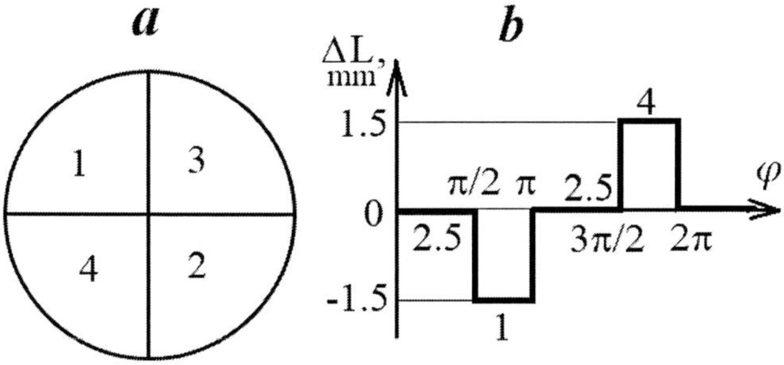

Figure 2.9-5. (a) The four-quadrant quartz chopper. (b) The modulation waveform. φ is the rotation angle. Numbers near the wave stand for L.

Different modulation units were built in Russia in the 1970-80s [2]. Unfortunately, the literature sources of these devices are not available to the English readers. For example, two static plates, shifted around the vertical axis by a small angle relative to each other (Fig. 2.9-6, panel a) or two achromatic prisms (panel b), were used instead of the oscillating plate. The light beams were alternately directed to the wavelength shifting elements by the splitters. The prisms, rotating with frequency 150 Hz, displaced the spectrum by 0.25 and 0.75 nm. Their combination produced the following modulation amplitudes: 0.25, 0.50, 0.75, and 1.00 nm. The differentiation photometer "KAMA", which was built in 1974, used the oscillating interference filter. The wavelength in the maximum filter transmission [20] is

$$\lambda_m \approx \lambda_{\theta=90^0}\sqrt{n^2 - sin^2(90^0 - \theta)}/n, \qquad (2.9-31)$$

where θ is the incident angle; and n is the refractive index of the filter. The angle oscillations produced the WM.

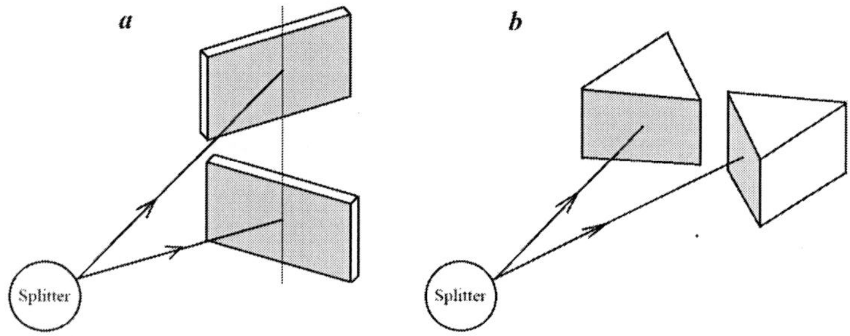

Figure 2.9-6. Wavelength shifting elements: static plates (a) and prisms (b).

The plate was undergoing small-amplitude oscillations excited by the electrostriction device (built on material that changes its shape under the application of an electric field) [13]. The galvanometer system produced the high-frequency modulation (1000 Hz) [21].

In some differentiation devices, the detector was a vidicon (Fig. 2.9-7) [22, 23]. Vidicon is mounted on the place of the exit slit of the monochromator. A light beam hits the vidicon target, which is a matrix (e.g., 1000 x 1000) of the silicon diodes. The electron gun, which consists of the cathode and two grids (control and accelerator), emits an electron beam. The electron optics includes focusing and alignment coils. The collimating mesh focuses the beam in a perpendicular direction to the target. The deflection coil scans the beam over the target diodes vertically and horizontally with high (50-100 kHz) and low (200 Hz) frequencies. During the scanning, diode charge neutrality, which was distorted by the light, is restored. A computer processes the charge current. WM is performed by the superposition of a high-frequency low-amplitude electrical wave (some kHz) on the horizontal scanning signal.

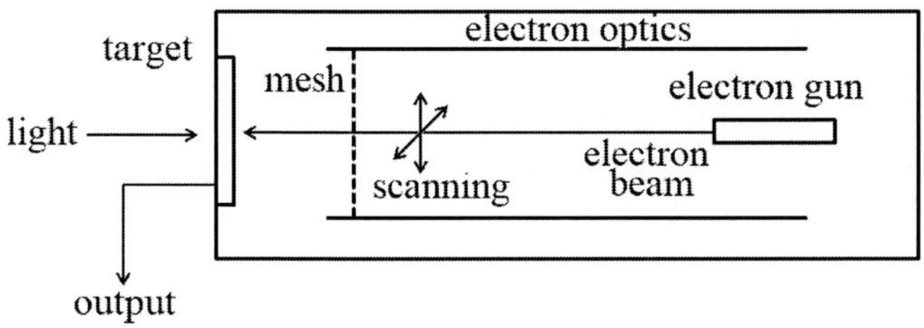

Figure 2.9-7. The vidicon modulation device.

If the modulation and the vertical scan frequencies are equal, then the modulation and the vertical sweep cycles will also be identical. It is similar to the mechanical modulation when a "narrow line is swept back and forth across the exit slit of a monochromator" [22].

The vidicon-based spectrometer generated the 1st-order derivative in the visible region over two intervals: 40 and 400 nm scanned during 100 ms [23]. The atomic emission and molecular absorption in the gas and liquids were studied.

One of the main disadvantages of the single-beam modulation technique is the existence of the blank sample channel's modulated signal. This conclusion was explained by modulating the radiation source's spectral distributions, the transmission of the optical system, and the detector's sensitivity [24]. To overcome this drawback, double-beam devices with two identical detectors [24, 25] or with a single sensor were used. The second method is preferable since selecting similar detectors is a very cumbersome task. The authors of the study [26] formulated requirements for designing the optical part of the wide-range WM spectrometer:

1. High-resolution monochromator.
2. Only using reflection mirrors with a constant reflection index.
3. Minimal incident angles on both the mirrors and the detector to decrease image distortions.
4. Symmetrical optical set-up with minimum components.
5. Focusing the two beams on the same area of the single sensor.

It was stated [26] that most of the optical differentiators, which were developed in the 1960-70s, had many disadvantages. This study described the modulation spectrometer, which measured the logarithmic derivative of the reflectivity of one sample. The 2nd-order derivative was estimated for two samples, which differ by a small amount in the parameter under study A slight composition variation of the ternary semiconductor alloy was detected.

A modulation device was based on the symmetrical monochromator (Fig. 2.9-8). Interested readers may satisfy the requirement '3' "in the mind's eye" by moving the grating and mirrors. The diffracted beam, which is modulated by the oscillating mirror (Mod, f=122Hz)), passes through the slit. Then, the beam is focused on the chopper (Ch) by the spherical mirror S_1. The transmitter chopper divided the beam into two parts in the vertical plane. The spherical mirror S_2 alternatively illuminates two objects (O_1 and O_2) that reflect the radiation. The spherical mirror S_3 directs these objects to the detector.

This device contains only three mirrors mutual to both channels, in addition to the monochromator. The transmitter chopper is the design's main advantage. But the channel intensities are not equal since different parts of the radiation source produce the beams. The electronic block must compensate for this drawback. The detected signal is amplified (1) and switched with frequency 8.4 Hz synchronously with the chopper. The differential amplifier (3) of the servomechanism (Sm) compares the direct voltage component (U_{dc}) with the reference signal (U_0). The Sm uses a large dynamic photocoupler, which is a combination of the diode and the photoresistor. The difference $U_{dc} - U_0$ is amplified and applied to the electro-luminescence diode (ED). The diode illuminates the photoresistor (PR), which is in the feedback circuit of amplifier 2. Therefore, the amplifier gain is

$$K = U_0/U_{dc} = const. \qquad (2.9-32)$$

The output signal of the amplifier 2 is

$$U_{out} = KU_{ac}, \qquad (2.9-33)$$

where U_{ac} is the alternating component.

The narrow-band differential amplifier, tuned to the modulation frequency, extracts the difference between two channels:

$$U_{dif} = U_{out}(1) - U_{out}(2). \qquad (2.9-34)$$

If the modulation amplitude is small, then the amplitude of the harmonic (with the modulation frequency 122 Hz) is proportional to the derivative. So,

$$U_{dc}(i) \approx I_i(\nu)R_i(\nu), \qquad (2.9-35)$$

$$K_i \approx U_0/U_{dc}(i) = U_0/[I_i(\nu)R_i(\nu)], \qquad (2.9-36)$$

$$U_{ac}(i) \approx [I_i(\nu)\, dR_i/d\nu + R_i(\nu)\, dI_i/d\nu]\Delta\nu, \qquad (2.9-37)$$

where $i = 1$ and 2 for the first and the second channel; $I_i(\nu)$ and $R_i(\nu)$ are the radiation intensities incident on and reflected by the sample O_i, respectively. By substituting Eq. (2.9-37) into Eq. (2.9-34), and using Eqs.(2.9-35) and (2.9-36) , we have

$$U_{dif} = K_i U_{dif} =$$

$$\{U_0/[I_1(\nu)R_1(\nu)]U_{ac}(1)\} - \{U_0/[I_2(\nu)R_2(\nu)]U_{ac}(2)\} \cong$$

$$U_0\Delta\nu\left\{\left[\frac{1}{R_1(\nu)}\frac{dR_1}{d\nu} - \frac{1}{R_2(\nu)}\frac{dR_2}{d\nu}\right] + \left[\frac{1}{I_1(\nu)}\frac{dI_1}{d\nu} - \frac{1}{I_2(\nu)}\frac{dI_2}{d\nu}\right]\right\}. \quad (2.9-38)$$

If the channels are identical, due to the special tuning, then the second term in the curly brackets is zero. Finally, we obtain

$$U_{dif} = U_0(\Delta R_1/R_1 - \Delta R_2/R_2), \quad (2.9-39)$$

where $\Delta R_i = \Delta\nu(dR_i/d\nu)$ is the differential of R_i corresponding to the increment $\Delta\nu$.

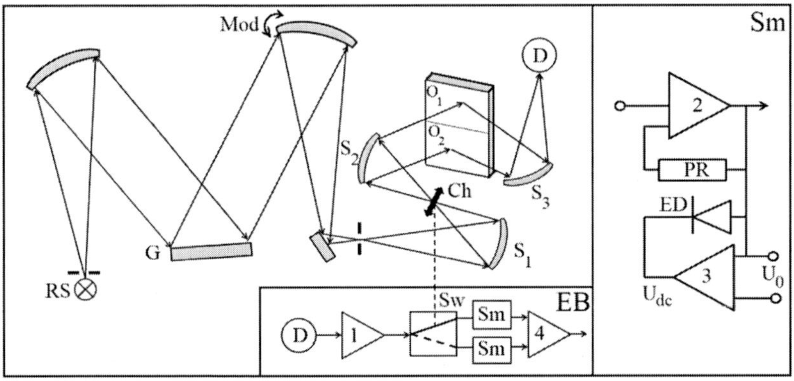

Figure 2.9-8. Setup of the double-beam modulation device (adapted from [26]): RS-radiation source, G-grating, Mod-modulator, Ch-light chopper, D-detector, EB-electronic block, Sw-switch, Sm-servomechanism, PR-photoresistor, ED-electroluminescent diode, and U_0- reference voltage.

Some elements of the device [26] were modified [27]. The feedback circuits, which were controlled by the photomultiplier tube's high voltage source, replaced the servomechanism (Sm). The conjunction with a sample-and-hold element provided a reflectivity measurement free of the intensive characteristic lines inherent to the arc lamp [27]. A high-speed chopper significantly increased the signal-to-noise ratio of the output data.

A motor-driven saw-tooth diaphragm, which was mounted before the entrance slit, compensated for the reference channel's alternating component [28].

A particular electronic block eliminated the dark current and the noises caused by the electrical power source [29]. This procedure was performed during the switching of the optical channels.

Since the modulation spectrometer simultaneously produces wavelength and amplitude modulations, the accurate frequency and phase proportions between these processes should be set [2, cross ref]. Also, for the nth-order differentiation, the electronic block must generate the modulation (ω_{mod}), the chopper (ω_{ch}), and combinations ($n\omega_{mod} \pm \omega_{ch}$) frequencies. The last frequency was the reference in the synchronous detector. The relation $\omega_{mod}/\omega_{ch} \geq 10$ provided a high signal-to-noise ratio. The tuning fork generator allowed researchers to satisfy the above requirements [2, cross ref]. The instability of ω_{mod}=1250 Hz was less than 10^{-4} %.

CHAPTER TEN

LASER MODULATION SPECTROSCOPY

Laser devices significantly expanded the technical opportunities of spectroscopy experiments due to their numerous advantages. Laser analyzers are powerful diagnostic tools widely used for the analytical control of technological processes.

More than 30 years ago, Prof. Zare [1] underlined the perspectives of laser analytical applications using multiphoton ionization (MPI) and laser fluorimetry (LF). In MPI, an atom or molecule absorbs more than one photon to cause the ejection of an electron. Resonant enhanced multiphoton ionization (REMPI) allows the following:

1. Elemental analysis and isotope ratio measurements.
2. High sensitive surface analysis.
3. Counting the individual atoms of nearly every element, one by one, in a gas sample or flow.

LF was used for measure the following:

1. Atomic constituents.
2. Velocity distributions and relative populations of excited state and fine structure levels of atoms sputtered from surface samples.

LF diagnostics were performed to check the quality of plasma devices and estimate the wall integrity of confined plasma sources. LF trace analysis of complex organic mixtures allowed researchers to detect a single molecule.

Raman spectroscopy [4] and IR absorption gas analysis widely use lasers [5].

Let us note some essential advantages of lasers [1-3]. First of all, they displace high-quality, expensive monochromators due to the narrow spectral width of the emitted radiation in the mid-IR (2-20 μm) (tunable diode lasers) and UV- VIS region (0.4-1.2 μm) (due lasers). Frequency-doubled lasers produce continuous UV laser emissions (200-400 nm). Secondly, researchers can analyze weak and strongly absorbing samples due to the high intensity of laser emission. Thirdly, high coherent beam directivity (the degree to which the laser emission is concentrated in a single direction) allows researchers to study remote objects and micro-samples. Finally, ultra-short powerful laser impulses significantly reduce the time required for analysis.

A reader unfamiliar with laser diodes' operational principles can familiarize himself with the simplified tutorial on this problem in Appendix K.

The wavelength or/and frequency modulation of emitted radiation significantly improved the efficiency of laser spectrometry. In semiconducting lasers, the injection current is modulated. In due lasers the modulation is performed by changing the resonator's length.

Theory

This paragraph is mainly based on article [6].

In the laser wavelength modulation (WM) spectroscopy, the modulated light emitted by a laser is slowly scanned through an analyte's absorption line. The injection current modulation in diode lasers produces a combined frequency (FM) and intensity modulation (IM) of the emitted beam with a phase difference between FM and IM.

A principal difference between the WM and FM techniques is the relationship between the modulation frequency (ν_m) and the width of the absorption line ($w = FWHM$). In the FM, $\nu_m \gg w$ is equal to several gigahertzes at atmospheric pressure, and the modulation amplitude $\Delta\nu \ll w$. Contrary, in the WM, ν_m is generally between some kilohertz and some tens of kilohertz and $\nu_m \ll w$.

The optical power (intensity) of the radiation, which was incident on the sample, measured in the center of the absorption line, is

$$I_0(x) = I_0(pwx + 1), \qquad (2.10-1)$$

where $x = (\nu - \nu_m)/w$ is the normalized frequency; $p(\nu)$ is the laser power.

According to the Beer-Lambert law, the light transmitted through the sample is

$$T_0(x) = I_0 exp[-\alpha(x)] \cong I_0[1 - \alpha(x)], \qquad (2.10-2)$$

where the absorbance of the sample $\alpha(x) \ll 1$.

In the pure FM [6],

$$x = x_0 - mcos(\omega t + \psi), \qquad (2.10-3)$$

where $m = \Delta\nu/w$ is the modulation index; the phase $\psi = 0$.

The output signal was represented as the infinite sum of harmonics,

$$S(x) = T_0(x_0 - mcos\omega t) = \sum_{n=0}^{\infty} s_n(x) cosn\omega t, \qquad (2.10-4)$$

where $s_n(x)$ was defined similar to Eq. (2.9-8) using the Lorentzian line shape.

For example,

$$s_2(x) = I_0\,\alpha(0)\Phi(x)/m^2, \qquad (2.10-5)$$

where $\Phi(x) = \{-4 + (\sqrt{2}/r)[(r + 1 - x^2)\sqrt{r + X} + 2|x|\sqrt{r - X}]\}$.

$X = 1 - x^2 + m^2;\ r = \sqrt{X^2 + 4x^2}$.

Reid and Labrie [7] obtained an expression similar to Eq. (2.10-5). They discussed the properties of the 2nd-order harmonic and showed that its experimental and theoretical shapes profiles were close to each other.

Figure 2.10-1 shows that for $m = 0.1$ the central part of the normalized 2nd -order harmonic has a minimum width. If m increases, then the harmonic broadens. The shape is constant for a small m. This case is referred to as derivative spectrometry characterized by a good resolution of the overlapping lines. However, for such an m, the signal-to-noise ratios are low. $m = 2.2$ was found to be the optimal value [6, 7].

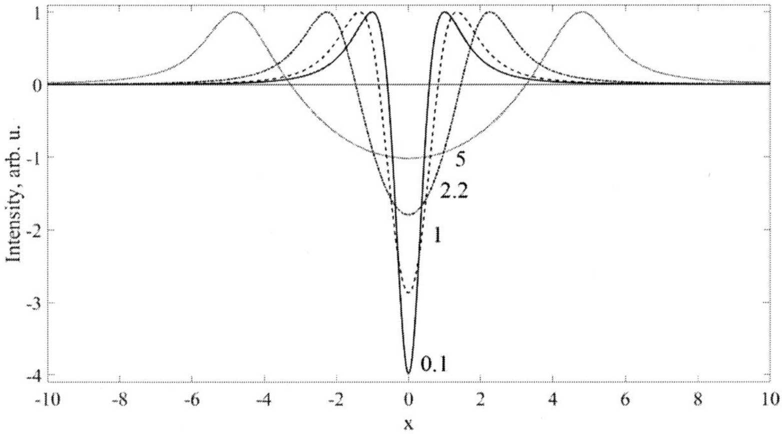

Figure 2.10-1. The normalized second harmonic (Eq. (2.10-5)). The numbers near the curves are m.

For the combined IM–FM [6],

$$I_0(x) = I_0(p_\Omega w x_0 + p_\omega w m cos\omega t + 1), \qquad (2.10-6)$$

where p_Ω and p_ω are the coefficients related to the low-frequency ramp (scanning the laser line through the absorption line) and the modulation frequency. For pure FM, $p_\Omega = p_\omega = 0$.

Eq. (2.10-4) is modified:

$$S_{IM-FM}(x) = \sum_{n=0}^{\infty} s_{np}(x) cosn\omega t - \sum_{n=0}^{\infty} s_{nq}(x) \sin n\omega t, \quad (2.10-7)$$

where s_{np} and s_{nq} are related to the in-phase (with the current modulation) and quadrature terms, respectively.

For example,

The second harmonic signals $S_{2v}(x)$ are

$$\begin{matrix} I_\Omega(x)cos2\psi\, s_2(x) - P_\omega[cos\psi\, s_1(x) + cos3\psi\, s_3(x)], v = p \\ I_\Omega(x)sin2\psi\, s_2(x) - P_\omega[sin\psi\, s_1(x) + sin3\psi\, s_3(x)],\ v = q \end{matrix} \quad (2.10-8)$$

where $P_\omega = p_\omega wm/2$; ψ is given in Eq. (2.10-3).

For the pure FM, $p_\omega = p_\Omega = 0$; $I_\Omega = 1$; therefore,

$$S_{2p}(x) = s_2(x)cos2\psi, S_{2q}(x) = s_2(x)sin2\psi. \quad (2.10-9)$$

Eq. (2.10-8) demonstrates that, for the small modulation amplitude, the combined IM-FM 2nd-order harmonic is not proportional to the corresponding derivative due to the impact of $s_1(x)$ and $s_3(x)$.

The theoretical model was found to be a good approximation of experimental data [6].

Di Rosa and Reiten [8] developed a new model [6]. They obtained very cumbersome formulas for the Voigt line shape without restrictions on the strength of absorption, the line-shape variables, or the FM parameters.

In nonlinear optical spectroscopy, the transmission of a weakly modulated beam, passed through a saturable-absorbing sample, is usually measured [9]. An intense pumping saturates the sample. The in-phase and in-quadrature components of the IM signal, which are measured by the square-law detector, are separated by the lock-in amplifier (Appendix J). The authors [9] studied IM spectra for ruby, alexandrite, and fluorescein in boric acid glass.

Conventional light sources of the high-resolution absorption spectrometers are characterized by thermal noise, which is independent of the light intensity [10]. However, in laser spectrometers, the source radiation is corrupted mainly by the flicker (shot) low-frequency noise, which is dependent on the light power incident on the detector [10, 11].

The author of the study [11] pointed out that the noise in the multipass cell laser devices is dependent on the number of reflections in a cell. They estimated the optimum absorption path length for the FM and WM diode laser absorption. It was shown that the multipass cell FM absorption spectrometers could improve the signal-to-noise ratio only by one order of magnitude.

High-order harmonics are usually detected using the lock-in amplifier. A mathematical theory of another method, which was based on the

Fourier analysis of the digitized light signal [12], was developed by Mei and Svanberg [13].

Techniques and applications of modulation laser spectroscopy

FM spectroscopy using the auxiliary sample modulation [14]

This is the most direct form of the laser FM spectroscopy in which an electro-optic phase modulator (EOM) adds radio frequency (RF) sidebands to a continuous wave laser (Fig. 2.10-2). The phase EOM controls the phase of light using an electrical signal, which modifies the refractive index of a nonlinear crystal in proportion to the strength of the electrical field produced by this signal [15]. This method is different from the light modulation performed by the absorbing sample disturbance (Chapter 2.9) (e.,g., the photochemical modulation). A brief description of this technique, which is used in transient spectroscopy, was reviewed by Gregory and North [14].

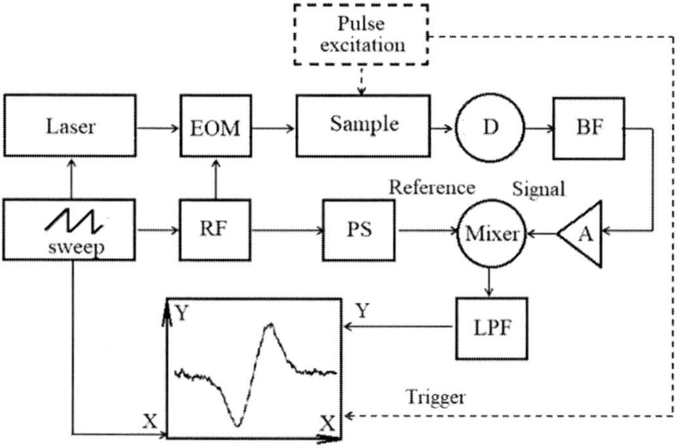

Figure 2.10-2. Block diagram of the auxiliary sample modulation device. EOM, electro-optic phase modulator; D, photo-detector (diode); BF, band-pass filter; A, amplifier; RF, radio frequency generator; PS, phase shifter; and LPF, low-pass filter (adapted from [14]).

In the experimental setup (Fig. 2.10-2), a double-balanced mixer performs the phase-sensitive detection of the photocurrent varying at the modulation frequency. A lowpass filter narrows the effective bandwidth of the detection system. The laser sweeps across an absorption line, which is scanned by recording the changing of the mixer's direct current output.

A transient sample's pulse excitation was used for kinetics and dynamics studies on the submicrosecond time scale. The transient waveforms of demodulated FM signals were processed on a sequence of fixed laser frequencies.

The disadvantage of the direct method is wavelength-dependent background signals. However, some improvements were made [14].

Weak absorption detection using a quantum-cascade laser [16]

One of the disadvantages of the semiconductor lasers is mode hopping (switching from one longitudinal mode to another), which occurs because their output spectrum depends strongly on case temperature and injection current. Quantum-cascade lasers (QCL), operated at room temperature, overcome this drawback. The QCL with a distributed feedback architecture (DFB) has no tuning gaps.

Figure 2.10-3 illustrates the detection of the 10% N_2O-90% N_2 mixture using the QC DFB laser. The ramp causes a weak temperature modulation of the laser. The sinusoidal current adds high-frequency (1.8 kHz) temperature modulation, which result in the WM. The derivative line shape of the 10% N_2O-90% N_2 mixture was observed. The study [16] demonstrated excellent sensitivity of the WM-based QC DFB laser gas detection.

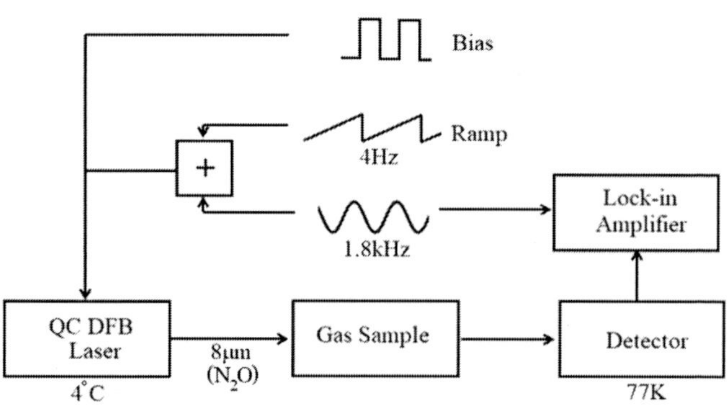

Figure 2.10-3. Experimental setup.(adapted from [16]).

Gas detection with continuous wave Optical Parametric Oscillator [17]

The authors of this study developed rapid detection, at the ppbv level sensitivity of gases, e.g., ethane, methane, and hydrogen cyanide. They used a fiber-amplified Distributed Bragg Reflector (DBR) diode laser to pump a continuous wave, singly resonant Optical Parametric Oscillator (OPO) (Appendix K) (Fig. 2.10-3). The input signal was modulated with 25 kHz modulation frequency on top of the 2 MHz applied to the DBR to improve the sensitivity. A lock-in amplifier demodulated the signal detected at the 1st-, 2nd-, and 4th-order harmonics.

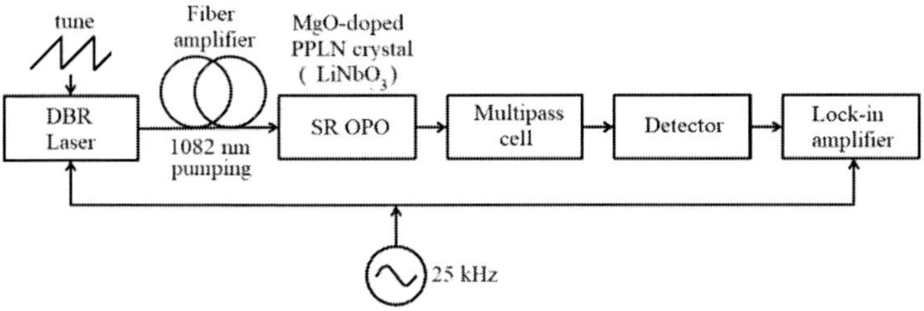

Figure 2.10-4. Simplified experimental setup (adapted from [17, 18]).

In the FM mode [18], the modulation was applied to the laser via the bias-tee unit (a diplexer that implements frequency-domain multiplexing). A double-balanced mixer demodulated the FM signal.

Readers, interested in an in-depth study of technique and some WM and FM spectroscopy applications, are directed to the reviews, oriented on laser specialists [14,19].

FM Fourier Transform spectroscopy [20]

The FM broad-band spectroscopy associates the advantages of the FM and high-resolution FT methods. The experimental pilot setup (Fig. 2.10-5) includes a fiber-coupled distributed feedback laser diode, whose current of which was modulated by a low-frequency ramp generator. The interferometer output light is phase-modulated by the electro-optical modulator (EOM). The InGaAs photodetector signal, which was mixed with the reference signal, was processed by the dual-phase lock-in amplifier (LIA). The LIA allowed detecting in-phase and in-quadrature components corresponding to the absorption and dispersion interferograms. The FT of these interferograms gave the final spectra.

The study in [21] showed that the FT modulation technique has distinct advantages compared with the traditional lock-in amplifiers technique: simple experimental setup, repidly available information on all the required harmonics, and high speed of data processing using FFT algorithm. A differential interpolation eliminated a systematic error (bias) of a signal parameters estimation using the Fourier coefficients interpolation method.

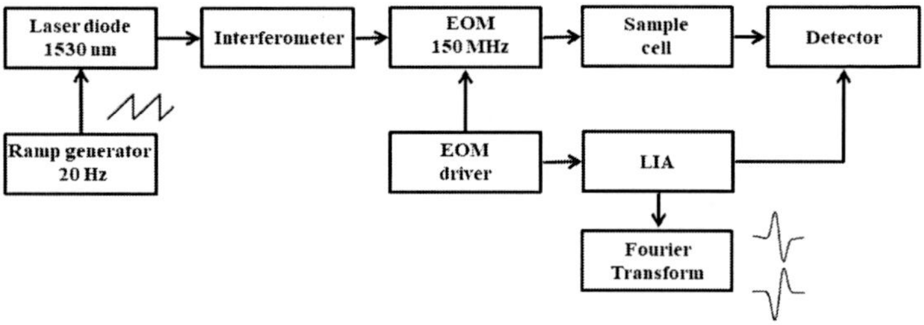

Figure 2.10-5. Experimental setup: EOM, electro-optic phase modulator; LIA, lock-in amplifier (adapted from [20]).

PART III:

ANALYSIS OF POORLY-RESOLVED SPECTRA STRUCTURE USING MATHEMATICAL DIFFERENTIATION

CHAPTER ONE

PEAK DETECTION

Among spectral peak parameters, the most important is the peak position, which is widely employed in theoretical and applied spectroscopy [3]. For example, identifying unknown elements and chemical compounds is usually performed by comparing the experimentally measured peak positions with those found in the databases. The peak maxima shifts are indicative of intra- and intermolecular interactions and changes in the external parameters (e.g., temperature). In this connection, the evaluation of the uncertainty of the peak maximum is of exceptional importance.

One should regard the existence of two types of errors that arise when solving this problem. The computational procedure [4] and instrumental factors [5] involve the first type of errors. Other errors are due to the component spectra's overlapping, which may cause apparent shifts [6]. The first type of errors can be easily eliminated by simple mathematical processing [4, 5] and careful recalibration of the spectrometer. In contrast to this, the resolution of overlapping peaks is a complicated task. Therefore, from a practical point of view, it seems reasonable to evaluate the upper limits of the apparent shifts and decide whether sophisticated computer methods should be further used in this particular case.

The peak must be detected before taking any measurements of the peak parameters. Statistical methods for detecting peaks involve testing the null hypothesis ("some samples of analytical signal (AS) belong to the peak") against the alternative hypothesis ("this is a noise sample") [7]. For this purpose, a baseline-free AS was processed by a digital filter. The result (the weighted sum of the AS values) was compared with a threshold. The drawback of this method is the broadening of the AS. Another one-point algorithm matched the 1st-order derivative of AS with some threshold. This algorithm does not require the correction of the constant baseline. However, since smoothing of the noisy AS is needed, the resolution of overlapping peaks decreases. Peak detection in the 2nd-order derivatives has some advantages due to removing the linear baseline and the resolution enhancement [7].

Zhang et al. [8] reviewed the peak detection algorithms used in Liquid-Chromatography-Mass Spectrometry (LC/MS) for protein identification and quantification. The LC elution profile of these objects demonstrated a structured asymmetrical shape that has some local maxima. Software packages for LC/MS analysis performed peak recognition using polynomial approximation and the wavelet transform. Peak shapes were Gaussians and the exponentially modified Gaussians.

The number of peaks in spectra and characteristic points of a spectral contour (maxima, shoulders, and inflection points) can be automatically estimated using the mathematical rules obtained by applying the derivatives in the investigation of functions [9] (Table 3.1-1, Fig. 3.1-1). Inflection points are not typical for elementary components (peaks) in spectra and chromatograms [10].

Unfortunately, the above rules do not take the noise present in spectra into account. Differentiation, especially of the high-orders, significantly increases this noise. A linear approximation of a small sign-changing curve segment around the zero point decreases the uncertainty when determining a position in which a transition over zero occurs.

Table 3.1-1. Using the derivatives in investigation of functions

$F^{(1)}$	$F^{(2)}$	$F^{(1)}$
$+ \to -$	-	Maximum
$- \to +$	+	Minimum
+	$+ \to -$	Left shoulder* (inflection point**)
-	$- \to +$	Right shoulder* (inflection point**)

Arrows point to the change of the sign.
* unresolved peaks; ** a single function.

The following conditions of the peak maximum (λ_m) involve high-order derivatives [11]:

$$F^{(2)}(x_m) < -\varepsilon_2; \;\; F^{(3)}(x_m) = 0 \; ; \;\; F^{(4)}(x_m) > \varepsilon_4, \qquad (3.1-1)$$

where $\varepsilon_n = \gamma_\alpha \sigma_n$; γ_α is defined by the confidence level α; and σ_n is the standard deviation of the nth-order derivative. The drawback of the criteria defined by Eq. (3.1-1) is the use of the noisy 3rd- and 4th-order derivatives. Also, the peak detection may be erroneous due to the false structure present in the derivatives.

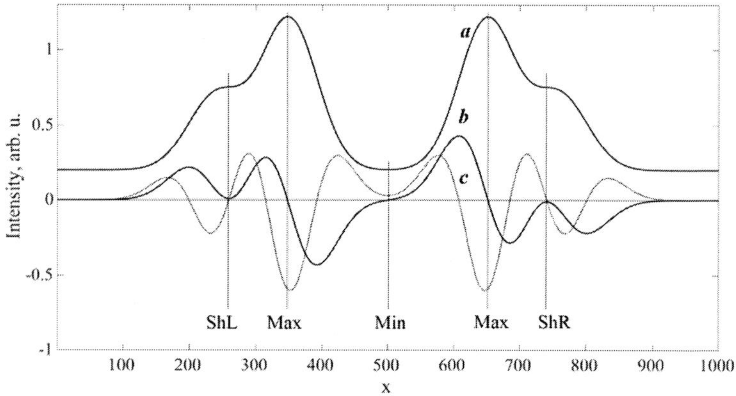

Figure 3.1-1. Identification of the characteristic points of contour (a) using the 1st-(b) and 2nd-order (c) derivatives. ShL and ShR stand for the left and right shoulders, respectively.

The false structure has random and systematic origins. In the first case, the structure occurs because of insufficient smoothing (noise peaks). In the second case, systematic instrumental factors, such as the Gibbs phenomenon (Appendix A), may produce false ripples. Also, these peaks are due to the shape features of the even-order derivative curves, including satellites. The following examples have an false structure (the illusion of the "super-resolution") (Fig. 3.1-2):

1. Overlapping of the central parts of the 2nd-order derivatives of the peaks; the maxima are close to each other but the widths are different, which produces two false peaks at the left and the right sides (panel A).
2. Overlapping of the satellites of the well-resolved 2nd- and 4th-order derivatives creates an incorrect central peak (panels B-D). The position of this peak does not depend on the doublet separation, but the intensity does. If the separation increases, then the top is split (panels C and D).
3. Weak positive satellites in the 4th-order derivatives look like real peaks. Their positions relative to the intensive central part do not depend on the doublet separation.

Figure 3.1-3 shows the Lorentz triplet and doublet (panels a and b, respectively):

$$F_{tr} = F_1 + F_2 + F_3, \quad F_{db} = F_1 + F_3, \qquad (3.1-2)$$

where $F_1 = 1/[1 + 4(x - 4.6)^2] + \eta_1$, $F_2 = 0.1/[1 + 4(x - 5.4)^2] + \eta_2$, and $F_3 = 0.7/[1 + 4(x - 6.2)^2] + \eta_3$; η_i is the normal noise with zero mean and the standard deviation $\sigma = 10^{-3}$. The 4th-order derivatives were obtained by the SG filter ($2t = 4, m = 50$).

The figure demonstrates that the false middle peak observed in the 4th-order derivative of the doublet (panel B) overlaps with the true derivative peak of the central triplet (panel A). So, the final decision of the middle peak's origin requires additional information, which can be obtained from the 2nd-order derivative (Fig. 3.1-4). However, for the accurate detection of an incorrect structure, one must perform a least-squares decomposition of the poorly-resolved contour into individual elementary peaks [12].

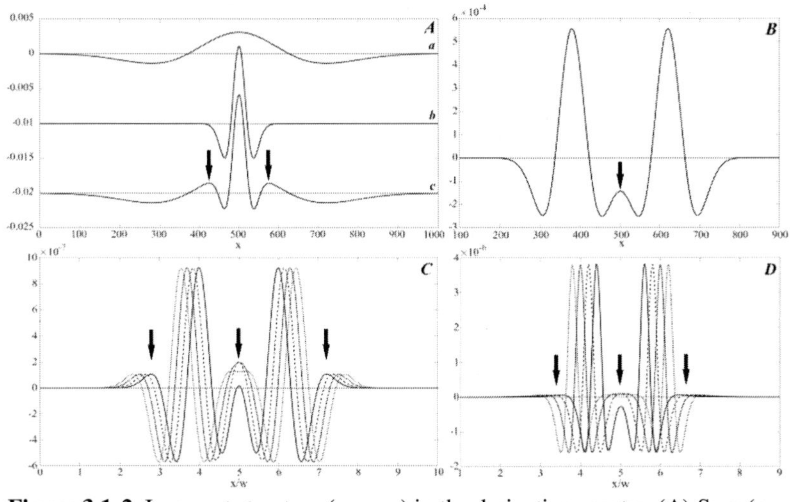

Figure 3.1-2. Incorrect structure (arrows) in the derivative spectra. (A) Sum (c = a+b) of the 2nd-order derivatives (equal peak positions, but different widths). (B) The well-resolved symmetrical doublet of the 2nd-(B) and 4th-order derivatives of (C) the Gaussian and (D) Lorentzian peaks: the relative peak separation: 2.00:0.30:2.90 (C) and 1.4:0.4:2.4 (D).

The informational content criterion of derivatives involved the expansion of the measured spectrum in the Taylor series (Eq. (1.1-4)) [13]. According to this criterion, the even-order derivatives (taken with the sign$(-1)^t$) added to the spectrum increase the intensity of its true structure components. However, the problem of choosing the empirical weights of the derivatives remains unsolved.

Recently, combining bad-resolved chromatographic peaks with their 1st- and 2nd-order derivatives allowed resolution enhancement [14]. This well-known variant of the deconvolution methods suffers from some drawbacks [15]. More mathematically rigorous results may be obtained using the least-squares decomposition procedures [12].

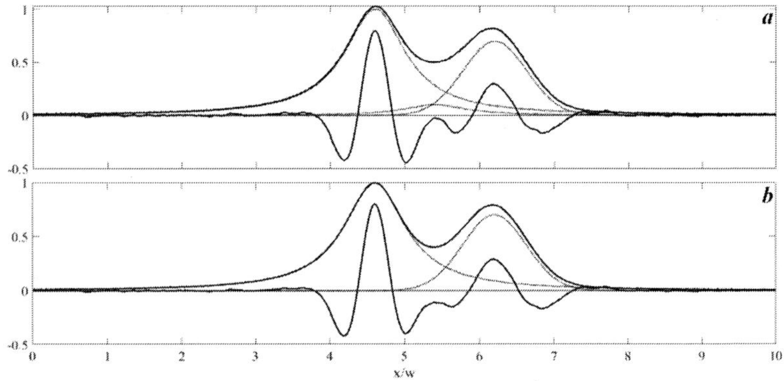

Figure 3.1-3. The origin of the middle peak in the 4th-order derivative of the Lorentzian triplet (a) (false + true peaks) and doublet (b) (false peak). The multiplet's components are plotted by the dotted curves.

The form parameter ($\varphi^{(n)}$, Table 1.2-2) may give a hint of the internal structure of the even-order derivatives of the unresolved doublet. For example, Figure 3.1-5a shows the noisy Lorentzian peak $(F_L(x))$ and the unresolved doublet $(F_{LD}(x))$:

$F_L(x) = 1/[1 + 4(x - 5)^2] + \eta_1$;

$$F_{LD} = 1/[1 + 4(x - 4.85)^2] + 0.5/[1 + 4(x - 5.15)^2] + \eta_2, \quad (3.1-3)$$

where η_i is the normal noise with zero mean and the standard deviation $\sigma = 0.01$. The negative 2nd-order derivatives of the peak and the doublet (normalized to the maximum intensity) are very similar. Still, the doublet form parameters, left $\varphi^{(2)}_{left}$ ≈0.40 and right $\varphi^{(2)}_{right}$ ≈0.37, are different, unlike both of the equal parameters of the single peak $\varphi^{(2)} \approx 0.40$. Unfortunately, this method only applies to the doublet wholly isolated from other structural components.

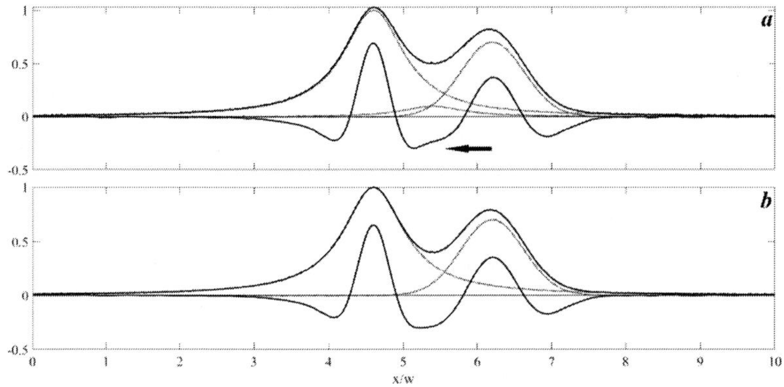

Figure 3.1-4. Identification of the possible false structure in the 4th-order derivative of the Lorentzian triplet (Fig. 3.1-3) using negative 2nd-order derivatives. The arrow points to the slight distortions of the valley between the central peaks.

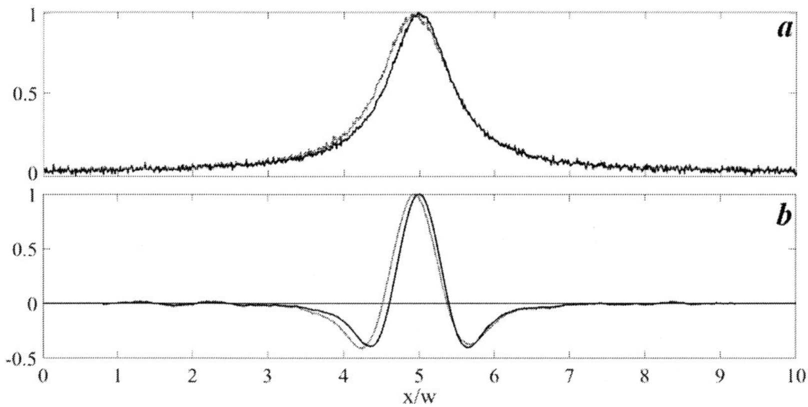

Figure 3.1-5. (a) Lorentzian peak and unresolved doublet; (b) their negative 2nd-order derivatives (solid and dotted curves, respectively) (Eq. (3.1-3)).

CHAPTER TWO

ERRORS OF THE PEAK MAXIMUM IDENTIFICATION IN THE 2ND-ORDER DERIVATIVE

The most-common estimation of the peak maximum positions in spectra is by using their even-order derivatives. There are two main reasons for this: the central (the most intensive) part of the derivatives has a shape that looks like the spectrum and it is noticeably narrower (Chapter 1.2). However, the satellites introduce shape distortions, and the signal-to-noise ratio (S/N) in derivatives decreases compared to the original spectra. The 2nd-order derivatives provide a compromise: a good resolving of overlapped peaks, a small decrease of the S/N relative to the higher-order derivatives, and only two satellites.

Let us consider the estimation of the peak positions, using the derivatives, from an informational-theoretical perspective [1]. Suppose that the maximum of a poor-resolved peak is located in the region [λ_1, λ_2] of a spectrum. Since this region is narrow, the probability density function of λ is uniform. Suppose that the random measurement errors of λ have a normal distribution with zero mean and the standard deviation σ_λ. Then the information quantity, which was acquired during the estimation of the peak position in the spectrum (Chapter 1.5), is

$$I^{(0)} = ln\left[(\lambda_2 - \lambda_1)/\left(\sigma_\lambda/\sqrt{2\pi e}\right)\right] - 0.5B_\lambda^2, \qquad (3.2-1)$$

where $B_\lambda = \delta_\lambda/\sigma_\lambda$; δ_λ is the systematic error of the λ determination.

In the derivative spectrum

$$I^{(n)} = ln\left[(\lambda_2 - \lambda_1)/\left(\sigma_{\lambda d}/\sqrt{2\pi e}\right)\right] - 0.5B_{\lambda d}^2, \qquad (3.2-2)$$

where subscript d stands for derivative.

From Eqs. (3.2-1) and (3.2-2) we have the following information change:

$$\Delta I = I^{(n)} - I^{(0)} = ln(\sigma_\lambda/\sigma_{\lambda \mathrm{d}}) - 0.5(B_{\lambda d}^2 - B_\lambda^2). \qquad (3.2-3)$$

We found the random errors σ_λ and $\sigma_{\lambda \mathrm{d}}$ in the maxima of the Gaussian and Lorentzian peaks:

$F_G = \exp[-4ln2(\lambda - 5)^2]$ and $F_L = 1/[1 + 4(\lambda - 5)^2]$, ($\lambda = [0.01\colon 0.01\colon 10]$), distorted by the normal noise with zero mean and the standard deviation σ_λ =0.01. The accurate maximum positions in the noisy peaks and their 2nd-order SG derivatives were calculated by the quadratic approximation of the data in the range [4.9:0.01:5.1] near $\lambda_{max} = 5$. The SG filter parameters were: $2t = 2$, $m =$ 30, 40, and 50. The standard deviations σ_λ and $\sigma_{\lambda d}$ were estimated from 10^5 repetitions. The ratio $\sigma_\lambda/\sigma_{\lambda d}$ =[0.94, 1.42, 3.36] and [1.08, 1.66, 2.08] for the Gaussian and Lorentzian peaks, respectively.

So, for the smoothed derivative (m is large), $\sigma_\lambda/\sigma_{\lambda d} > 1$. Therefore, the first term of Eq. (3.2-3) is positive. The second term is proportional to $[(\sigma_\lambda/\sigma_{\lambda d})(\delta_{\lambda d}/\delta_\lambda)]^2 - 1$. Due to the resolution improvement in the derivative, the ratio is $R_\delta = \delta_{\lambda d}/\delta_\lambda < 1$ for the poor-resolved peak, but R_δ depends on the smoothing factor [2]. We conclude that information gains when determining the overlapped peaks maxima's locations, using derivative spectra, can be achieved through the trade-off between noise suppression and resolution enhancement. Further, we will estimate the peak location uncertainties in the 2nd-order derivatives [2].

Models

Consider the doublet, which is defined as the function of a dimensionless argument,

$$D(x) = F(\Delta_x^+) + RF(\Delta_x^-), \qquad (3.2-4)$$

where $\Delta_x^+ = x + x_0$; $\Delta_x^- = (x - x_0)/r$; $x = \beta i/w$; β is the peak shape parameter; $i = -M, -M+1, \ldots, 0, \ldots, M-1, M$; $w = FWHM$; $x_0 = \beta\delta/2$; $\delta = \Delta/w$ is the relative separation of the doublet components; $\Delta = 2i_0$ is the absolute separation; $\pm i_0$ are the peak maxima; R and r are the relative intensity and the relative width of the second peak, respectively.

The 2nd-order derivatives of the doublet (Eq. (3.2-4), composed of the Gaussian $\left(F_G = \exp(-x^2), \beta_G = 2\sqrt{ln2}\right)$ and Lorentzian $(F_L = 1/(1 + x^2), \beta_L = 2)$, functions, are

$$D_G''(x) = 2T_G(\Delta_x^+)F_G(\Delta_x^+) + Z * T_G(\Delta_x^-)F_G(\Delta_x^-), \qquad (3.2-5)$$

$$D_L''(x) = 2T_L(\Delta_x^+)F_L(\Delta_x^+) + Z * T_L(\Delta_x^-)F_L(\Delta_x^-), \qquad (3.2-6)$$

where $T_G(t) = 2t^2 - 1$; $Z = 2R/r^2$; $T_L(t) = (3t^2 - 1)/(1 + t^2)$.

Relative shift

The relative shift of the peak position (ξ) is usually measured relative to the peak width. However, the widths of overlapping peaks can be evaluated very approximately [3]. Therefore, we chose to calculate ξ as a proportion of the doublet separation, which can be readily measured visually. In this case, for the ith-peak of the resolved doublet,

$$\xi_i = (x_{mi} - (-1)^i x_0)/2\,x_0), \qquad (3.2-7)$$

where x_{mi} is the zero point of the derivatives of Eqs. (3.2-5) and (3.2-6):

$$D'''(x_{mi}) = 0. \qquad (3.2-8)$$

The $D'''(x)$ changes sign when it passes through point x_{mi} of the well-resolved peak.

At and below the resolution limit of the 2nd-order derivative of the symmetrical doublet ($R = 1, r = 1$), the condition, defined by Eq. (3.2-8) is valid in the central point $x = 0$. This point may be considered as an apparent position of the peak maxima. Therefore, the absolute value of the relative shift of the point x_0 from the zero point, $|\xi| = x_0/(2x_0) = 0.5$, does not depend on the doublet components' separation.

Since the analytical solution of Eq. (3.2-8) is too complicated, we solved this equation numerically.

The shifts of a non-ideal derivative peak also depend on the parameters of the differentiation procedure [3]. Smoothing of noisy spectra causes the broadening of the derivative spectra. Therefore, the resolution decreases, and the shifts will be different from those obtained for perfect derivatives. These effects are only significant if the separation is near the resolution limit, where the displacements are substantial [3]. In such poorly-resolved derivatives, precise peak identification leads to erroneous results.

Results of Computer Modelling and Discussion

Equal-width peaks ($r = 1$)

Figure 3.2-1 represents the dependencies of the relative shifts of the 2nd-order derivative peaks on the parameter x_0 for the Gaussian and Lorentzian doublets $\left(\xi(x_0)\right)$. On the strength of symmetry,

$$\xi_1^{(r=1)}(R, x_0) = -\xi_2^{(r=1)}(1/R\,, x_0). \qquad (3.2-9)$$

The plots of the shift dependencies for the first component of the Gaussian ($\xi_{1G}^{(r=1)}(R, x_0), R = 1/3 \div 1$) and Lorentzian ($\xi_{1L}^{(r=1)}(R, x_0)$,

$R = 0.2 \div 1$) doublets pass through the intersection points, 0.612 and 0.500, respectively.

According to the symmetry rule (Eq. (3.2-9)), the plots related to the second peak also pass through these intersection points. The plots are cut off at the resolution limit, for the rest of the values R, and do not reach the intersection points. The shifts are zero at the intersection points, where the 2nd-order derivative minima coincide with the satellite maxima of the interfering component (Fig. 3.2-2). The sign of the slope of the interfering peak changes when the curve passes the intersection point. The shifts' signs are opposite on the left and the right from this point.

On the right of the intersection point, the shift values for the Lorentzian derivatives are smaller than those for the Gaussian derivatives. On the left, the Gaussian peaks are not resolved, except for $R = 1$.

Although the more intense second component (larger R values) causes more significant shifts of the first component, its shifts decrease. Thus, if $R_1 > R_2$, then

$$\begin{matrix} \left|\xi_1^{(r=1)}(R_1,x_0)\right| > \left|\xi_1^{(r=1)}(R_2,x_0)\right| \\ \left|\xi_2^{(r=1)}(R_1,x_0)\right| < \left|\xi_2^{(r=1)}(R_2,x_0)\right| \end{matrix}. \qquad (3.2-10)$$

A high-degree polynomial well approximates the shifts' dependencies on the doublet components' separation relative to the variable $A = 1/x_0^2$:

$$\xi_i(R,x_0) = \sum_{p=0}^{5} a_{pi}(R)A^p, \ i = 1,2. \qquad (3.2-11)$$

Figure 3.2-3 represents the dependencies of the polynomial coefficients on the relative intensity of the second doublet peak $\left(a_{p2}(R)\right)$ (Eq. (3.2-11)). The plots for the Gaussian and Lorentzian peaks ($a_{p2}^G(R)$ and $a_{p2}^L(R)$) demonstrate different patterns.

For the large R values, the shifts increase inversely proportional to the two highest degrees of x_0 (Eq. (3.2-11)), namely, 10 and 8 for the Lorentzian, and 8 and 4 for the Gaussian doublets. Fortunately, for these R values, the absolute shifts of the Lorentzian peaks are small, but not in the case of the Gaussians, where the steep slopes of function $\xi_i(x_0)$ may cause noticeable sensitivity of the peak maxima to the noise.

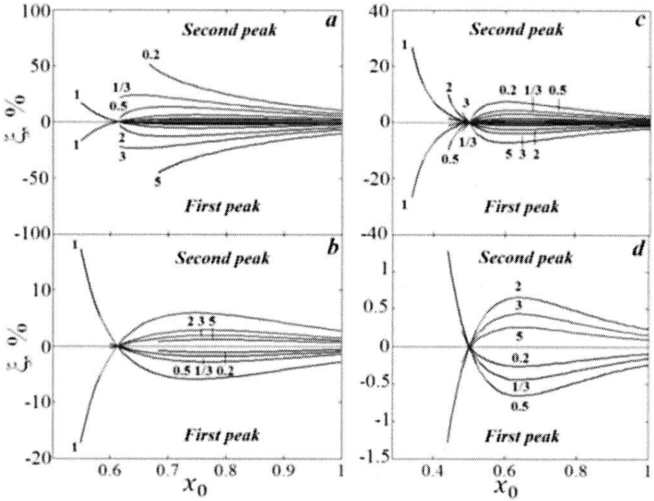

Figure 3.2-1 [2]. The relative shifts of the 2nd-order derivatives of the Gaussian (a, b) and Lorentzian (c, d) doublets, consisting of equal-width peaks. The numbers near the curves are R values.

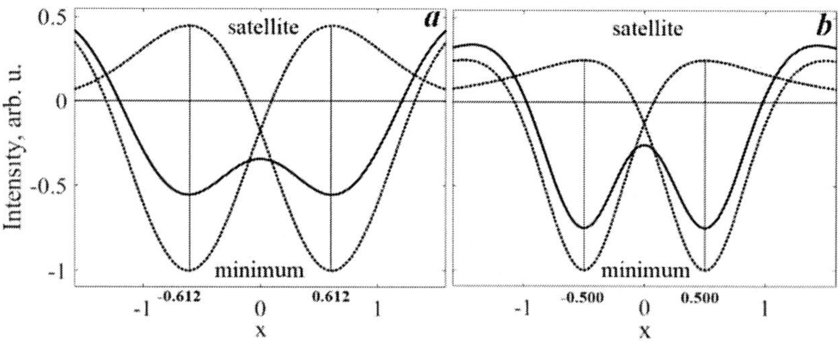

Figure 3.2-2 [2]. The zero-shift points in the 2nd-order derivatives of the Gaussian (a) and Lorentzian (b) doublets.

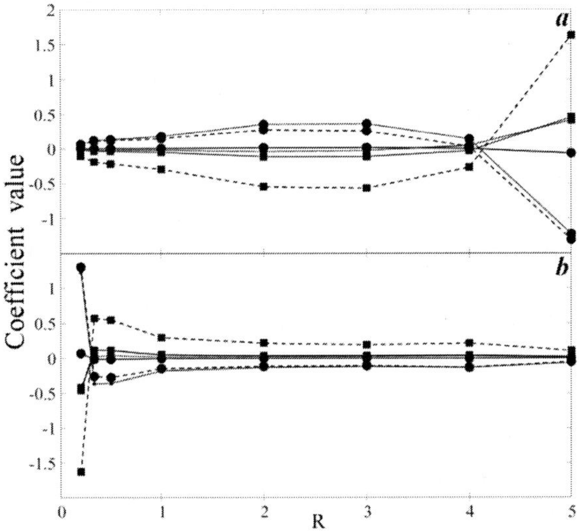

Figure 3.2-3 [2]. The polynomial coefficients (for the second component) of the 1st - (a) and the 2nd-order (b) derivatives of the Gaussian doublet peaks: a_5 (circles) and a_4 (squares); a_3 (circles) and a_2 (squares); a_1 (circles) and a_0 (squares) (solid, dotted, and dashed curves, respectively).

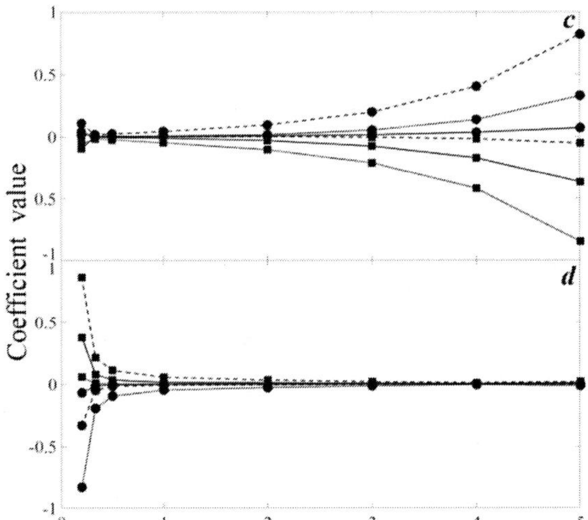

Figure 3.2-3 [2] (continued). The same data for the Lorentzian peaks.

Unequal-width peaks $(r \neq 1)$

The shifts of the second peak, which is broader than the first one $(\xi_2^{(r>1)})$, are substantial, especially for the Gaussian doublets (Figs. 3.2-4 and 3.2-5). $\xi_2^{(r>1)}$ significantly increases with the increase of r (Figs. 3.2-8 - 3.2-14, R = const) because the intensity of the 2nd-order derivative spectrum is inversely proportional to the squared peak width [3]. The shift of the second peak is only less than 0.1, if its intensity is large enough ($R \geq 2$), and the doublet components are well separated (Table 3.2-2). The shifts of the first peak are an increasing function of intensity R. If the second peak is narrower than the first peak ($r < 1$), then the negative shifts of the first peak become very large, since the intensity of the 2nd-order derivative increases (Figs. 3.2-6 and 3.2-7).

The plots, which correspond to the second component of the Gaussian $\left(\xi_{2G}^{(r<1)}(R, x_0)\right)$ and Lorentzian $\left(\xi_{2L}^{(r<1)}(R, x_0)\right)$ doublets, pass through the intersection points (0.612 and 0.500, respectively) (Figs. 3.2-6 and 3.2-7, panels c and d). Figure 3.2-2 shows that the shifts are zeros at these points, where the narrow 2nd-order derivative peaks coincide with the satellite maxima. If $r > 1$, then the broad derivative peaks are shifted to the region outside the satellite, and thus no intersection points can appear.

For R = const, if r grows from 1/3 to 1, then, for the Gaussian doublets, the absolute values of the first peak' negative shifts increase (Figs. 3.2-8 - 3.2-14, panels a). A further broadening of the second peak ($r = 2$) results in decreasing the shift and changing its sign to positive. For the Lorentzian doublets, the signs and the ordinates of the $\xi_{1L}^{(R=const)}(r, x_0)$ plots (Figs. 3.2-8 - 3.2-14, panels b) vary in a complicated manner depending on the location of the intersection points.

In the 2nd-order derivative, the overlapping of the second peak with the satellite of the first peak may generate a "super-resolution" pattern even at a very low separation of the doublet components (Fig. 3.2-15, a). Near this separation ($x_0^{(min)}$), the shifts of the weak second peak increase very quickly and may be more than ten times as large as the $x_0^{(min)}$ for the Gaussian shapes (Fig. 3.2-5, a). The symmetrical false "resolved" peak (denoted by an arrow on the left of Fig. 3.2-15, a) indicates that the right-hand component is wrong.

The positions of the real and the false peaks are the same, only if $x_0 = 1$ (Fig. 3.2-15, panel b). This "super-resolution" gives rise to significant errors in analysis.

Acceptable relative shifts ($\xi < 0.1$) for $r > 1$ and $r < 1$ for the first and the second peaks, respectively, are sometimes observed for smaller separations (δ) of the Gaussian doublets than those of the Lorentzian ones (marked in bold in the Table 3.2-2). In other words, for a given separation value, the peak of the narrow, intense 2nd-order derivative of the Gaussian doublet may be less shifted from its actual position than that of the Lorentzian doublet. These different shifts are accounted for by different slopes of the interfering derivatives of the Gaussian and Lorentzian peaks.

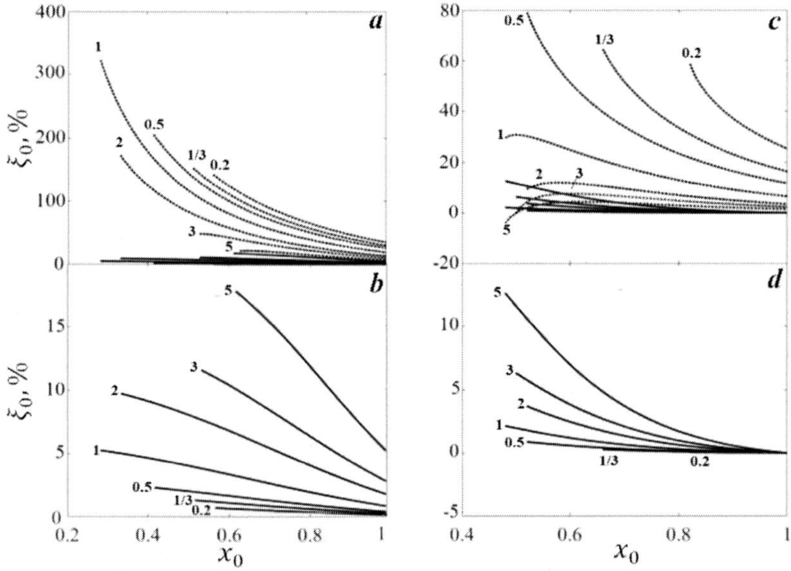

Figure 3.2-4 [2]. The relative shift of the first and the second peaks (solid and dotted curves, respectively) of the 2nd-order derivatives of the Gaussian(a, b) and Lorentzian (c, d) doublets. $r = 2$. The numbers near the curves are R.

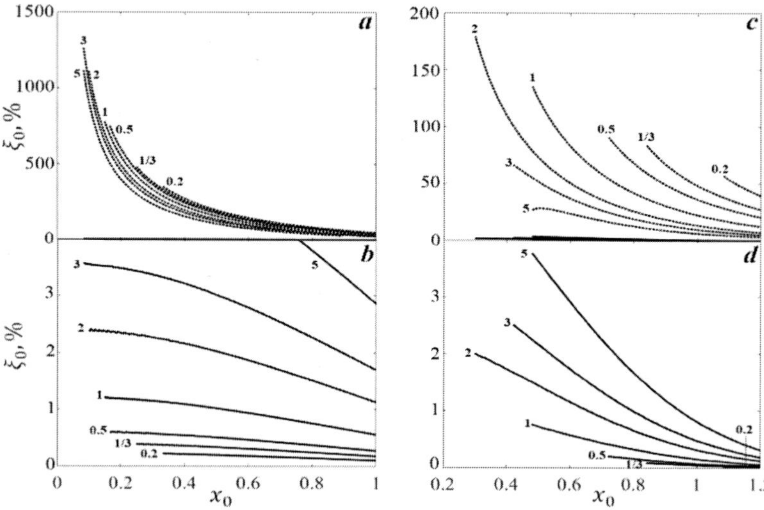

Figure 3.2-5 [2]. The relative shift of the first and the second peaks (solid and dotted curves, respectively) of the 2nd-order derivatives of the Gaussian (a, b) and Lorentzian (c, d) doublets. $r = 3$. The numbers near the curves are R.

We conclude that the correctness of the peak locations measured in the 2nd-order derivative spectrum should be checked, in each particular case, by computer modelling the overlapping peaks. The shifts, connected with the changes of the physicochemical parameters of the sample under study, must be differentiated from apparent shifts, which may be caused by changes of the peak's shape, width, and degree of overlapping. For this reason, correlating the peak shifts in the 2nd-order derivative spectrum with physicochemical parameters may lead to erroneous conclusions.

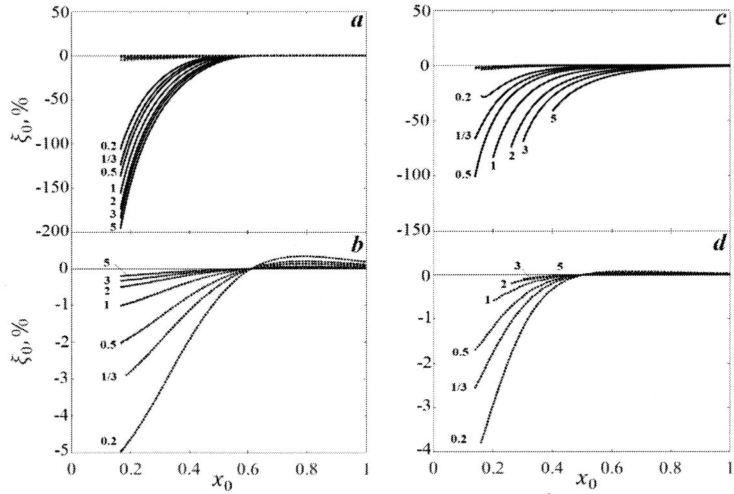

Figure 3.2-6 [2]. The relative shifts of the 1st and the 2nd peaks (solid and dotted curves, respectively) of the 2nd-order derivatives of the Gaussian (a, b) and Lorentzian (c, d) doublets. $r = 1/3$. The numbers near the curves are R.

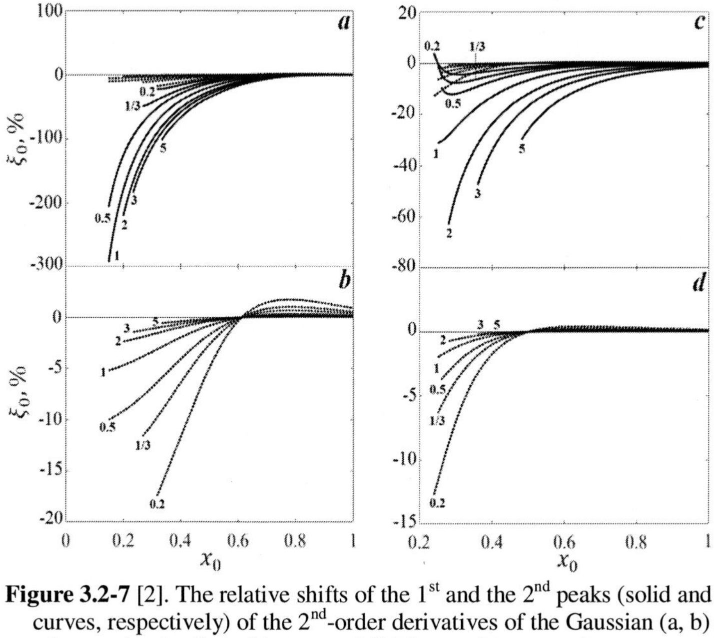

Figure 3.2-7 [2]. The relative shifts of the 1st and the 2nd peaks (solid and dotted curves, respectively) of the 2nd-order derivatives of the Gaussian (a, b) and Lorentzian (c, d) doublets. $r = 0.5$. The numbers near the curves are R.

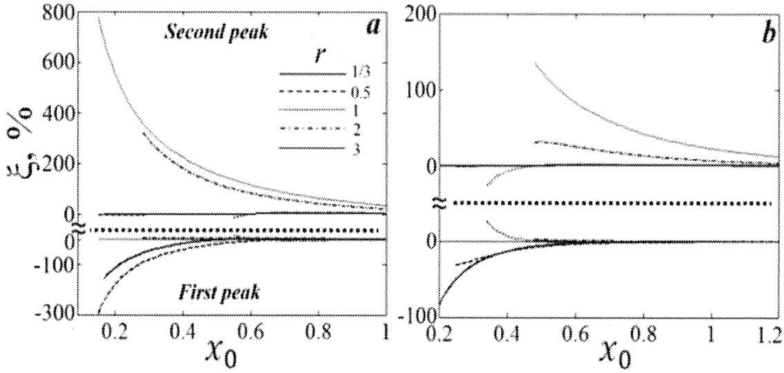

Figure 3.2-8 [2]. The relative shift of the equal-intensity ($R = 1$) Gaussian and Lorentzian peaks (panels a and b, respectively). The bold dotted lines separate the data between the first and the second doublet peaks.

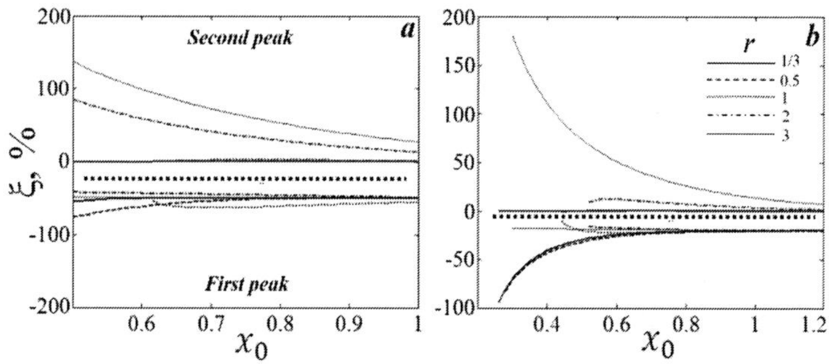

Figure 3.2-9 [2]. The relative shifts of the Gaussian (a) and Lorentzian (b) peaks. $R = 2$. The plots of the first peak are shifted down by 50% (a) and 20% (b) The bold dotted lines separate data between doublet peaks.

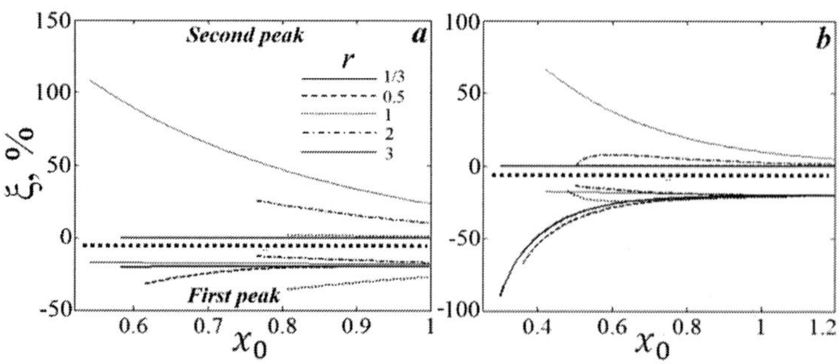

Figure 3.2-10 [2]. The relative shifts of the Gaussian (a) and Lorentzian (b) peaks. $R = 3$. The plots of the first peak are shifted down by 20%. The bold dotted lines separate data between doublet peaks.

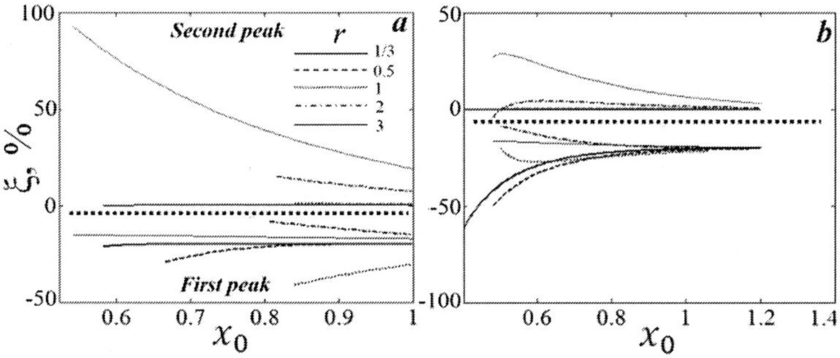

Figure 3.2-11 [2]. The relative shifts of the Gaussian (a) and Lorentzian (b) peaks. $R = 5$. The plots of the first peak are shifted down by 20%. The bold dotted lines separate the data between the doublet peaks.

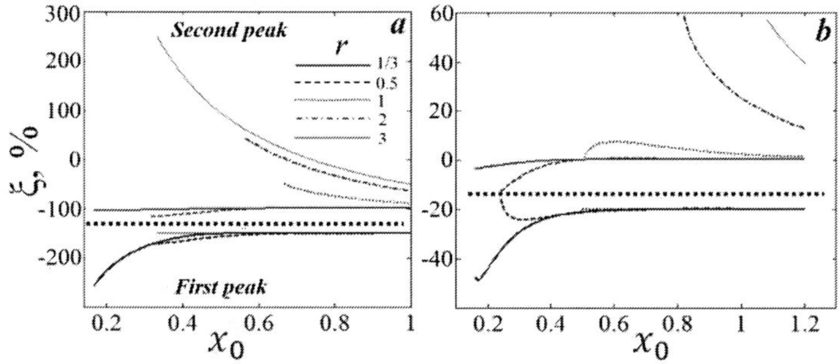

Figure 3.2-12 [2]. The relative shifts of the Gaussian (a) and Lorentzian (b) peaks. $R = 3$. .The plots of the first peak are shifted down by 50% (a) and 20% (b).The bold dotted lines separate the data between the doublet peaks.

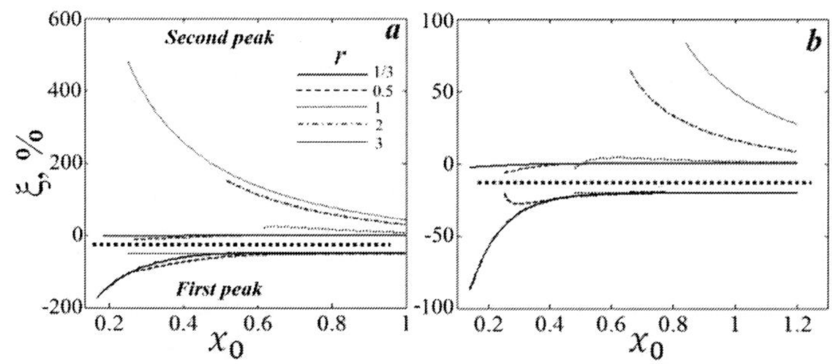

Figure 3.2-13 [2]. The relative shifts of the Gaussian (a) and Lorentzian (b) peaks. $R = 1/3$. The plots of the first peak are shifted down by 50% (a) and 20% (b).The bold dotted lines separate the data between the doublet peaks.

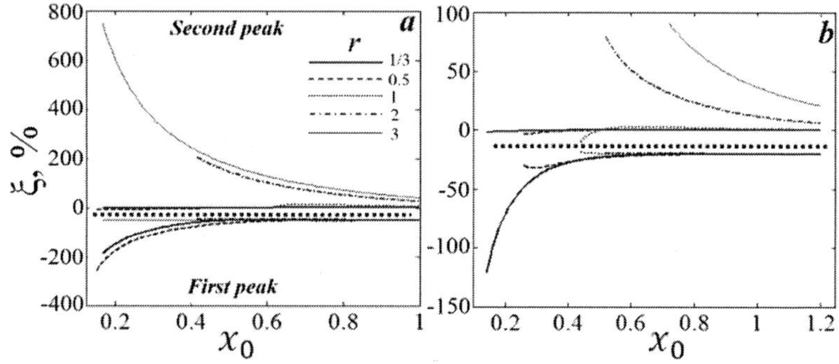

Figure 3.2-14 [2]. The relative shifts of the Gaussian (a) and (b) Lorentzian peaks. $R = 0.5$. The plots of the first peak are shifted down by 50% (a) and 20% (b).The bold dotted lines separate the data between the doublet peaks.

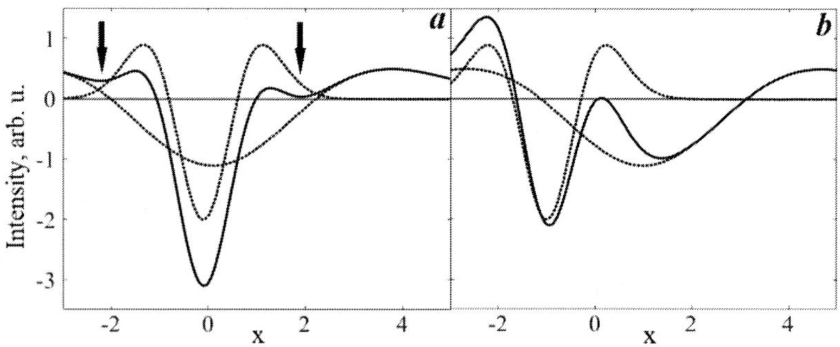

Figure 3.2-15 [2]. False (designated by arrows) and accurate peaks in the 2nd-order derivatives of the Gaussian doublets. $R = 5, r = 3$. $x_0 = 0.1$ (a) and 1 (b). The dotted curves represent the doublet components.

Table 3.2-2 [2]. Minimum relative separation values where relative shifts do not exceed 0.1

R \ r	1/3		0.5		1		2		3	
	G	L	G	L	G	L	G	L	G	L
0.2	0.46	0.30	0.56	0.24	0.80	0.50	**0.68**	**0.82**	**0.40**	**1.1**
	0.20	0.16	0.56	0.27	>1	0.50	>1	>1	>1	>1.2
1/3	0.48	0.34	0.61	0.25	0.74	0.48	**0.62**	**0.66**	**0.30**	**0.84**
	0.22	0.14	0.38	0.25	>1	0.48	>1	>1	>1	>1.2
0.5	0.50	0.37	0.64	0.26	0.74	0.44	**0.50**	**0.52**	**0.20**	**0.72**
	0.20	0.14	**0.18**	**0.26**	0.74	0.44	>1	>1	>1	>1.2
1	0.53	0.43	0.69	0.44	0.68	0.39	**0.34**	**0.48**	**0.18**	**0.48**
	0.20	0.20	**0.17**	**0.25**	0.68	0.39	>1	0.88	>1	>1.2
2	0.55	0.50	0.73	0.53	0.74	0.44	**0.40**	**0.52**	**0.12**	**0.30**
	0.20	**0.26**	**0.24**	**0.28**	0.74	0.44	>1	0.52	>1	1.1
3	0.56	0.54	0.76	0.58	>1	0.48	0.75	0.50	**0.10**	**0.42**
	0.20	**0.30**	**0.28**	**0.36**	0.74	0.48	>1	0.50	>1	1.0
5	0.58	0.59	0.79	0.65	>1	0.50	>1	0.54	**0.10**	**0.48**
	0.20	**0.40**	**0.40**	**0.48**	0.82	0.50	>1	0.48	>1	0.88

The relative separations of the first and second components (upper and lower rows) of the Gaussian (G) and Lorentzian (L) doublets. The values in bold correspond to $\delta_L > \delta_G$.

CHAPTER THREE

DETERMINATION OF THE PEAK INTENSITIES AND WIDTHS IN THE 2ND-ORDER DERIVATIVE

Peak intensity measurements

From the previous chapter, one can see that mathematical methods of estimating of the resolved peak positions in spectra and their even-order derivative spectra are similar. Generally speaking, the direct evaluation of the peak intensities using derivatives is cumbersome since they are inversely proportional to the peak width in terms of a power that is equal to the differentiation order (Chapter 1.2). The measurements of the poorly-resolved peak widths suffer from significant errors [1]. Therefore, the precise correction of the derivative peak intensity for the width is impossible. Also, four satellites in the 4th-order derivative produce false negative and positive peaks, which distort intensities of the true spectra components.

Based on the above, we studied uncertainties only when determining the relative peak intensities of the 2nd-order derivatives of the Gaussian and Lorentzian doublets consisting of the equal peak width:

$$D(x) = F(\Delta_x^+) + RF(\Delta_x^-), \qquad (3.3-1)$$

where $\Delta_x^+ = x - x_c + x_0$; $\Delta_x^- = x - x_c - x_0$; $x = \beta i/w$; β is the peak shape parameter; $i = 1{:}1000$; $w = 100$ is the *FWHM*; $x_c = i_c\beta/w$ is the central point; $i_c = 500$; $x_0 = \beta\delta/2$; $\delta = \Delta/w$ is the relative separation of the doublet components; Δ is the absolute separation; $x_c \pm x_0$ are the doublet peaks positions; and R is the relative intensity of the second peak. The SG derivative filter parameters were: $2t = 2$, $m = 2$ and 30 (precise and smoothed derivatives, respectively).

Figure 3.3-1 illustrates the zero-line $(R_{zl} = I_2/I_1)$ and baseline $\left(R_{bl} = (I_2 + \Delta_1)/(I_1 + \Delta_2)\right)$ measurements of the relative intensity. Figures 3.3-2 and 3.3-3 represent dependences of the relative errors (ξ_{meas}) on the relative separation (δ) for the Gaussian and Lorentzian doublet components using R_{zl} and R_{bl} data.

$$\xi_{meas} = R_{meas}/R - 1. \qquad (3.3-2)$$

The dependences $\xi_{meas}(\delta)$ allowed us to draw the following conlclusions:

1. For the zero-line measurements, the relative errors are negative. For the baseline measurements, the errors change the sign from negative to positive at some point δ_s, while the separation increases. Near δ_s, $|\xi_{bl}(\delta)| \ll |\xi_{zl}(\delta)|$.
2. For the large separations, our experiments showed that the false structure elements in the derivatives should not be confused with true peaks when calculating the value of R_{meas}.
3. If R increases, then ξ_{meas} decreases.
4. In all cases, the smoothing ($m = 30$) increases the errors.
5. For small doublet separations, the relative errors may be tens of percent.

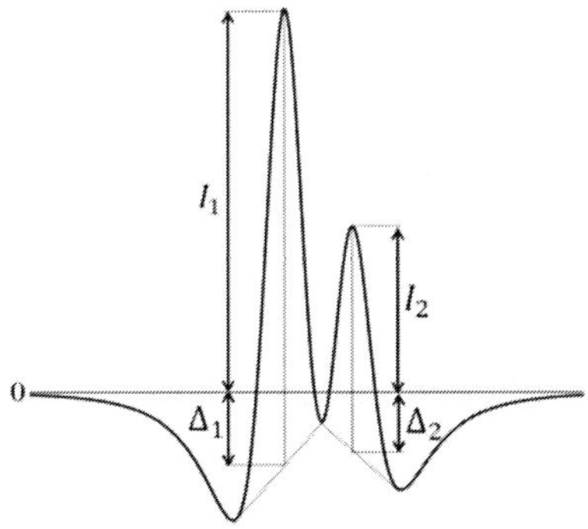

Figure 3.3-1. Measurements of the relative intensity of the 2nd doublet peak in the negative 2nd-order derivative from the zero line: $R_{zl} = I_2/I_1$ and the baselines:$R_{bl} = (I_2 + \Delta_1)/(I_1 + \Delta_2)$.

Peak width measurements

Direct method

The peak width (w) is estimated by the measuring of the width of its 2nd-order derivative on the two horizontal levels (a-b and c-d) of the central part (Fig. 3.3-4).

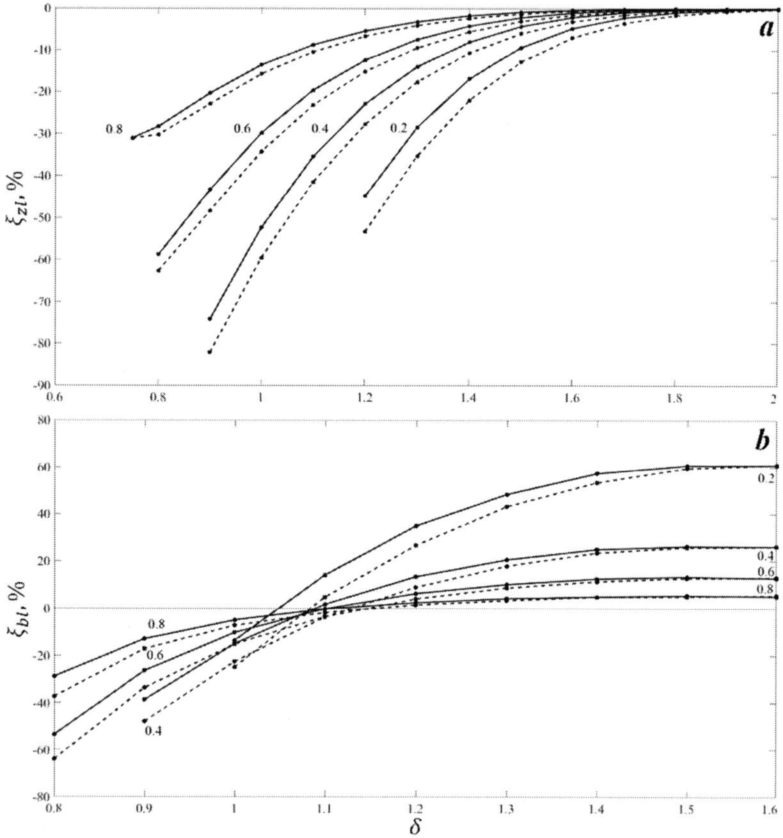

Figure 3.3-2. The relative measurement errors of the relative intensity of the second doublet peak in the 2nd-order derivative of the Gaussian doublets using the zero-line (a) and the baseline (b) methods. The width of the SG filter $m = 2$ and 30 (solid and dashed lines, respectively).

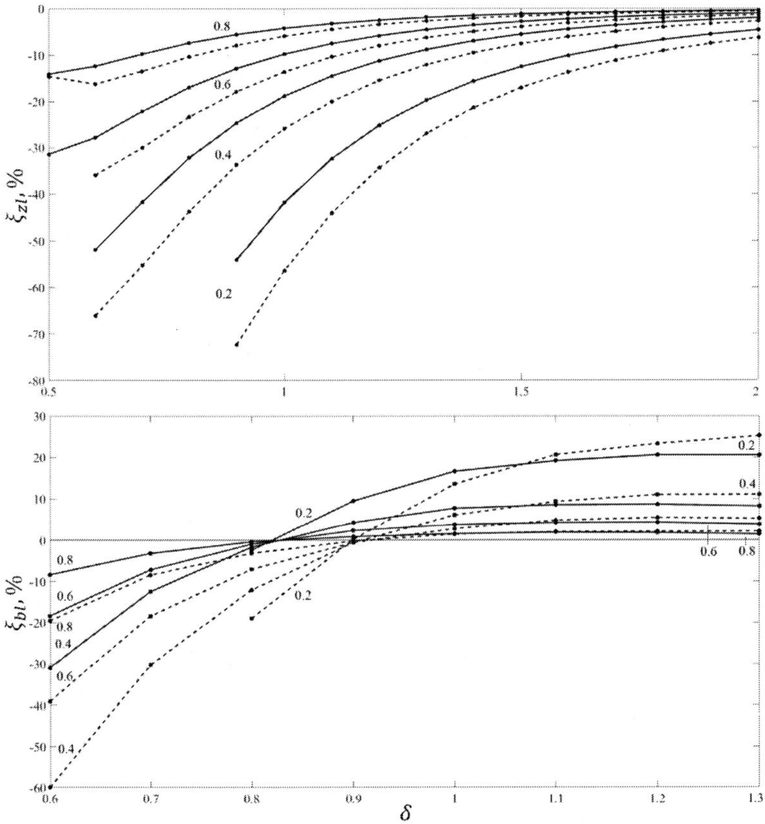

Figure 3.3-3. The relative measurement errors of the relative intensity of the second doublet peak in the 2nd-order derivative of the Lorentzian doublets using the zero-line and baseline methods (top and bottom panels, respectively). The width of the SG filter $m = 2$ and 30 (solid and dashed lines, respectively).

Given the Gaussian peak, defined in the dimensionless abscissa scale $x = w_x i/w$ (Eq. (3.3-1)):

$$F_G(x) = exp(-x^2), \quad (3.3-3)$$

where $w_x = 2\sqrt{ln2}$ is the peak FWHM in the x scale (it is the peak shape parameter β).

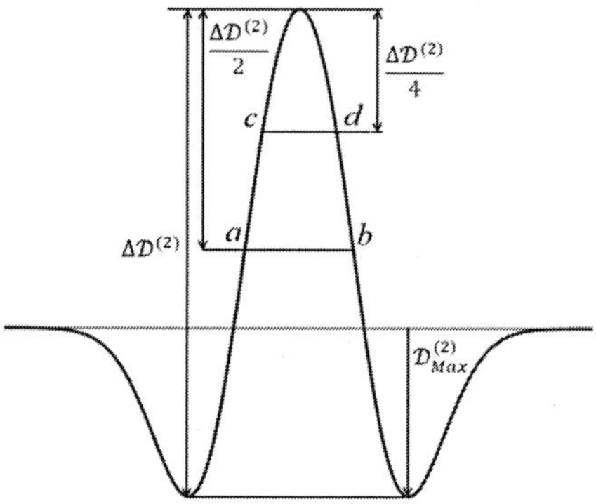

Figure 3.3-4. The width measurements in the negative 2nd-order derivative peak on the two horizontal levels (a-b and c-d) of the central part. The units of the figure parameters are w^2/F_0.

The 2nd-order derivative of Eq. (3.3-3) with respect to x, is

$$F_G^{(2)}(x) = (4x^2 - 2)F_G(x). \quad (3.3-4)$$

According to Table 1.2-2 and Eqs. (1.2-15)-(1.2-16), the characteristic parameters of Eq. (3.3-4) (Fig. 3.3-4), measured in the units of w^2/F_0, are

$$\Delta\mathcal{D}_G^{(2)} = 8.0198, \mathcal{D}_{Max}^{(2)} = 2.4746. \quad (3.3-5)$$

In the i-scale, used for a numerical calculation, these values must be divided by β^2. According to Figure 3.3-4, for $F_0 = 1$, the heights of the levels "ab" and "cd" are $(0.5\mathcal{D}_G^{(2)} - \mathcal{D}_{Max}^{(2)})/\beta^2 = 0.5537$ and $\left(0.75\mathcal{D}_G^{(2)} - \mathcal{D}_{Max}^{(2)}\right)/\beta^2 = 1.2769$, relative to the zero-line.

The relationships between the Gaussian peak width and the widths, measured in its 2nd-order derivative are:

$$w_G = K_{ab}w_{ab} = K_{cd}w_{cd}, \quad (3.3-6)$$

where $K_{ab} = 1{,}4928$, $K_{cd} = 2.2681$.

There was similar data for the Lorentzian peak:

$$F_L(x) = 1/(1 + x^2), \quad (3.3-7)$$

where $x = w_x i/w$; $w_x = 2$.

$$F_L^{(2)}(x) = (6x^2 - 2)/(1 + x^2)^3. \quad (3.3-8)$$

$$\Delta\mathcal{D}_L^{(2)} = 10;\ \mathcal{D}_{Max}^{(2)} = 2. \quad (3.3-9)$$

The levels "ab" and "cd" are on the heights $(0.5\mathcal{D}_G^{(2)} - \mathcal{D}_{Max}^{(2)})/2^2 = 0.75$ and $\left(0.75\mathcal{D}_G^{(2)} - \mathcal{D}_{Max}^{(2)}\right)/2^2 = 1.375$, relative to the zero-line.

The relationships between the Lorentzian peak width and the widths measured in its 2nd-order derivative are

$$w_L = K_{ab}w_{ab} = K_{cd}w_{cd}, \qquad (3.3-10)$$

where $K_{ab} = 2.6226$, $K_{cd} = 4.0733$.

The drawbacks of the width measurements (Fig. 3.3-4) are:

1. Significant dependence of the coefficients K_{ab} and K_{cd} on the peak shape (Eqs. (3.3-6) and (3.3-10)).
2. Peak shape distortions due to the smoothing of the derivative spectra and overlapping with adjacent peaks.

The broadening of the 2nd-order differentiation SG filter increases the derivatives' width and decreases the coefficients K_{ab} and K_{cd} (Table 3.3-1). This effect is the most pronounced for the Lorentzian peak.

Table 3.3-1. Impact of the width of the SG filter on the coefficients K_{ab} and K_{cd}

Peak	m								
	2			100			200		
	$\frac{w_x}{m}$	K_{ab}	K_{cd}	$\frac{w_x}{m}$	K_{ab}	K_{cd}	$\frac{w_x}{m}$	K_{ab}	K_{cd}
Gauss	833	1.495	2.268	16.6	1.486	2.257	8.3	1.468	2.237
Lorentz	1000	2.625	4.082	20.0	2.558	3.984	10.0	2.398	3.690

However, the primary factor of the uncertainties in determining peak width using derivatives is the overlapping with the adjacent peaks. We calculated the relative errors of the width estimation of the first doublet peak disturbed by the second one. The measurements were performed only on the left side of the component under study, thus partly eliminating the overlapping (Fig. 3.3-5). The results (Fig. 3.3-6) demonstrate that the absolute values of the errors may be more than 70% and 15% for the Gaussian and Lorentzian shapes, respectively. However, if a peak is disturbed from both sides (left and right), the errors may be unpredictable significant. Therefore, this measurement procedure has a very limiting application.

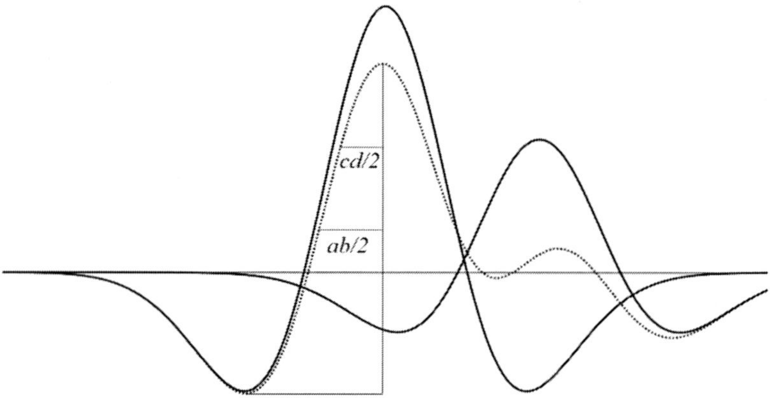

Figure 3.3-5. Width measurements of the first doublet peak using the 2nd -order derivative (dotted line) at the levels "ab" and "cd"'. Solid lines represent doublet peaks.

Use of the 2nd -order derivatives of the logarithm of a spectral curve

For the width measurements, we used the 2nd-order derivatives of the logarithm of the peak intensity:

$$W(i) = d^2 \ln F(i)/di^2. \qquad (3.3-11)$$

For the Gaussian and Lorentzian peaks, from Eqs. (3.3-3) and (3.3-7) we have

$$W_G = -8ln2/w^2, \qquad (3.3-12)$$

$$W_L(i) = -(8/w^2)[(1-4i/w^2)/(1+4i/w^2)]. \qquad (3.3-13)$$

In the peak maximum: $W_L(0) = -8/w^2$.

From Eqs. (3.3-11) - (3.3-13) we obtain, in the peak maximum:

$$w = C\sqrt{|d^2\, ln\, F(i)/di^2|}, \qquad (3.3-14)$$

where $C =$ 2.3548 and 2.8284 for the Gaussians and Lorentzians, respectively.

The critical advantage of this method is that the constant C weakly dependents on the peak shape. Also, the peak intensity has no direct impact on the calculations, except for an indirect influence of the signal-to-noise ratio. However, the differentiation does not suppress interfering signals (e., g., a constant baseline) since the logarithmic function is not additive (logarithm of a sum is not equal to the sum of logarithms).

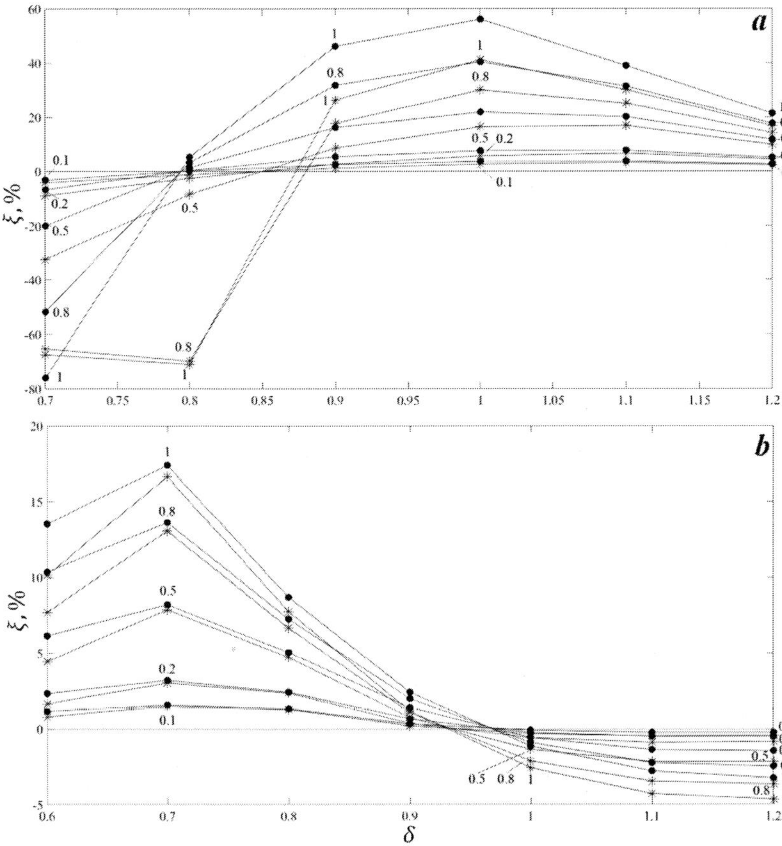

Figure 3.3-6. The width measurements of the first Gaussian (a) and Lorentzian (b) doublet peaks using the 2nd-order derivative at the levels "ab" (asterisks) and "cd" (circles).

Figure 3.3-7 illustrates the measurements at the maximum of the left satellite (point "a") and at the minimum (point "b") of the 2nd-order derivatives of the Gaussian and Lorentzian doublets:

$$D_G = exp(-x^2) + Rexp[-(x - \Delta x)^2], \quad (3.3 - 15)$$

$$D_L = 1/(1 + x^2) + R/[1 + (x - \Delta x)^2], \quad (3.3 - 16)$$

where $x = (\beta/w)[-2{:}\,0.001{:}\,2]$; $\beta_G = 2\sqrt{ln2}$; $\beta_L = 2$; $\Delta x = (\beta/w)\delta$; δ is the relative doublet separation; $w = 1$; R is the relative intensity of the second peak. R =[0.1, 0.2, 0.5, 0.8, 1.0]; δ =[0.7:0.1:1.2] and [0.6:0.1:1.2] for the Gaussian and Lorentzian doublets, respectively. As an example, the dotted lines (Fig. 3.3-7) represent the case: R=0.5 and δ=1. The SG differentiation filter parameters are: $2t = 2$ and $m = 20$.

The solid curves (Fig. 3.3-7) represent the 2nd-order derivative of the logarithmic functions (Eq. (3.3-11)):

$$W_{R=0.5,\delta}(x) = d^2 \ln D(x)/di^2. \qquad (3.3-17)$$

If the doublet separation δ increases, then, following Eqs. (3.3-12) and (3.3-13), $W_G(a)$ and $W_G(b) \rightarrow 8ln2$, and $W_L(b) \rightarrow 8$. The difference between the curves at the point "a" is significantly less than that at the point "b". That is, the dependence of the separation on the measure $W(a)$ is small. Figure 3.3-8a proves this conclusion for the Gaussian doublet. The relative errors of the width measurements at the point "a" is less than 5% compared to the 50% for the measurements at the point 'b". However, the worse signal-to-noise ratio at the point "a" may increase the errors.

For the Lorentzian peaks, $W_L(a)$=0 (Fig. 3.3-7a). In principle, the measurements may be performed to the left of the satellite maximum (e.g., in the maximum of the 3rd-order derivative); but the intensity of W_L in this region is small.

The major disadvantage of the width estimation based on Eq. (3.3-14) is the influence of the peak overlapping and distortions due to the background. The side regions of a peak, free of these effects exist in rare practical cases. Therefore, Eq. (3.3-14) is only useful for extracting approximate data about the peak's width.

“

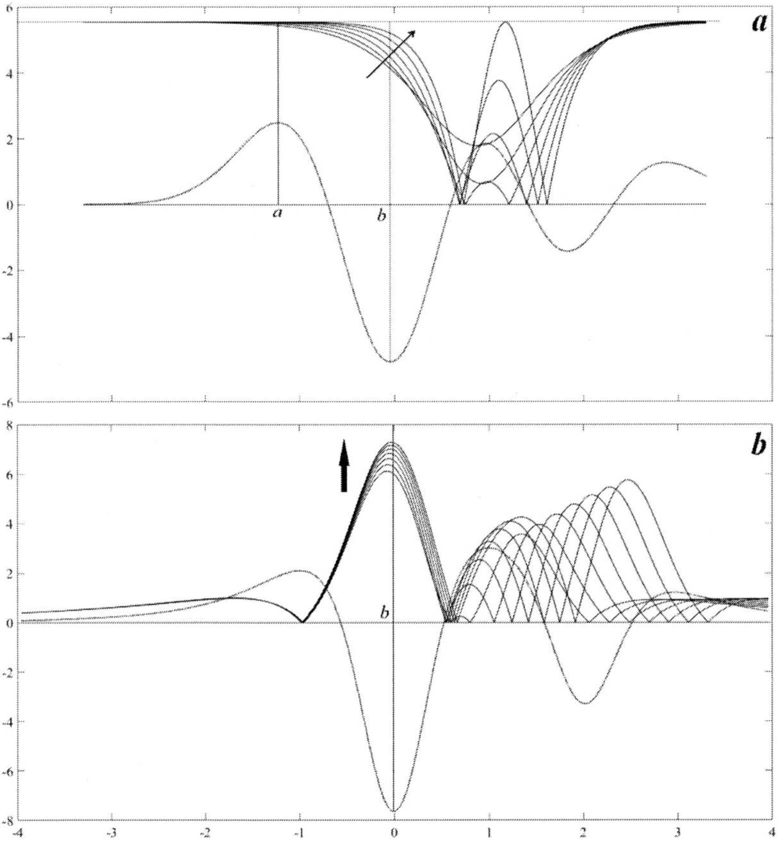

Figure 3.3-7. The width measurements of the 1st Gaussian (a) and Lorentzian (b) doublet peaks using the 2nd-order derivative of the logarithmic functions (Eq. (3.3-14, solid curves) at the maximum of the left satellite (point “a”), and at the minimum (point “b”) of the derivatives (dotted line). The arrows point at an increase of $\delta = 0.7{:}0.1{:}1.2$ (a) and 0.6:0.1:1.2 (b). $R = 0.5, m = 20$.

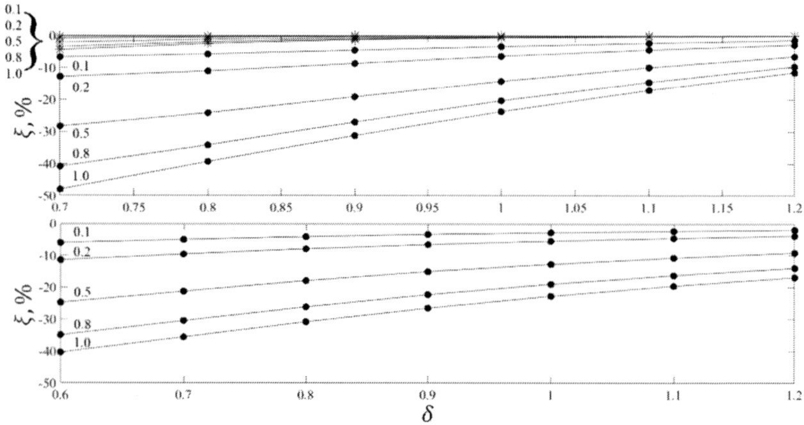

Figure 3.3-8. The width measurements of the 1st Gaussian and Lorentzian (top and bottom panels, respectively) doublet peaks using the 2nd-order derivative of the logarithmic functions (Eq. (3.3-14)). The asterisks and circles correspond to the measurements performed at the points "a" and "b", respectively. The numbers near the curves stand for R.

Use of the fractional derivatives

We estimated uncertainty in determining the ratio of the peak widths of the Gaussian and Lorentzian doublets using the fraction derivatives (FD) (q =1.6:0.1:2.4) (Eq. (1.2-36)):

$$D(x) = F(\Delta_x^+) + RF(\Delta_x^-), \qquad (3.3-18)$$

where $\Delta_x^+ = x - x_c + x_0$; $\Delta_x^- = (x - x_c - x_0)/r$; $x = \beta i/w$; β is the peak shape parameter; $i = 1{:}\,10000$; $w_G = 500$ and $w_L = 200$ are the FWHMs (to eliminate the Gibbs phenomenon (Appendix A), $w_L < w_G$); $x_c = i_c\beta/w$; $i_c = 4000$; $x_0 = \beta\delta/2$; $\delta = \Delta/w$ is the relative separation of the doublet components; Δ is the absolute separation; $i_c \pm i_0$ are the doublet peaks positions ; R and r are the relative intensity and the relative width of the second peak.

Figure 3.3-9 shows that the absolute errors in determining the ratio of the peak widths for the Lorentzian doublets are less than for the Gaussians. If the doublet separation increases, then the errors increase from negative to the positive values for $R = 0.5 - 1.0$.

We explain this increase, as follows. For small separations, the errors are due to the overlap of central parts of the derivatives' components (Fig. 3.3-10a). For large δ, the overlap of the central part of the first peak's derivative with the satellite of the second peak (Fig. 3.3-10c) is the essential factor of the errors. Only for $\delta = 4$, when the overlap is small, the relative error decreases to 2.2%.

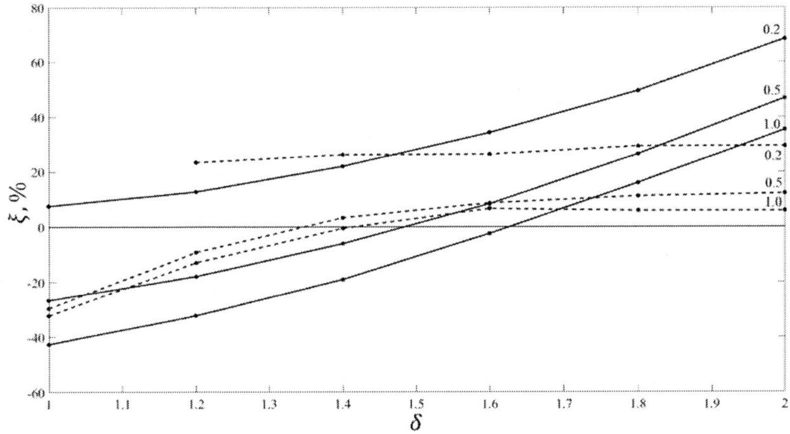

Figure 3.3-9. The peak widths ratio ($r = 3$) measurements for the Gaussian and Lorentzian doublets (solid and dotted lines, respectively) using the fraction derivatives (Eq. (1.2-36)). The numbers near the curves stand for R.

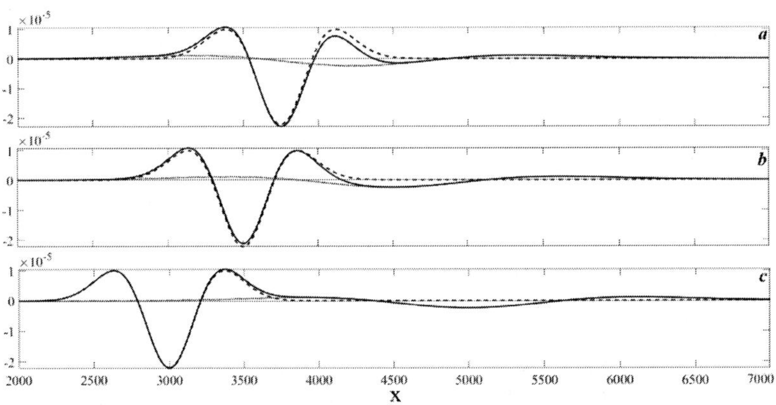

Figure 3.3-10. The 2nd-order derivatives of the Gaussian doublet and doublet components (Eq. (3.3-18)) (solid, dashed, and dotted curves, respectively). $R = 1$, $\delta = 1$, 2 and 4 (panels a-c, respectively).

Dependences of the zero-crossing and extreme values of the FDs of the Lorentzian peak on the differentiation order were approximated by parabolas (Fig.1.2-7) [2]:

$$P_0(q) = c_2q^2 + c_1q + c_0, \quad (3.3-19)$$
$$P_e(q) = d_2q^2 + d_1q + d_0. \quad (3.3-20)$$

We found that the linear approximation of the $lnP_e(q)$ (Eq. (1.2-32)) for the first satellite is more precise than Eq. (3.3-20):

$$lnP_e(q) = f_1q + f_0. \quad (3.3-21)$$

The left zero-points of the first three derivatives of the Lorentzian peak are readily obtained from Eq. (1.2-10):

$$y_0^{(1)} = 0;\ y_0^{(2)} = \sqrt{1/3};\ y_0^{(3)} = 1, \quad (3.3-22)$$

where $y = 2(\lambda - \lambda_0)/w$. Therefore,

$$P_0(1) = \lambda_0;\ \ P_0(2) = \lambda_0 - w/(2\sqrt{3});\ P_0(3) = \lambda_0 - w/2. \quad (3.3-23)$$

Substituting $P_0(1)$ and $P_0(2)$ into Eq. (3.3-19), we have two estimators:

$$\hat{\lambda}_0 = c_2 + c_1 + c_0, \quad (3.3-24)$$
$$\widehat{w} = -2\sqrt{3}(3c_2 + c_1). \quad (3.3-25)$$

Exercise 3.3-1

Readers are invited to calculate the estimator, similar to Eqs. (3.3-24) and (3.3-25) and based on the zero points $P_0(2)$ and $P_0(3)$.

The left extrema of the first three derivatives of the Lorentzian peak are (Table 1.2-2 (Chapter 1.2)):

$$P_e(1) = 1.2990F_0/w;\ \ P_e(2) = 2F_0/w^2;\ \ P_e(3) = 3.3677F_0/w^3. \quad (3.3-26)$$

Substituting the pairs $[P_e(1),\ P_e(2)]$ and $[P_e(2),\ P_e(3)]$ into Eq. (3.3-20), one can obtain two different sets of estimators for $\widehat{w}$ and $\widehat{F}_0$ [2]. Readers who love tedious algebra can do this. For example, in the first case:

$$\widehat{w} = 1.539 * Num/Den, \quad (3.3-27)$$
$$\widehat{F}_0 = 1.1852 * Num^2/Den, \quad (3.3-28)$$

where $Num = d_2 + d_1 + d_0;\ Den = 4d_2 + 2d_1 + d_0$.

For the logarithmic approximation (Eq. (3.3-21)), using $lnP_e(1)$ and $lnP_e(2)$ (Eq. (3.3-26)), we have

$$\widehat{w} = (K_m/K_w)exp(-f_1), \quad (3.3-29)$$
$$\widehat{F}_0 = \left(K_m/{K_m}^2\right)exp(f_2), \quad (3.3-30)$$

where $K_m = P_e(2)w^2/F_0 = 2; K_w = P_e(1)w/F_0 = 1.2990$.

Computerized studies showed amazingly accurate results, which were obtained by the decomposition of some simulated sets of overlapping Lorentzian peaks [2] and experimental Laser-Induced Breakdown spectra

[3]. In the last case, the authors performed background correction and data denoising.

Prof M. Kh. Salakhov' team (Russia) [4] first suggested this FD-based method, which was successfully developed by Japan researchers [2, 3].

To illustrate the potential possibilities of the method, we estimated uncertainties in determining the positions and widths (by Eq.(3.3-24) and (3.3-25)) of the Lorentzian peaks distorted by the interfering peaks of the same shape. Eq. (3.3-18) is slightly changed so that the first peak is identical to the following model [2]:

$$D(x) = F(x - x_c) + RF\{(x - x_c - 2\delta)/r\} + \eta, \qquad (3.3-31)$$

where η is the normal noise with zero mean and standard deviation $\sigma = 0.01$ (the maximum signal-to-noise ratio is 100).

Figures 3.3-11 and 3.3-12 show dependences of the relative errors in the determining of the peak positions and widths, and corresponding standard deviations, on the doublet separation (δ) for noise-free and noisy data, respectively. The plots of Figure 3.3-12 (panels signed by a and b with indexes 1-3) are the mean values of 1000 repetitions. The denoising of the derivatives was performed in the Fourier domain (Chapter 2.4). The smoothing filter's parameters were: $t = 2$; $\alpha = 3 * 10^{-8}, 10^{-8}$, and $8 * 10^{-9}$ for $r = 1$, 0.5, and 2, respectively. These parameters were chosen empirically based on a trade-off between the mean errors and the standard deviations.

The Lorentz peak shape manifests itself as the constant $2\sqrt{3}$ in Eq. (3.3-25). Therefore, small shape distortions, such as smoothing, that change this constant, may cause significant systematic errors of the peak width measurements.

As opposed to this parameter, the expression of the peak's maximum data, which was obtained by the least-squares regression (Eq. (3.3-24)), does not directly depend on the peak's shape, just like $y_0^{(1)} = 0$ (Eq. (3.3-22)).

Figure 3.3-13 (panel a) demonstrates shifts of the zero-points (λ_z) in the 2nd-order derivatives of the Lorentzian peak smoothed by the FFT method. The dependence of these shifts on the smoothing parameter α allowed us to modify Eq. (3.3-25) by introducing the correcting coefficient $K_\alpha = w/\left(2\sqrt{3}|\lambda_z - \lambda_0|\right)$:

$$\widehat{w} = -K_\alpha\left[2\sqrt{3}(3c_2 + c_1)\right]. \qquad (3.3-32)$$

The linear dependence (Fig. 3.3-13, panel b), calculated in the range $[5, 50] * 10^{-9}$, is

$$K_\alpha = -0.0878 * ln\,\alpha - 0.8820. \qquad (3.3-33)$$

The following conclusions can be drawn from the graphical data:
1. The errors increase while R increases and δ decreases.
2. The errors reduce while the relative peak width (r) changes from 1 to 0.5 or 2.
3. The accuracy of the peak's maximum estimation is better than that of the peak width.
4. The smoothing increases both types of errors (compare Figs.3.3-11 and 3.3-12). However, all dependencies $\xi_w(\delta)$ rise from zero-line by more than 30% due to the distortion of the Lorentzian peak shape. The above numerical experiment explained this result.
5. The plots (Fig. 3.3-14) which were corrected according to Eqs. (3.3-32) and (3.3-33), exclude systematic shifts observed in Figure 3.3-12.

Figure 3.3-15 shows the dependences of the relative errors in the determining of the peak maxima and widths on the doublet separation for the noisy model (Eq. (3.3-31)). The curves were plotted using Eqs. (3.3-29) and (3.3-30). We modified these equations, taking the fact that the constants K_w and K_m are functions of the smoothing parameter α into account Two sets of the discrete values $K_w(\alpha_i) = P_e(1) * w/F_0$ and $K_m(\alpha_i) = P_e(2) * w^2 \ /F_0$ were obtained by numerical estimations of the peak maxima $P_e(1)[\alpha_i]$ and $P_e(2)[\alpha_i]$ in the 1st- and 2nd-order Fourier derivatives, respectively. Then $K_w(\alpha_i)$ and $K_m(\alpha_i)$ were well approximated by the 3rd-order polynomials used in Eqs. (3.3-29) and (3.3-30).

Let us briefly consider some studies in which information about the approximated peak parameters was extracted from the derivatives.

The models were the Gaussian, Lorentzian, and Student peaks, and a combination of the first and second peak shapes [5, 6]. The peaks' maximum intensities and widths were approximately estimated using cumbersome expressions included the 2nd- and 4th-order derivatives.

The low signal-to-noise ratio of the 4th-order derivatives and four satellites are the major disadvantages of this method. One of the pioneering computer programs, which is based on the 2nd-order derivative, refined the widths of the Gaussian peaks in the decomposition procedure [7]. A software package based on high-order differentiation was designed to reconstruct the parameters of the unresolved doublets consisting of the Gaussian, Lorentzian, and Voigt peaks [8]. The calculated profiles of the atomic spectral lines well coincided with the experimental ones.

Conclusion

Based on the literature data and our experience, this chapter showed that the derivatives of the poorly-resolved spectral contours allow the extraction useful information about peak parameters using many smart algorithms. However, each algorithm suffers from specific drawbacks, including the negative impact of noise and background, and errors occur because of unknown correct peak shapes. Attempts to develop a "panacea" to the problem of overlapping peaks is doomed to failure. Only a combination of the rigorous mathematical methods of decomposing unresolved spectra based on spectral curves modelling and the derivative techniques for estimating the approximated initial peak parameters allows an analyst to obtain realistic preliminary data. However, without physical-chemical data justification, the mathematically-obtained results may be erroneous.

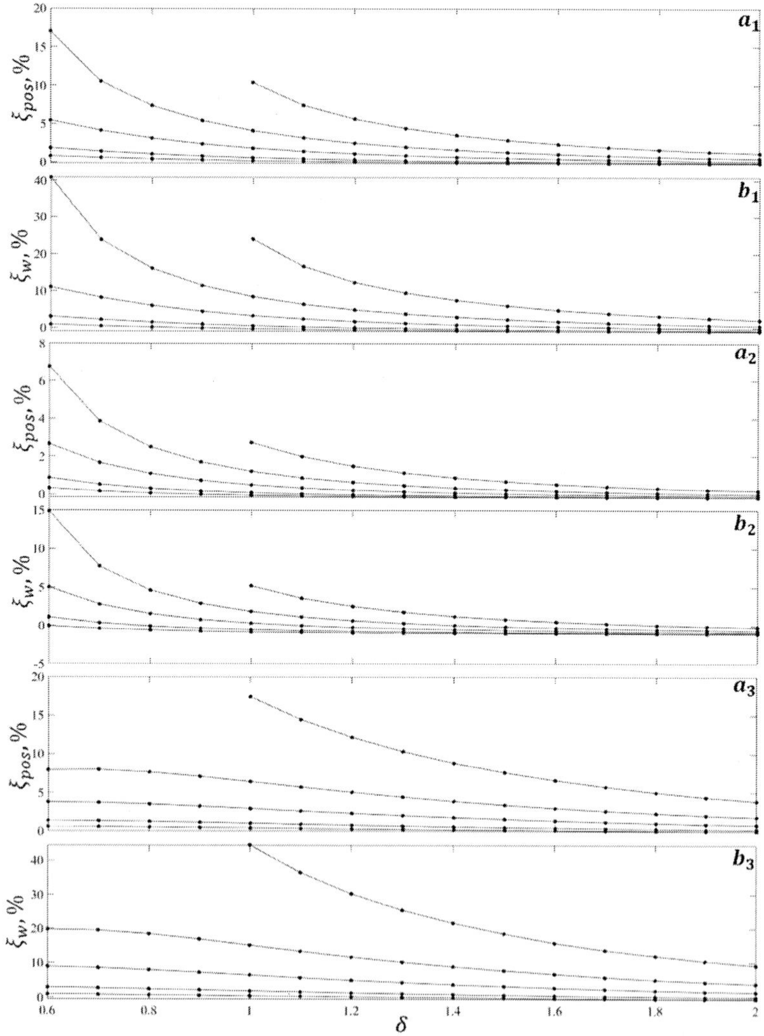

Figure 3.3-11. Errors in determining positions (a) and widths (b) the noise-free Lorentzian peaks distorted by the interfering peaks of the same shape. Subscripts 1-3 correspond to r = 1, 0.5, 2, respectively. R =0.1, 0.2, 0.5, 1, 2 from the lower to upper curves.

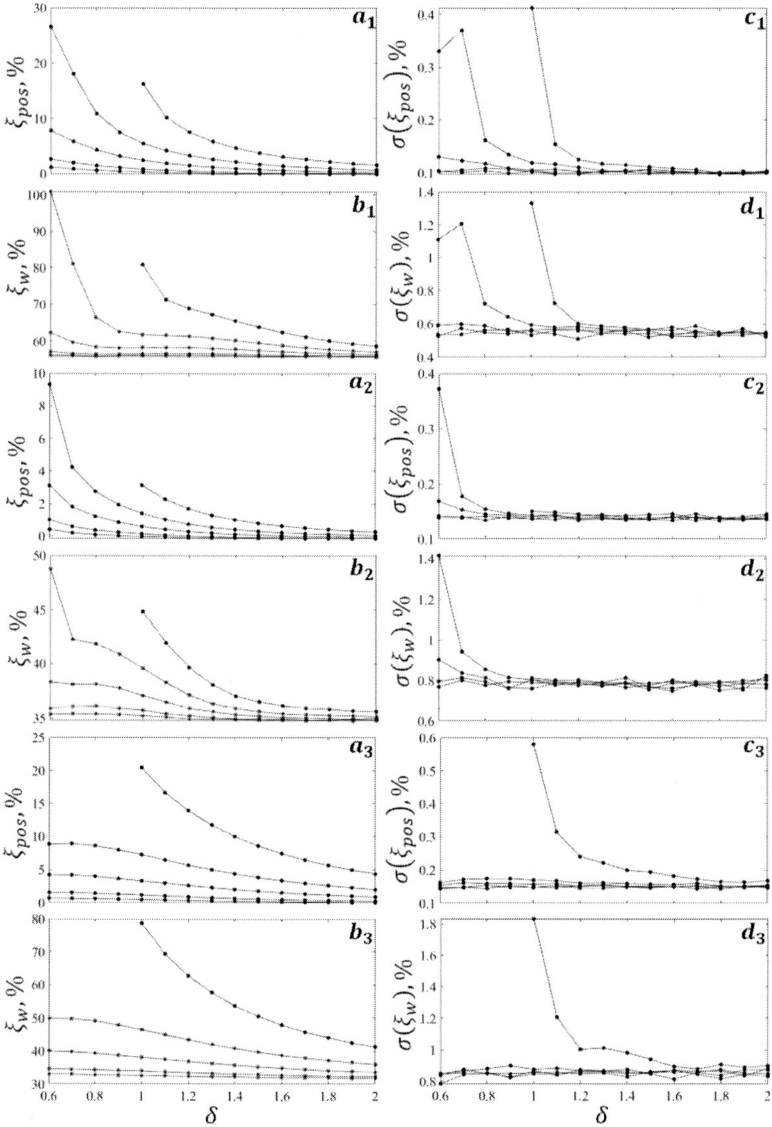

Figure 3.3-12. The mean errors and standard deviations found in determining the positions (a, c) and widths (b, d) of the noisy Lorentzian peaks distorted by the peaks of the same shape. Subscripts 1-3 correspond to $r = 1, 0.5, 2$; $\alpha = [3, 1, 0.8] * 10^{-8}$, respectively. R =0.1, 0.2, 0.5, 1, 2 from the lower to upper curves.

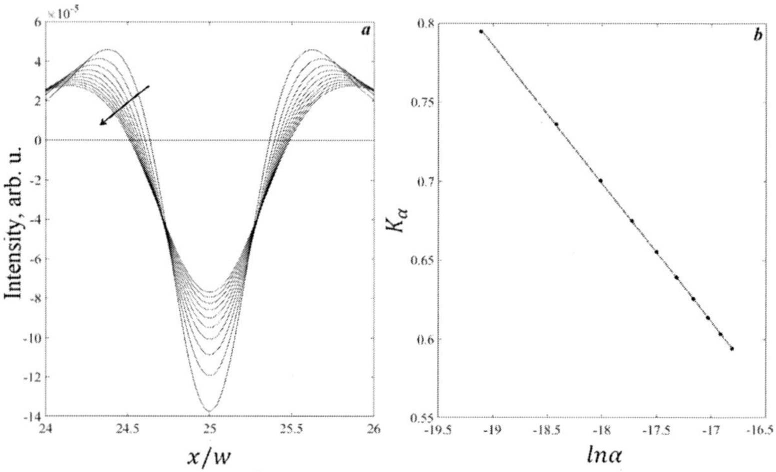

Figure 3.3-13. (a) The 2nd-order FFT derivatives of the Lorentzian peak ($2t = 4$). The arrow points to $\alpha = (5:5:50) * 10^{-9}$. (b) Dependence $K_\alpha(\ln\alpha)$.

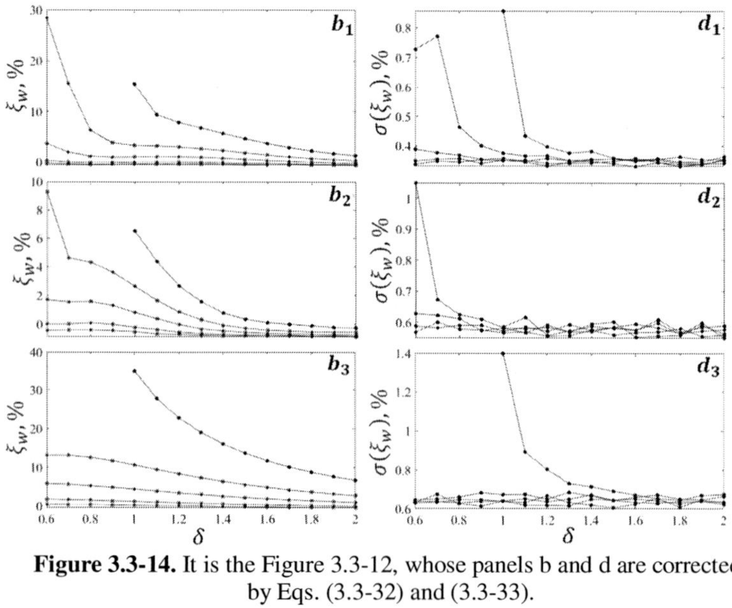

Figure 3.3-14. It is the Figure 3.3-12, whose panels b and d are corrected by Eqs. (3.3-32) and (3.3-33).

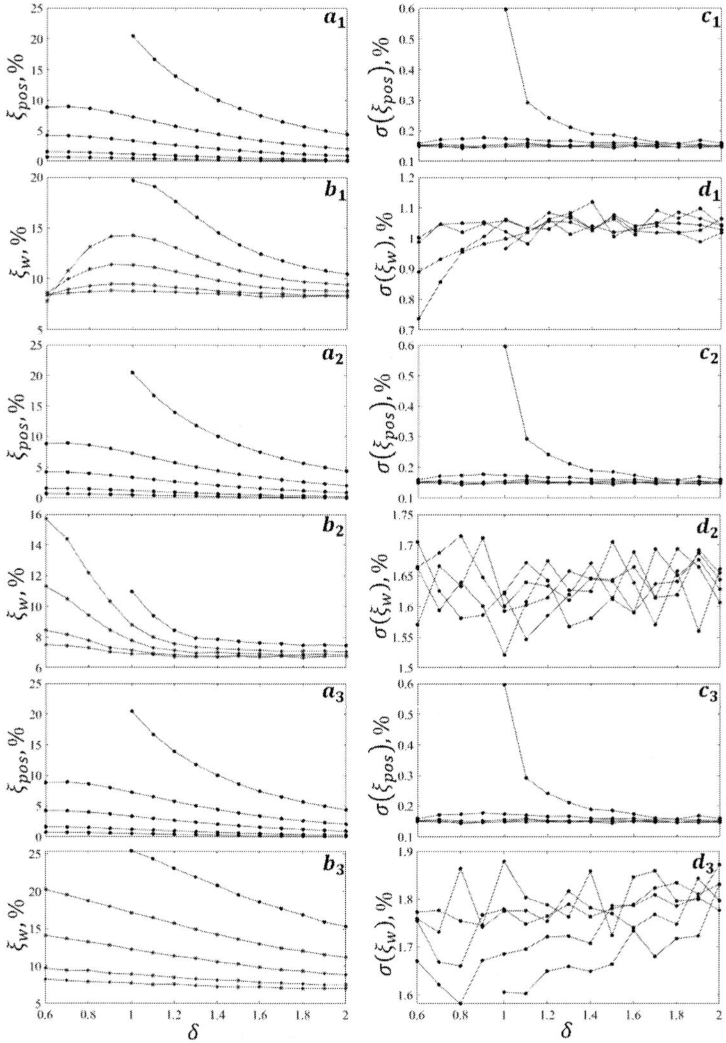

Figure 3.3-15. The mean errors and standard deviations in determining maxima (a, c) and widths (b, d) of the noisy Lorentzian peaks distorted by the peaks of the same shape. Subscripts 1-3 correspond to $r = 1, 0.5, 2$; $\alpha = [3, 1, 0.8] * 10^{-8}$, respectively. R =0.1, 0.2, 0.5, 1, 2 from the lower to upper curves.

CHAPTER FOUR

DETERMINATION OF DOUBLET PARAMETERS USING THE EMPIRICAL COEFFICIENTS OF THE 2ND-ORDER DERIVATIVE

According to the previous chapters, the parameters of the poorly-resolved overlapping peaks, which were estimated using their derivatives, suffer from significant errors. Therefore, generally speaking, these estimates, being approximated values of the peak parameters, may be useful as initial data for the decomposition of spectra into elementary components. Let us consider this problem in more detail.

Usually, the decomposition is performed by searching for the parameters (Prm) that corresponded to the minimum Euclidian distance (the objective function) between a measured spectrum (F_{ms}) and the model (F_{mod}) [1]:

$$Prm = ArgMin\left\{\sqrt{\sum_{i=1}^{N}[F_{ms}(i) - F_{mod}(i, Prm)]^2}\right\}. \qquad (3.4-1)$$

For an in-depth look into the problem, we recommend a book [2].

Introducing computerized tools in the 1960-70s allowed analysts to perform extensive use of the decomposition procedures: e.g., the authors of the pioneering study in analytical spectroscopy in [3] compared errors when determining peak parameters using different numerical algorithms. Now, there are many commercial software products designed for decomposition [1].

The main problem of minimizing the objective function (Eq. (3.4-1)) is that it may have multiple local minima in the interval. In other words, many sets of parameters that best fit to the model are possible. Finding all the sets is a cumbersome mathematical task [4]. Another widely used method requires guesses (initial values) of possible peak parameters. However, a searching algorithm may converge to different sets due to small changes in the guesses. From all data, one must select parameters that are most appropriate to the spectrum under decomposition. The correct choice of initial model parameters may help to solve this challenging problem [1].

To illustrate some difficulties when choosing these parameters, consider the multiplet's normalized form consisting of peaks with a similar shape (F):

$$Mult(x) = \sum_{t=1}^{p} R_t F(\Delta_{xt}), \qquad (3.4-2)$$

where $\Delta_{xt} = x - x_c + x_{0t}$; $x = \beta i/w_t$; $i = 1,2,\dots N$; β is the peak shape parameter; $w_t = wr_t$ is the peak width; w is the width-scale parameter; $x_c = i_c\beta/w_t$ is the reference (central) point; $x_{0t} = \beta i_{0t}/w_t$; $\delta_t = \Delta_t\beta/w_t$ is the relative shift of the peak from the i_c; Δ_t is the absolute shift; $x_c - x_{0t}$ is the peak position ; R_t is the normalized peak intensity. The pair $\{i_c, w\}$ and the set of p triples $\{\delta_t, R_t, r_t\}$ define the multiplet with symmetrical peaks.

Figure 3.4-1 demonstrates triplet and quartet spectra and their negative 2nd- order derivatives. Combining analysis of the well-resolved peaks and their derivatives allows us to estimate the initial peak parameters for further decomposition. However, the origin of the weak hump marked with arrows is doubtful. The hump (panel b) and its combinations with correct peaks (panels a and c) are false peaks.

Exercise 3.4-1

Readers are invited to inspect some triplets and quartets, plotted by the program plotQuartets.m, and to try to guess the number of the peaks and their parameters visually. Pay attention to the possibility of a false structure in the 2nd-order derivative.

Figure 3.4-2 illustrates two possible ways of initial parameters estimation. The left side of the diagram describes the traditional method in which the parameters are extracted from the spectrum using some preprocessing algorithm combined with the iterative procedure of the decomposition. If the results are unsatisfactory, then the parameters, which were obtained in a given step, are corrected, and the process continues to fit the model to the experimental data properly. Also, the model (peaks' number and their shapes) may be changed.

Prior information about the sample, which spectrum is to be studied, helps clarify the parameters' values. This information includes the theoretically-obtained spectral data.

In another approach suggested in the 1980s [5] (the right side of Fig. 3.4-2), we searched for the unknown parameters of the spectral segment (doublet) in the database containing a broad set of the parameters that occur in practice. For this goal, we generated the 2nd- order SG derivatives of 600 pairs of the Gaussian and Lorentzian doublets by varying the peak parameters in broad ranges:

$$D(x) = F[\beta(i + i_0)/w] + RF[\beta(i - i_0)/rw], \quad (3.4-3)$$

where $i_0 = \delta w/2$; δ is the relative peak separation. Other parameters are defined in Eq. (3.4-2).

The empirical coding coefficients of derivatives were:

$$\alpha = \bar{\Delta}/\bar{w}_1\,, \beta = \bar{w}_2/\bar{w}_1\,, \gamma = \bar{I}_2/\bar{I}_1 \le 1. \quad (3.4-4)$$

Figure 3.4-3 defines parameters of the negative 2nd-order derivative used for the calculation of these coefficients for the poorly- (a) and well-resolved Gaussian doublets (b). The last panel shows the false peak. The coefficients, α, β and γ, correspond the relative separation, width, and intensity of the doublet peaks, respectively. The triples $\{\alpha, \beta, \gamma\}$ were mapped to the triples of the doublet parameters $\{\delta, R, r\}$. The sets of both triples were saved in the database. The use of the coefficients (Eq. (3.4-4)), instead of the full derivative spectra, allowed the significant compresion of the data.

The computer program scanned the database and searched for the triple $\{\alpha, \beta, \gamma\}$ closest to that estimated for the doublet under study. The database was small because of the insufficient calculation power of the Russian minicomputer available in our laboratory. Therefore, we used this method only to estimate relative values $\{\delta, R, r\}$. Also, unique initial parameters were not obtained in all cases.

To overcome these drawbacks, we took the multiplet model (Eq. (3.4-2)) instead of the old one (Eq. (3.4-4)):

$$D(x) = R_1F(\Delta_{x1}) + R_2F(\Delta_{x2}), \quad (3.4-5)$$

where $\Delta_{x1} = (x - x_c + \beta\delta)/r_1$; $\Delta_{x2} = (x - x_c - \beta\delta)/r_2$; $r_1 = w_1/w$; $r_2 = w_2/w$; $\delta = \Delta/w$; Δ is the peak shifts from the central point i_c.

The coefficients β and γ (Eq. (3.4-4)) were also split into two pairs:

$$\beta_1 = \bar{w}_1/(\bar{w}_1 + \bar{w}_2)\,;\ \beta_2 = \bar{w}_2/(\bar{w}_1 + \bar{w}_2)\,;\ \gamma_1 = \bar{I}_1/(\bar{I}_1 + \bar{I}_2)\,;$$

$$\gamma_2 = \bar{I}_2/(\bar{I}_1 + \bar{I}_2). \quad (3.4-6)$$

It follows from Eqs. (3.4-4) and (3.4-6) that $\beta = \beta_2/\beta_1$ and $\gamma = \gamma_2/\gamma_1$.

The quintet $\{\alpha, \beta_1, \beta_2, \gamma_1, \gamma_2\}$ does not depend on the x and y scales. However, a large sampling interval of the x-axis and smoothing of the noisy data disturb the data.

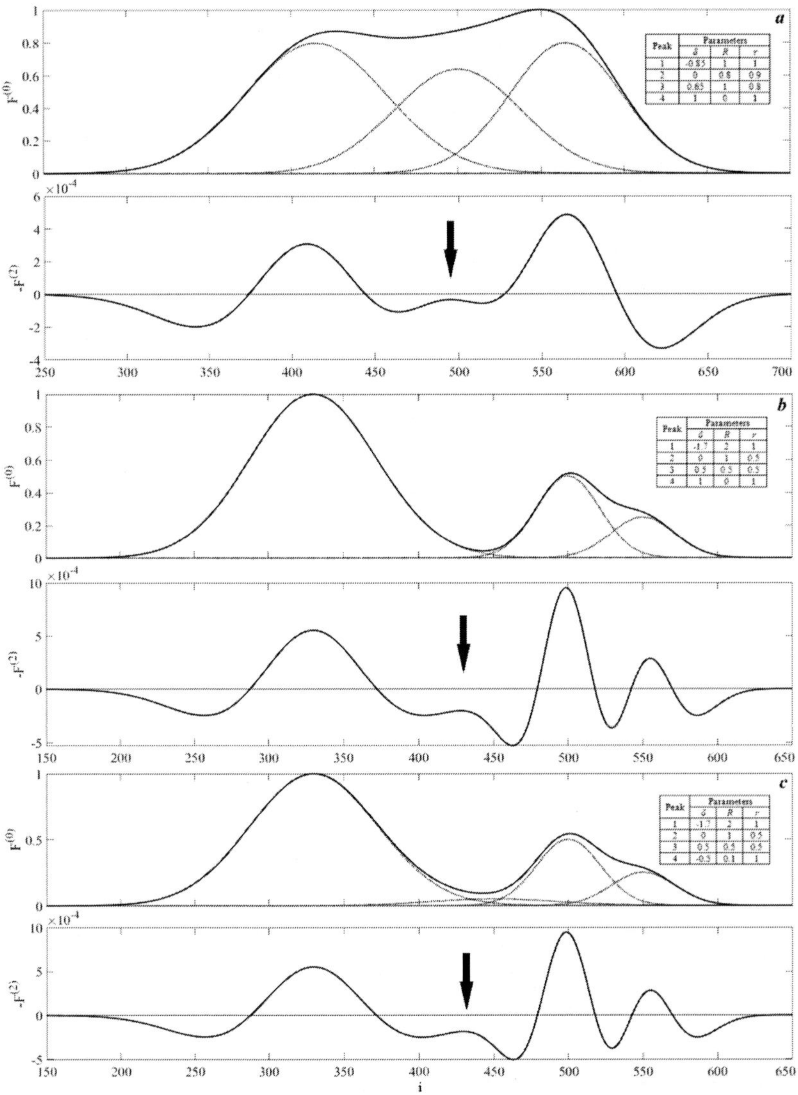

Figure 3.4-1. The Gaussian triplets (a, b), quartet (c), and their negative 2nd-order derivatives (upper and lower panels, respectively). The dotted curves are the multiplet's components. The arrows point to "doubtful" peaks.

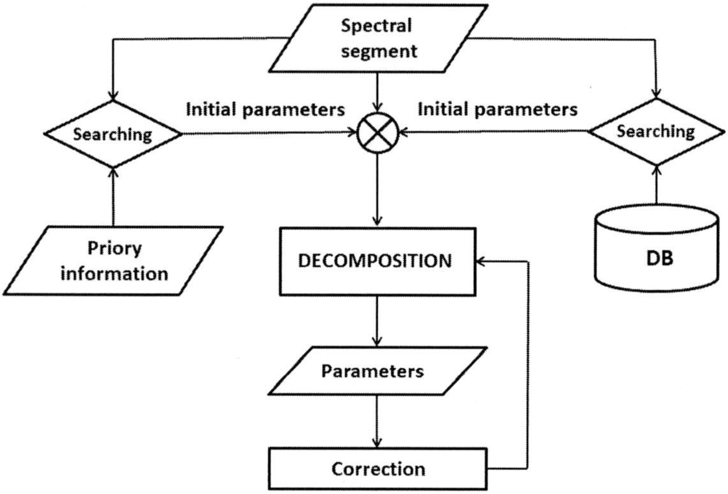

Figure 3.4-2. Diagram of the spectral segment decomposition. DB stands for a database.

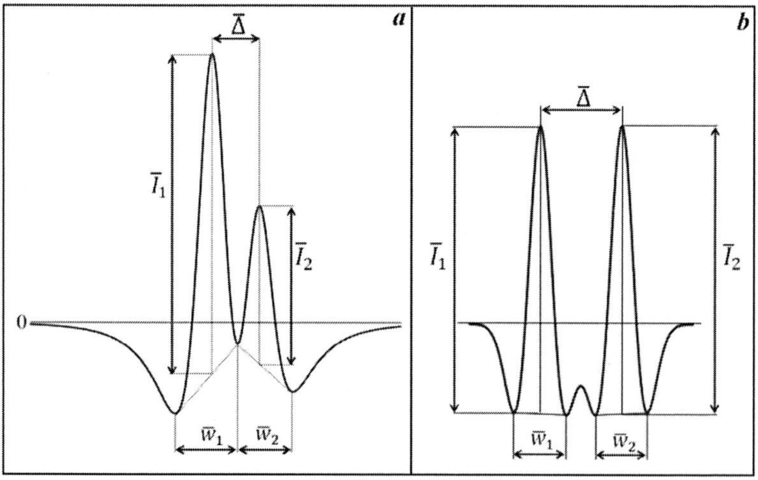

Figure 3.4-3. Estimation of the empirical coefficients (Eq. (3.4-4)) for poor (a) and well-resolved 2nd-order derivatives of the Gauss doublets (b). For simplicity, the derivatives have a negative sign. The last panel shows the false peak.

Relationships between $\{\alpha, \beta_1, \beta_2, \gamma_1, \gamma_2\}$ and the doublet parameters $\{w, R_1, R_2, r_1, r_2, \delta\}$ cannot be theoretically calculated. As an example, Figure 3.4-4 shows dependences $\alpha(\delta)$ obtained numerically. There is only one way to build a two-column lookup table with the following rows: $\{w, R_1, R_2, r_1, r_2, \delta\}, \{\alpha, \beta_1, \beta_2, \gamma_1, \gamma_2\}$. The second row depends on w. Since $\beta_1 + \beta_2 = 1$ and $\gamma_1 + \gamma_2 = 1$, only one term of these pairs must be saved. Another is calculated. However, below we show why both coefficients are needed.

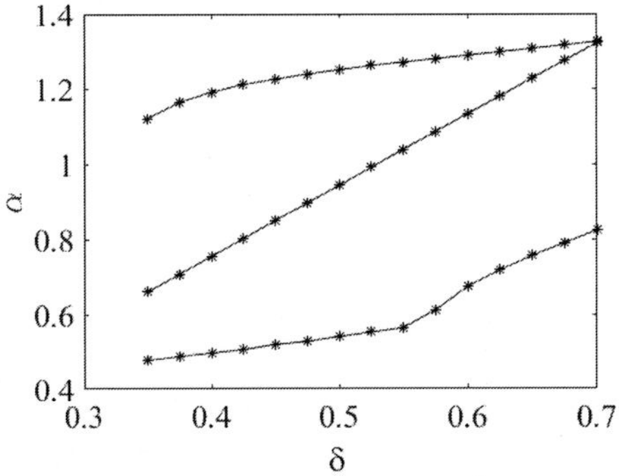

Figure 3.4-4. Dependence $\alpha(\delta)$ for the 2nd-order derivative of the Gaussian doublet. $R_1 = R_2 = 1, r_1 = r_2 = [0.1, 0.5, 1]$ from upper to lower curves, respectively.

The algorithm of thc cstimation of the initial parameters consists of the following steps (Fig. 3.4-5):

1. Loading the database file (DB), the doublet under analysis, and the parameter θ_{max}. θ_{max} is the applicable relative discrepancy between the coefficients $C_k = \{\alpha, \beta_1, \beta_2, \gamma_1, \gamma_2\}$ and those found in DB $\left(C_{k,DB}\right)$, that is,

$$\left|C_k / C_{k,DB} - 1\right| \leq \theta_{max}. \quad (3.4-7)$$

The same relative errors of the terms β_1, β_2 and γ_1, γ_2 only occur if they equal to 0.5. E.g., suppose that $\beta_{1,DB} - \beta_1 = \varepsilon$, then

$$|\beta_1/\beta_{1,DB} - 1| = \left|\left(\beta_{1,DB} - \varepsilon\right)/\beta_{1,DB} - 1\right| = |\varepsilon|/\beta_{1,DB}. \quad (3.4-8)$$

$$|(1-\beta_1)/\Delta\beta_{1,DB} - 1| = \left|\left(\Delta\beta_{1,DB} + \varepsilon\right)/\Delta\beta_{1,DB} - 1\right| = |\varepsilon|/\Delta\beta_{1,DB}, \quad (3.4-9)$$

where $\Delta\beta_{1,DB} = 1 - \beta_{1,DB}$.

Eqs. (3.4-8) and (3.4-9) are equal, if $\beta_{1,DB} = 0.5$.

2. Searching for the sets $\{\alpha, \beta_1, \beta_2, \gamma_1, \gamma_2\}$ that are satisfied Eq. (3.4-7), and mapping the doublet parameters $\{w, R_1, R_2, r_1, r_2, \delta\}$ to these sets. The rows of the matrix $\boldsymbol{M}_{param}$ contain these parameters.

3. Calculating the model doublets F_{mod} (Eq. (3.4-5)), which corresponded to matrix $\boldsymbol{M}_{param}$. The width w has the most significant impact on the accuracy of the solution. Therefore, for each row of the matrix $\boldsymbol{M}_{param}$, this parameter is varied in the range given by DB. The set of the approximately best-fit models of initial parameters are found using Eq. (3.4-1). From these models, one selects the set (named $\alpha\beta\gamma$-best-fit doublet parameters), which has the minimum fitting error. owever, a number of the competing models may be used as candidates for the initial parameters in the least-squares (LS) fitting.

The choice of relative discrepancy θ_{max} (Eq. $(3.4-7)$) is critical. If θ_{max} decreases, then the size of the matrix $\boldsymbol{M}_{param}$ decreases, and the program running time significantly reduces. However, the condition, defined by Eq. $(3.4-7)$, must be satisfied. Therefore, a compromise is needed.

As an example, we built DB, combined 6,293,700 sets of the doublet parameters $\{w, R_1, R_2, r_1, r_2, \delta\}$, which varied in the following ranges: $\{20:5:200;\ 0.1:0.1:1; 0.1:0.1:1; 0.2:0.1:1; 0.2:0.1:1;\ 0.3:0.05:1.3\}$.

To reduce the dimension and the size of DB, data do not include a scaling factor along y-axis. For the scaling, the best initial parameters $R_1$and R_2, found in DB, were multiplied by the factor

$$f_s = max(D_x) / max(D_{est}), \qquad (3.4-10)$$

where D_x and D_{est} are doublets; the first is under study, and the second is built, using the $\alpha\beta\gamma$-best-fit parameters.

The parameters of the SG 2nd-order derivative filters were: $2t = 2$, $m = w/2 = \{10:2:100\}$. The smoothing, with a factor that was constant for basic widths w, took the possible noise in the spectra into account. However, if w was small, then the peaks with small relative widths r, might be oversmoothed. The $w_1 < 2m$ and $w_2 < 2m$ cases were eliminated from the calculations. Therefore, the number of combinations decreased to 5,501,202. The Matlab program running time on the laptop (Intel(R) Core (TM) i7-4700HQ CPU @ 2.40 GB, RAM 16.0 GB) was approximately 143s.

Figure 3.4-6 illustrates the decomposition process of the three Gaussian doublets. The $\alpha\beta\gamma$-best-fit coefficients served as the initial parameters for the LS-fitting. For this goal, we used the nlinfit Matlab function. Table 3.4-1 includes the precise doublet parameters and their estimates obtained by the $\alpha\beta\gamma$ method and the LS-fitting.

The laptop running time of the Matlab program (Fig.3.4-5), which included DB loading, searching, and LS-fitting, was less than 3-5s.

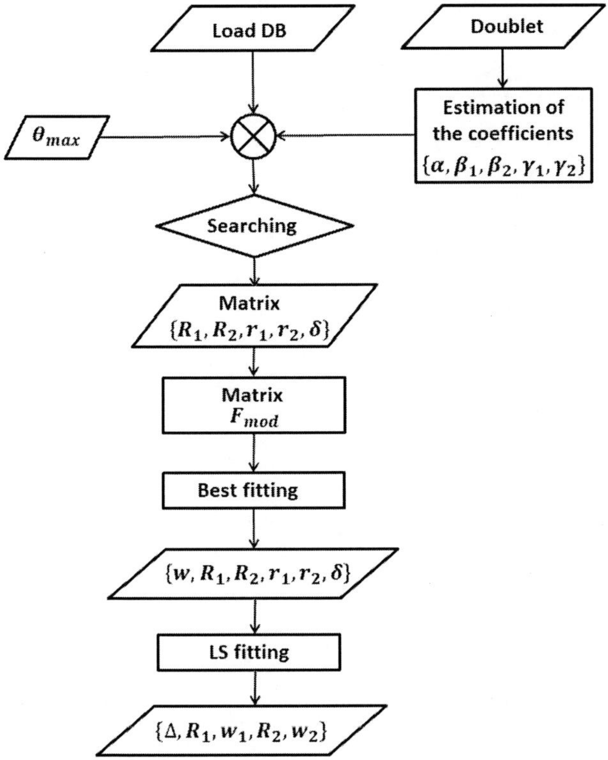

Figure 3.4-5. Doublet parameters estimation.

From these data, we conclude the following:

1. Parameters estimated by the $\alpha\beta\gamma$-method are in good accordance with their precise values. The exception for the width w_2 of the model 3 is given in bold in Table 3.4-1.
2. In some cases, the 95% confidence intervals of the LS-estimates (given in bold in Table 3.4-1) do not include the doublet parameters' precise value. However, the figures demonstrate visually similar curves of the best-fitting models and doublets.

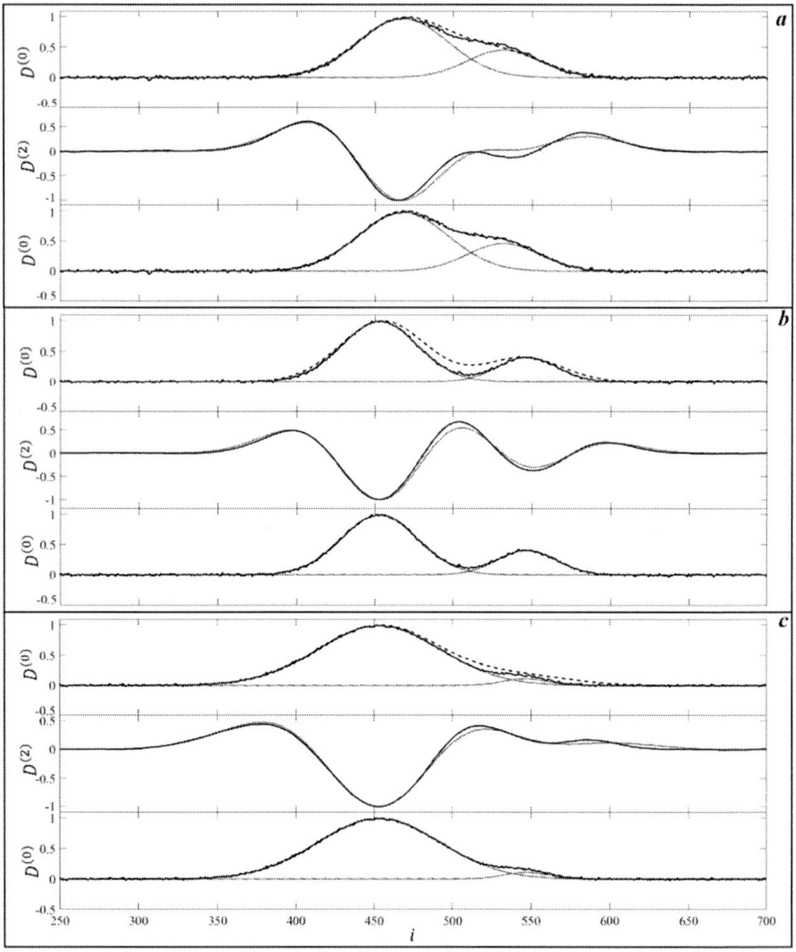

Figure 3.4-6. Decomposition of the doublets (Table 3.4-1). In the first section of each panel, solid, dotted, and dashed curves designate the model, its components, and the doublet calculated using $\alpha\beta\gamma$-best-fit parameters. The second section represents the 2nd-order derivatives of the model and the $\alpha\beta\gamma$-best-fit doublet (solid and dotted lines, respectively). The last section illustrates the LS-best-fit decomposition. The curve designation is similar to the first section.

To illustrate this phenomenon, we repeated calculations 1000 times for the model 1 distorted by the newly generated noise. Statistically invalid results were observed in 934, 893, and 967 cases for the standard deviation of the normal noise σ =0.01, 0.005, and 0.001, respectively.

These findings proved the well-known fact that searching of the best-fitting regression models is not a trivial task [1]. Blind use of commercial software, which has been designed to decompose of overlapping peaks, may give unpredictable results. Some research has demonstrated forests of peaks found in unresolved contours without any statistical data.

Finally, we conclude that even the most straightforward doublet decomposition requires special attention.

Table 3.4.-1. Data obtained by decomposition of the model Gaussian doublets

<table>
<tr><th rowspan="2">Model</th><th rowspan="2">Type of parameters</th><th colspan="5">Parameters</th></tr>
<tr><th>Δ</th><th>R_1</th><th>w_1</th><th>R_2</th><th>w_2</th></tr>
<tr><td rowspan="3">δ =0.32
R_1 =0.62
r_1 =0.67
R_2 =0.29
r_2 =0.54</td><td>Precise</td><td>32.0</td><td>0.62</td><td>67.0</td><td>0.29</td><td>54.0</td></tr>
<tr><td>Estimated by $\alpha\beta\gamma$-method</td><td>31.5</td><td>0.62</td><td>70.0</td><td>0.25</td><td>63.0</td></tr>
<tr><td>Estimated by LS-fitting</td><td>32.1± 0.16</td><td>**0.52± 0.10**</td><td>**66.1± 0.54**</td><td>**0.24± 0.047**</td><td>53.4± 0.99</td></tr>
<tr><td rowspan="3">δ =0.47
R_1 =0.78
r_1 =0.57
R_2 =0.32
r_2 =0.44</td><td>Precise</td><td>47.0</td><td>0.78</td><td>57.0</td><td>0.32</td><td>44.0</td></tr>
<tr><td>Estimated by $\alpha\beta\gamma$-method</td><td>44.8</td><td>0.79</td><td>64.0</td><td>0.32</td><td>57.6</td></tr>
<tr><td>Estimated by LS-fitting</td><td>46.9± 0.12</td><td>0.75± 0.30</td><td>56.6± 0.26</td><td>**0.30± 0.12**</td><td>44.0± 0.69</td></tr>
<tr><td rowspan="3">δ =0.47
R_1 =0.99
r_1 =0.92
R_2 =0.11
r_2 =0.34</td><td>Precise</td><td>47.0</td><td>0.99</td><td>92.0</td><td>0.11</td><td>34.0</td></tr>
<tr><td>Estimated by $\alpha\beta\gamma$-method</td><td>46.0</td><td>1.00</td><td>92.0</td><td>0.14</td><td>**73.6**</td></tr>
<tr><td>Estimated by LS-fitting</td><td>46.8± 0.18</td><td>1.25± 0.44</td><td>**90.8± 0.43**</td><td>**0.14± 0.050**</td><td>36.2± 2.3</td></tr>
</table>

The LS-estimates are the 95% confidence intervals.

PART IV:

QUANTITATIVE DERIVATIVE SPECTROMETRY

INTRODUCTION

Derivative-based quantitative analysis is similar to the conventional method since the differentiation is the linear mathematical operation. If the analyte's pure analytical signal is proportional to the amount (concentration) of the analyte in the sample, then the derivatives of the signal, obtained by precise linear methods, will have this property. However, in practice, the assumption of linearity may be violated. Also, as in conventional procedures, one must consider the background (including random noise) and interfering signals that disturb the accurate instrumental response to an analyte. Denoising the derivative spectrum is one of the essential problems since the differentiation significantly decreases the signal-to-noise ratio.

In the first chapter, we will consider the intensity measurements in the derivative spectra and the relationships between the intensity and the analyte concentration for different differentiation methods. The following chapters' subjects are the selectivity, sensitivity, and informational concepts of quantitative derivative spectrometry.

CHAPTER ONE

INTENSITY MEASUREMENTS: RELATIONSHIPS BETWEEN ANALYTE CONCENTRATION AND INTENSITY

The intensity of the derivatives of the analytical signals is usually measured from a base- and zero-lines (Fig. 4.1-1, panels (a, b) and (c-g), respectively [1, 2]). Consider the more detailed classification of these baseline and zero-line methods [2].

The main advantage of the first group of methods is that the measurements from a horizontal baseline do not depend on the constant shifts along the y-axis. Therefore, the linear backgrounds do not disturb the analytical signal in the 1st- and the 2nd-order derivatives, but only the parabolic in the last case. The non-horizontal baseline corrects the residual linear background in the 1st-order derivative.

The errors of the background approximation by the non-horizontal baseline decrease in the derivative spectrum. It follows from the fact that the approximation interval narrows due to the narrowing of the derivative peaks, thus, improving the background correction. New extremal points appear in the derivatives. Drawing of the baselines between these points reduces the uncertainties of the baseline endpoints locations along the x-axis.

In the null-measurement methods, the analytical points of the analyte (x_0) are the points of the intersection of the interfering curves with the x-axis (panels c, d). Unfortunately, the extrema positions in the derivatives and the points x_0 match each other in rare cases. Therefore, the measurements at the steep branches of the peak may be subject to significant errors due to the random shifts along the x-axis. The authors of one of the pioneering studies in derivative spectrometry [3] adjusted the extremal and the null-points by varying the monochromator slit's width. The authors of the study [4] estimated both the dependence of the distance between these points on the slit width and the optimal width for a given peak separation. The null-point was shifted to the extremum of the analytical derivative peak by varying the modulation amplitude of the optical differentiator and the slit width. We achieved the same results by changing the RC differentiator parameters [2].

Panel d simulates a hypothetic case in which extrema of the derivative coincide with two null-points located on the two sides of the interfering peak. Since the slopes of these sides have opposite signs, the shifts of the abscissa are partly compensated. Therefore, the concentration measure, which equals the y-distance between extrema, becomes less sensitive to the x scale's drifts.

The absolute value of the ratio $|F_1^{(1)}/F_2^{(1)}|$ (panel f) in the poor-resolved 1st-order derivative spectra, was the concentration measure of the analyte in the binary mixture [5].

Figure 4.1-2a demonstrates the interesting example of the unresolved 1st-order derivative of the Gaussian doublet consisted of strongly overlapped peaks:

$$D(i) = R_1 exp\left\{-\beta\left[\frac{i-i_c+\Delta}{w_1}\right]^2\right\} + R_2 exp\left\{-\beta\left[\frac{i-i_c-\Delta}{w_2}\right]^2\right\}, \qquad (4.1-1)$$

where R and w are the peak intensity and width, respectively; $\beta = 4ln2$; Δ is the peak shift from the central point i_c.

The extrema of the derivatives of the doublet components have opposite signs at the point i_c=250. The total intensity is $F_1^{(1)} = F_{12}^{(1)} - F_{11}^{(1)}$, where the second subscript stands for the component number. The right minimum $F_2^{(1)} \cong F_{22}^{(1)}$. Therefore, since $F_{12}^{(1)} = |F_{22}^{(1)}|$, the ratio is (Fig. 4.1-2b)

$$|F_1^{(1)}/F_2^{(1)}| = \left|\left(F_{12}^{(1)} - F_{11}^{(1)}\right)/F_{22}^{(1)}\right| = 1 - F_{11}^{(1)}/F_{22}^{(1)} = 1 - kR_1, \qquad (4.1-2)$$

where $k = w_2/(R_2 w_1)$ is the slope.

If the intensity R_1 linearly depends on the amount of the first component in the binary mixture, then the ratio $|F_1^{(1)}/F_2^{(1)}|$ may be used in the quantitative analysis.

Unfortunately, the above pattern (Fig. 4.1-2a) is rare in practice. Usually, the appropriate analytical points do not coincide with the extrema. Therefore, the shifts along the x-axis cause additional errors similar to those of the null-method.

The pioneering study in [6] used areas under peaks in the 1st- and 2nd-order derivatives for quantitative purposes. Later, Kalmanovskyi [7] received a patent for the chromatographic analysis based on this method. He constructed a new RC differentiator, which generated the unipolar signal (absolute value of the derivatives).

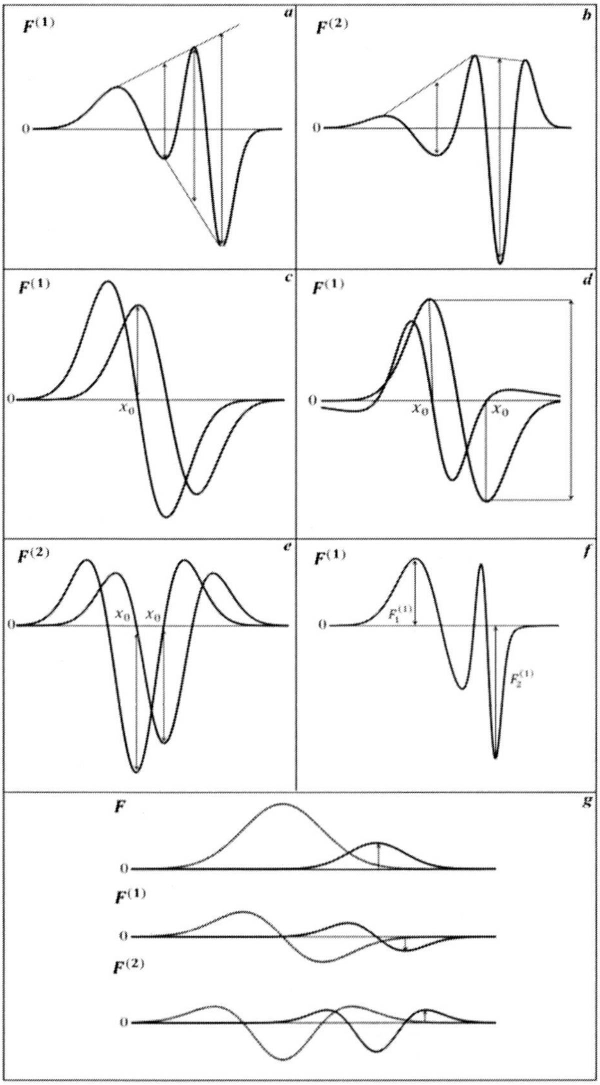

Figure 4.1-1 (adapted from [2]). Intensity measurements in the derivative spectra: from a baseline (a, b), in the zero points of the interfering components (c-e), by the ratio $F_1^{(1)}/F_2^{(1)}$(f), and at the points of a weak distortion of the analyte signal (g).

Exercise 4.1-1.

Readers are invited to check the validity of Eq. (4.1-2) using Figure 4.1-2b.

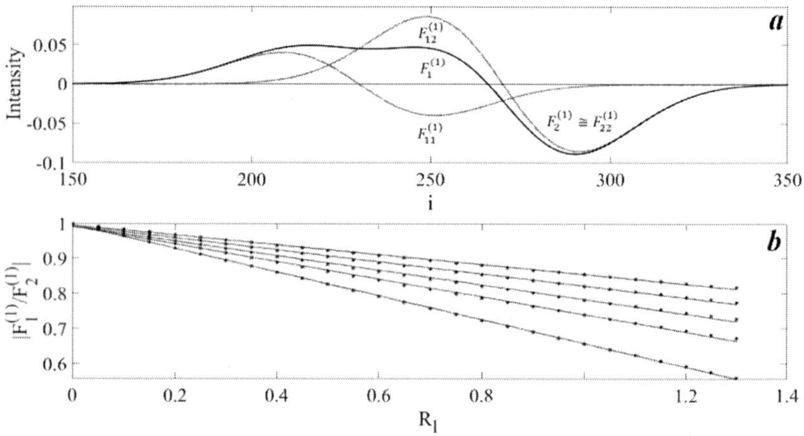

Figure 4.1-2. (a) The 1st-order derivative spectra of the symmetrical Gaussian doublet and its components (solid and dotted curves, respectively ($R_1 = 1.4$, $w_1 = 50$). (b) Dependence of the ratio $\left|F_1^{(1)}/F_2^{(1)}\right|$ of R_1. $w_1 = 50{:}25{:}150$ from the lower to the upper lines. In both panels: $R_2 = 3, w_2 = 50$, $\delta = 0.4$.

For example, consider two overlapping Gaussian peaks and their 2nd-order derivatives (Fig. 4.1-3), representing a model chromatogram of a binary mixture. Eq. (1.2-18) showed that the areas under the positive and negative branches of any order derivatives are equals. Therefore, the central peak and satellite of the 2nd-order derivative include 50% and 25% of the whole area, respectively. The areas of the darkened humps in the left and right sides of the panel b are free of the overlapping.

These areas are accurate measures of 75% and 25% of the amounts of the first and second mixture components, respectively. The corresponding darkened segments of the peaks also do not approximately overlap. However, while the humps endpoints in the derivative chromatogram are zeros, the end- and start-points of the first and second segments, respectively, lie on the peaks' steep slopes. Small shifts of these points cause significant errors in the quantitative analysis. Figure 4.1-3 represents the patterns, which are typical for nearly entirely resolved peaks. For these peaks, reducing the area significantly increases the uncertainties in the analysis [8]. The correction method [7] improves accuracy. However, in the case of a substantial overlap, this method is not applicable.

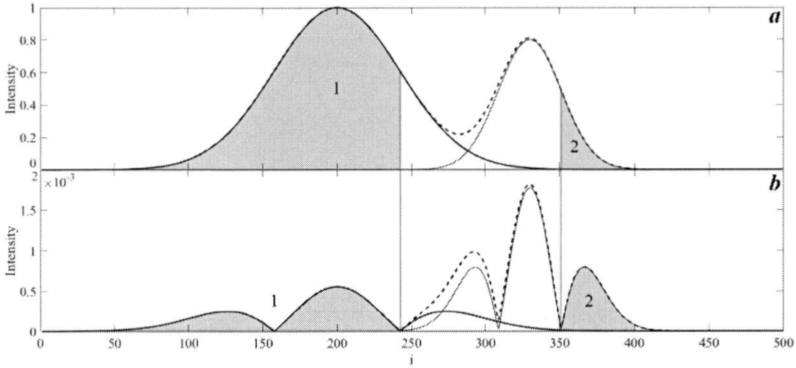

Figure 4.1-3. The area measurements in the Gaussian doublet (a), and the unipolar 2nd-order derivatives (b). Dashed and solid curves stand for the doublet and its components, respectively.

Currently, chromatographists are trying to resolve the overlapped peaks by the least-squares decomposition method.

The errors in the quantitative derivative spectrometry depend on the degree of the peak overlap, the method of measuring the intensity of the analytical signal, and the noise. These errors are related to the uncertainties in determining the analytical data in derivatives (Chapter 3.3). In this chapter, we estimated the relative systematic errors for the zero- and baseline measurement methods of the relative intensity of the second doublet peak in the 2nd-order derivative of the Gaussian and Lorentzian doublets.

In the first piece of research on this issue [1], O'Haver simulated quantitative analysis using the Gaussian doublet and its 1st- and 2nd-orders derivatives. He showed that,

1. Null and baseline measurements of the intensity in the 2nd-order derivatives demonstrate minimum errors in determining the concentration ratio of the binary mixture components.
2. Systematic errors of the baseline methods are minimal if the analytical peak is narrower than the interfering one. The null-method is the best in two cases: (a) equal-width components; and (b) constant parameters of the interfering peak.
3. The measurements using the 2nd-order derivatives are, as a rule, better than in 1st-order derivatives.

4. Differentiation has an advantage over unprocessed data only in the case of poor-resolved peaks.

Since the differentiation significantly reduces the signal-to-noise ratio, the noise must be removed to a large extent by smoothing (denoising). Therefore, one obtains distorted derivative peaks. They broaden, and the spectral resolution decreases. While intense smoothing hampers the structure analysis of spectra, the estimation of the mixture concentrations may successfully involve inaccurate derivatives distorted by the denoising [2]. However, the parameters of the differentiators must be unchanged in all studies of the quantitative analysis.

One of the significant advantages of nth-order derivative spectrometry is the suppression of the polynomial background, which degree $t < n$. In this case, the differentiation may reduce the quantitative analysis' systematic errors that appear in the presence of an uncorrected background.

We showed that (a) the background correction by the differentiation is a partial case of algebraic correction; and (b) from a signal processing view, the latter method is better than the derivative technique for the background suppression [9]. The algebraic correction involves the polynomial approximation of the background. The polynomial coefficients and unknown mixture concentrations are variables of a system of equations. The solution of this system suffers from many errors. Therefore, background correction is now performed in the chemometrics framework, but it requires multidimensional data [10].

Due to Chapter 2.1, the output voltage of the differentiator is proportional to the concentration of the analyte; the analytical signal of which is not disturbed by interfering mixture components. This conclusion is valid for inaccurate derivatives of the absorbance or any quantity related linearly to the concentration. The above is also valid for linear numerical differentiation.

Consider the following expression of absorbance:

$$A = -log_{10}T = log_{10}I_0 - log_{10}I, \qquad (4.1-3)$$

where $T = I/I_0$ is the transmittance of the sample; I and I_0 are the radiation intensities received and transmitted by this sample.

If the first term (Eq. (4.1-3)) is approximately constant, then the nth-order derivative of the output signal of the single-beam spectrometer will be as follows:

$$A^{(n)} \cong -(log_{10}I)^{(n)}. \qquad (4.1-4)$$

If $A \ll 1$, then $I/I_0 = 10^{-A} \sim (1 - A)$. We stated above that $I_0^{(n)} = 0$, therefore,

$$I^{(n)} \sim -A^{(n)} \qquad (4.1-5)$$

This method was used for the analysis of weak-absorbing objects, e.g., gases.

Suppose that the Beer-Lambert law is valid:

$$A = \varepsilon_\lambda lc, \qquad (4.1-6)$$

where ε_λ is the absorptivity of the sample; λ is the abscissa of the spectrum (e.g., the wavelength); l is the optical path length; c is the analyte concentration. Then, from Eq. (4.1-3) we have

$$\frac{1}{I}\frac{dI}{d\lambda} = \frac{1}{I_0}\frac{dI_0}{d\lambda} - lck\frac{d\varepsilon_\lambda}{d\lambda}, \qquad (4.1-7)$$

where $k = ln10$.

For the derivative of Eq. (4.1-7)

$$\frac{1}{I}\frac{d^2I}{d\lambda^2} - \left(\frac{1}{I}\frac{dI}{d\lambda}\right)^2 = \frac{1}{I_0}\frac{d^2I_0}{d\lambda^2} - \left(\frac{1}{I_0}\frac{dI_0}{d\lambda}\right)^2 - lck\frac{d^2\varepsilon_\lambda}{d\lambda^2}$$, we have

$$\frac{1}{I}\frac{d^2I}{d\lambda^2} = \frac{1}{I_0}\frac{d^2I_0}{d\lambda^2} + \left(\frac{1}{I}\frac{dI}{d\lambda} - \frac{1}{I_0}\frac{dI_0}{d\lambda}\right)\left(\left(\frac{1}{I}\frac{dI}{d\lambda} + \frac{1}{I_0}\frac{dI_0}{d\lambda}\right)\right) - lck\frac{d^2\varepsilon_\lambda}{d\lambda^2}. \qquad (4.1-8)$$

By substituting Eq. (4.1-7) in the first brackets of Eq. (4.1-8) and taking the fact that in the peak maximum $d\varepsilon_\lambda/d\lambda$=0 into account, we obtain

$$\frac{1}{I}\frac{d^2I}{d\lambda^2} = \frac{1}{I_0}\frac{d^2I_0}{d\lambda^2} - lck\frac{d^2\varepsilon_\lambda}{d\lambda^2}. \qquad (4.1-9)$$

Exercise 4.1-2.

Readers are invited to obtain the expressions of the higher-order derivatives, which contain quadratic terms of concentration.

Eqs. (4.1-7) and (4.1-9) describe the derivatives obtained by optical modulation differentiators (Chapter 2.9). In some cases, it is not easy to use these devices for quantitative analysis: e.g., the 2nd-order derivative is only proportional to the analyte concentration in the peak maximum. Therefore, baseline measurements are impossible.

Quantitative laser analysis of the gas mixtures using frequency modulation technique

Before studying this section, we recommend revisiting the materials provided in Chapter 2.10.

In the lack of saturation, the power of the monochromatic laser radiation, at frequency ν, passed through the uniform absorption layer of the width l [11]:

$$P(\nu) = \chi P_0(\nu) exp[K(\nu)l] = \sum_{k=0}^{\infty} \mathcal{F}_k \cos(k\omega_m t), \qquad (4.1-10)$$

where χ is the loss factor independent of absorbance; $P_0(\nu)$ stands for the power of the incident radiation and for the jth line of ith mixture component (Eq.(4.1-6)):

$$K(\nu) = \sum_{i,j} \varepsilon_{i,j}(\nu) c_i; \qquad (4.1-11)$$

$\mathcal{F}_k$ is the amplitude of the Fourier transform; ω_m is the modulation frequency.

Dependence $P(\nu)\{c_i\}$ is nonlinear. Therefore, the estimation of the concentration, using the calibration method, is a cumbersome task. The linearization makes it possible to simplify the problem only if $K(\nu)l \ll 1$.

The quantitative analysis of gas mixture used combinations of signal harmonics, which were obtained by small modulation amplitudes (compared to peak widths) [11]:

$$A_1 = \mathcal{F}_1/\mathcal{F}_0 = l \sum_{i,j} R_{ij}^{(1)} c_i; \qquad (4.1-12)$$

$$A_2 = 4\mathcal{F}_2/\mathcal{F}_0 - A_1^2 = l \sum_{i,j} R_{ij}^{(2)} c_i, \qquad (4.1-13)$$

where the coefficients $R_{ij}^{(1)}$ and $R_{ij}^{(2)}$, obtained for the Lorentzian shape, are complicated functions of the peak intensity, width, and modulation amplitude. Their estimation involved a cumbersome calibration.

Equations, similar to Eqs. (4.1-12) and (4.1-13), obtained for the Voigt line shape, included amplitudes of the zero-, 2nd-, and 4th-order Fourier harmonics [12].

C. Weitkamp [13] performed the calibration using a single analytical line. The short reference cell contained the high-concentrated gas under analysis. The following equations describe the estimation of the unknown concentration of the sample gas (c_s).

Suppose that I_0 is a constant and the analytical line shape is the Gaussian. Then, from Eqs. (4.1.3) and (4.1-6), we have the following in the line center:

$$I_{max}/I_0 = exp(-cl\varepsilon_0) = (U_{max} - U_{dip})/U_{max}, \qquad (4.1-14)$$

where U_{max} is the maximum intensity in the line center; U_{dip} is the dip in this point caused by the saturation. From Eqs. (4.1-9) and (4.1-14), we obtain the ratio of the 2nd-order derivatives, generated in the sample (s) and reference (r) channels:

$$S_s^{(2)}/S_r^{(2)} = q_s c_s l_s exp(-c_s l_s \varepsilon_0)/q_r c_r l_r exp(-c_r l_r \varepsilon_0), \qquad (4.1-15)$$

where q_s and q_r are constants estimated in the regions free of the analytical line. Eq. (4.1-14) defines ε_0. The unknown concentration c_s was estimated from Eq. (4.1-15) [13].

J. A. Mucha [14, 15] used the standard addition technique for quantitative trace gas analysis. He measured the modulation-broadened Gaussian linewidths and absorptivities in the 2nd-order derivative spectra of the infrared transitions of the water vapor. The use of effective absorptivities allowed for the correction of the nonlinear correlation between derivative data and absorber concentration.

The research team in [16] developed the calibration-free method applicable to the absorbance less than 5.0%. In this case, the first-order Taylor series well approximates the exponential term of Eq. (4.1-10):

$$exp[K(\nu)l] \approx 1 - lK(\nu) = 1 - l\sum_{k=0}^{\infty} H_k \cos(k\omega_m t), \qquad (4.1-16)$$

where $H_0 = (1/2\pi)\int_{-\pi}^{\pi} K(\nu)d\varphi$;

$H_k = (1/\pi)\int_{-\pi}^{\pi} K(\nu)cos(k\varphi)d\varphi \ (k = 1, 2, 3, \dots)$ according to the Fourier transform properties (Appendix A).

By comparing Eq. (4.1-10) and Eq. (4.1-16), we obtain the zero and the kth-order signal harmonics:

$$A_0 \sim 1 - lH_0; \ A_k \sim -lH_k, k = 1, 2, 3, \dots \qquad (4.1-17)$$

The 2nd-order Taylor series reduced the approximation errors for the absorbance of more than 5.0% [16]. This method significantly improved the accuracy when determining the ammonia's mole fractions.

Multicomponent derivative spectroscopy. General approach

We showed that the quantitative derivative of spectrometry is a particular case of standard spectrochemical analysis [17]. In this context, a question arises: "Does the derivative spectrometry improve accuracy (trueness) and precision of the analytical procedures?" Assume that,

(1) The random measurement errors of experimental spectra are unbiased (that is, the errors have zero mathematical expectation (mean)).
(2) The pure component spectra of the mixture, whose exact composition is known, are linearly independent.

Then, according to the general principles of statistics [18], no linear transform (e.g., differentiation) of the experimental data can decrease the dispersion of the unbiased estimates of the amounts (concentrations) of the mixture components. This statement is also valid in the case of the linearly dependent components when the least squares estimates are obtained using the overdetermined system of the linear equations.

However, the biased estimation can decrease the total error consisted of random and systematic components. The trade-off between these components is the optimization principle, which is widely used in signal processing.

For simplicity, consider the molecular absorption spectra of the additive m-component mixture. For this mixture, the unknown concentrations (c_k) are found from the solution of the system of linear equations (similar to Eq. (4.1-11)):

$$F_i = \sum_{k=1}^{m} \varepsilon_{i,k} c_k, \quad (4.1-18)$$

where $i \in [0, L]$ is the serial number of the analytical point; the absorption pathlength is taken to be unity.

The total number of the analytical points is $L+1 \geq m$. However, the system is usually overdetermined, that is, $L \gg m$. Solutions for overdetermined systems are given in monograph [18].

Suppose that the spectrum contains the normal noise with zero mean and variance σ^2. The orthogonal polynomial expansion of $\varepsilon_k(i)$ on the interval $[1, L]$ gives (similar to the Savitzky-Goley (SG) method, Chapter 2.3):

$$\varepsilon_{i,k} = \sum_{\alpha=0}^{L} \theta_{\alpha,k} \varphi_\alpha(i), \quad (4.1-19)$$

where $\theta_{\alpha,k}=M_\alpha^{-1} \sum_{i=0}^{L} \varepsilon_{i,k} \varphi_\alpha(i)$ are the coefficients of the polynomial φ_α; $M_\alpha = \sum_{i=0}^{L} \varphi_\alpha^2(i)$.

Let us perform the following operations:

1. Substituting Eq. (4.1-19) into Eq. (4.1-18).
2. Multiplying both sides of the new equality by $\varphi_i(i)$.
3. Summing the result in the interval $[1, L]$.

Finally, taking the orthogonality of the polynomials into account, we have

$$M_q^{-1} \sum_{i=0}^{L} F_i \varphi_q(i) = \sum_{k=0}^{m} \theta_{q,k} c_k, \quad (4.1-20)$$

where $q \in [0, L]$.

According to Chapter 2.3, the left-hand side of Eq. (4.1-20) resembles the qth-order derivative $F_r^{(q)}$ at the point $r \in [0, L]$. If $L = 2N$, $r = N$, and the sampling interval $h = const$, this term represents the SG differentiation filter (the full width is $2N+1$). This filter only generates the derivative in the central point r. However, the derivatives in the rest points are not accurate; they are pseudo-derivatives.

The following numerical experiment illustrates the quantitative analysis of a binary mixture based on the orthogonal transformation method (OTM). The component concentrations are $c_1 = 0.1$, $c_2 = 0.9$.

The model spectrum is the Gaussian doublet:

$$D(i) = c_1 F_1(i) + c_2 F_2(i) + \eta, \qquad (4.1-21)$$

where $i = 1{:}25$; $F_1(i) = 1.5exp\{-\beta[(i - i_{max1})/6]^2\}$; $\beta = 4ln2$; $i_{max1} = 12 - \Delta$; $F_2(i) = exp\{-\beta[(i - i_{max2})/w_2]^2\}$; $i_{max2} = 12 + \Delta$; η is the normal noise with zero mean and the variance $\sigma^2 = 10^{-4}$.Table 4.1-1 represents the numerical values of Δ and w_2.

In the standard method, the mixture concentrations are estimated from the matrix equation:

$$\begin{vmatrix} \varepsilon_{11} & \varepsilon_{12} \\ \varepsilon_{21} & \varepsilon_{22} \end{vmatrix} * \begin{vmatrix} c_1 \\ c_2 \end{vmatrix} = \begin{vmatrix} D(i_{max1}) \\ D(i_{max2}) \end{vmatrix}, \qquad (4.1-22)$$

where $\varepsilon_{i,k} = F_k(i_{maxi})$.

The solution to Eq. (4.1-22) is

$$\hat{c}_1 = det1/det, \;\; \hat{c}_2 = det2/det, \qquad (4.1-23)$$

where the determinants, $det1 = D(i_{max1})\varepsilon_{22} - D(i_{max2})\varepsilon_{12}$; $det2= \varepsilon_{11}D(i_{max2}) - \varepsilon_{21}D(i_{max1})$; and $det = \varepsilon_{11}\varepsilon_{22} - \varepsilon_{12}\varepsilon_{21} \neq 0$.

In the OTM, the mixture concentrations are estimated from the matrix equation:

$$\begin{vmatrix} \bar{F}_1 & \bar{F}_2 \\ |F^{(2)}_{1max}| & |F^{(2)}_{2max}| \end{vmatrix} * \begin{vmatrix} c_1 \\ c_2 \end{vmatrix} = \begin{vmatrix} \bar{D} \\ |D^{(2)}_{max}| \end{vmatrix}, \qquad (4.1-24)$$

where the bar under the variables stands for the mean value, which is the zero-order coefficient $\theta_{0,k}$ in Eq. (4.1-19). The 1st-order coefficient ($\theta_{1,k}$) equals the negligible value of the 1st-order derivative in the middle point 12. Therefore, we used $\theta_{2,k}=|F^{(2)}_{k,max}|$.

According to Eq. (4.1-20), the full width of the 2nd-order SG differentiation filter is equal to the spectral segment (25 points). So, the derivative plot shows only a single point-peak in the central point (Fig. 4.1- 4, b).

The solution of Eq. (4.1-24) is similar to that of Eq. (4.1-22).

The mean relative errors of the binary mixture analysis and their standard deviations (Table 4.1-1) were estimated from 10,000 independent repetitions of the numerical experiment. Table data show that the narrowing of the second peak (Fig. 4.1- 4, c) or increasing the doublet peaks separation (panels d and e) reduce uncertainties in determining concentrations by the standard method. The OTM has better metrological characteristics in the case of a substantial peak overlap (in bold in the Table).

Exercise 4.1-3.

Readers are invited to estimate the errors in determining the mixture concentrations by varying the model parameters, including the total number of analytical points (Eq. (4.1-21)), in a wide range.

Table 4.1-1. Relative errors of the binary mixture analysis

Parameters		δ_{c_1}, %		δ_{c_2}, %	
Δ	w_2	spectrum	OTM	spectrum	OTM
1	**18**	**0.0±31.9**	**0.0±10.0**	**0.0±4.7**	**0.0±0.79**
1	**10**	-0.2±26.1	-0.5±43.8	-0.0±4.0	0.1±4.7
2	**18**	0.3±11.8	-0.1±12.3	-0.0±1.6	0.0±0.92
3	**18**	-0.0 ± 8.6	-0.1±17.4	0.0±1.2	0.0±0.1.2

The mean values ± standard deviations.

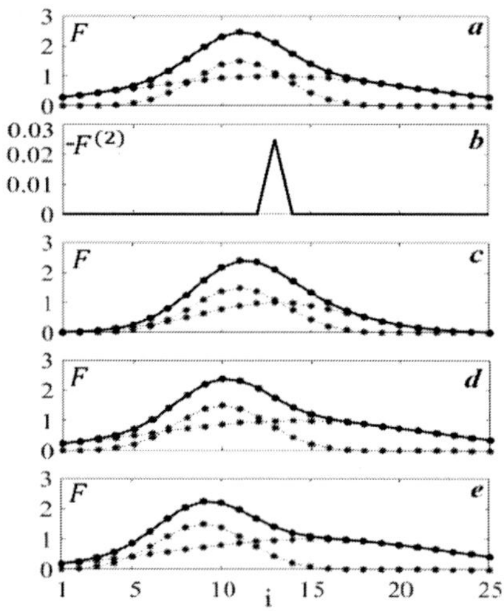

Figure 4.1-4. (a, c-e) The Gaussian doublet and its components (solid and dotted lines, respectively). $\{\Delta, w_2\} = \{1,18;\ 1,10;\ 2,18,3,18\}$ for panels a, c-e, respectively. (b) The negative 2nd-order derivative of "a" at point $i = 12$.

The mathematical q^{th}-order derivative of Eq. (4.1-18) at the point i:

$$F_i^{(q)} = \sum\nolimits_{k=1}^{m} \varepsilon_{i,k}^{(q)} c_k. \qquad (4.1-25)$$

For sufficiently small h , the solution of Eq. (4.1-24) is close to that of Eq. (4.1-25). The segment width in the abscissa scale $([0, L] * h)$ must be small. The difference between the derivatives, obtained by analogous or optical devices, and the accurate values (Eq. (4.1-25)) depends on the differentiator's accuracy. The linear transformation, represented by Eq. (4.1-19), is a partial case of the Linear Transform Coordinates Methods (LTCM) or the Generalized Fourier Transform (series) (GFT) [19]. So, the GFT (in particular, the orthogonal polynomial transforms) of the standard system of linear equations (Eq. (4.1-18)) is used for quantitative multicomponent analysis [20].

The inverse transform from Eq. (4.1-25) to Eq. (4.1-24) and, finally, to Eq. (4.1-18) is readily performed.

It is common to only solve one of the q equations of the system, represented by Eq. (4.1-25). Usually, this equation only involves the analyte signal in the single point, i. The impact of the other mixture components at this point must be negligible. However, this simple single-point derivative analysis suffers from serious errors (see discussion below).

The use of the pseudo-derivatives (Eq. (4.1-20)) of the noisy spectra is usually limited by 2nd-order differentiation. 4th-order differentiation requires a very high signal-to-noise ratio in the spectrum. In the best case, this method only applies to the quantitative analysis of a maximum of 3-5 component mixture.

Suppose that we have systems of linear equations (Eqs. (4.1-20) and (4.1-25)) whose number of unknown concentrations (m) equals the number of analytical points ($L + 1$). Then, rejecting g equations, allows us to find concentrations of $m - g$ mixture components using only a single analytical point in each equation. However, the systematic errors increase if the signals of these g components disturb the measurements. Now, suppose that the number of equations in the overdetermined systems is reduced, but $L + 1 - g \geq m$. In this case, the variances of the concentration increase according to the general principles of the least squares method [18].

The dimension of the systems (Eqs. (4.1-20) and (4.1-25)) may also decrease because of the fast convergence of the orthogonal polynomial series (Eq. (4.1-19)). That is, for some $\alpha < L - 1$, the coefficients are $\theta_{\alpha+1,k} \ll \theta_{\alpha,k}$. Therefore, the corresponding terms may be neglected.

Let us consider the general case of the multicomponent analysis using the derivatives.

For a given q, Eq. (4.1-26), written in the matrix form, represents the full-rank model:

$$\boldsymbol{F}^{(q)} = \boldsymbol{E}^{(q)} * \boldsymbol{C} + \boldsymbol{\eta}^{(q)}, \qquad (4.1-26)$$

where $\boldsymbol{F}^{(q)} = \begin{vmatrix} F_1^{(q)} \\ F_2^{(q)} \\ \vdots \\ F_L^{(q)} \end{vmatrix}$; $\boldsymbol{E}^{(q)} = \begin{vmatrix} \varepsilon_{11}^{(q)} & \varepsilon_{12}^{(q)} & \cdots & \varepsilon_{1m}^{(q)} \\ \varepsilon_{21}^{(q)} & \varepsilon_{22}^{(q)} & \cdots & \varepsilon_{2m}^{(q)} \\ \vdots & \vdots & \vdots & \vdots \\ \varepsilon_{L1}^{(q)} & \varepsilon_{L2}^{(q)} & \cdots & \varepsilon_{Lm}^{(q)} \end{vmatrix}$; $\boldsymbol{C} = \begin{vmatrix} c_1 \\ c_2 \\ \vdots \\ c_m \end{vmatrix}$;

$$\boldsymbol{\eta}^{(q)} = \begin{vmatrix} \eta_1^{(q)} \\ \eta_2^{(q)} \\ \vdots \\ \eta_L^{(q)} \end{vmatrix}.$$

In statistics, $\boldsymbol{F}^{(q)}$ and $\boldsymbol{E}^{(q)}$ are named the vectors of the response variables and the regression matrix [18]. It is supposed, that the elements of the regression matrix are *a priori* known with negligible errors. $\boldsymbol{\eta}^{(q)}$ is the vector of the measurement errors of $\boldsymbol{F}^{(q)}$. The mean of $\boldsymbol{\eta}^{(q)}$ is zero, and the variance is $\sigma^2\boldsymbol{I}$, where $\boldsymbol{I}$ is the identity matrix (q x q). The number of analytical points must not be less than the number of mixture components ($L \geq m$). The rank of matrix $\boldsymbol{E}^{(q)}$ equals m. In other words, all rows and/or columns of this matrix are linearly independent.

The least squares estimate of the vector of unknown concentration is

$$\widehat{\boldsymbol{C}} = \left\{\left(\boldsymbol{E}^{(q)}\right)^T \boldsymbol{E}^{(q)}\right\}^{-1} \left(\boldsymbol{E}^{(q)}\right)^T \boldsymbol{F}^{(q)}, \qquad (4.1-27)$$

where superscripts T and -1 denote the matrix transpose and inverse, respectively.

Above, we supposed that the derivatives measured in different analytical points are uncorrelated. To satisfy this condition, the time interval between the points is

$$\Delta t \geq (3 \div 5)\tau_m, \qquad (4.1-28)$$

where τ_m is the time constant of the measurement system, including the differentiator.

The analytical points lying inside the width of the digital differentiator filter are correlated.

The model represented by Eq. (4.1-26) has two significant drawbacks. First of all, the spectra of pure chemical mixture components needed for estimation of the matrix $\boldsymbol{D}$ may be unavailable, or some constituents are unknown. Also, differentiation significantly reduces the signal-to-noise ratio.

Chemometrics took a revolutionary step toward improving the metrological characteristics of multicomponent analysis by introducing multivariate calibration [21]. However, this method requires large analytical datasets obtained from the multicomponent samples, whose qualitative and quantitative compositions are identified by auxiliary analytical techniques (e.g., chromatography). Unidentified small impurities impair analysis accuracy by producing a weak background. Chemometricians widely used 1st-order derivatives to suppress the constant background (e.g., [22]).

Mathematics and software for the multivariate calibration are not evident for analysts, who are non-professionals in statistics and signal processing. Therefore, traditional single-point derivative-based spectrometry measurements are widespread in routine analysis, especially in pharmaceutical laboratories (Chapter 5.1). Researchers obtain the derivative ratio spectra to increase the number of analytical points undistorted by the interfering components. Unfortunately, this method suffers from many drawbacks [23].

An in-depth informational analysis of the quantitative derivative spectrometry, based on the theoretical concepts of sensitivity and selectivity, will be given in Chapters 4.3 and 4.4.

CHAPTER TWO

APPLICATIONS OF QUANTITATIVE DERIVATIVE SPECTROMETRY IN THE INDUSTRIAL LABORATORY

The previous chapter showed that, although the quantitative derivative analysis is quasi-optimal, it allows researchers to significantly simplify calculations by the following measurements:

1. The null-points (Fig. 4.1-1, c-e).
2. The points of the weak interfering signals in the 1st-order derivative, and at the maximum of the 2nd-order derivative satellite (Fig. 4.1-1, g).
3. The presence of a low-order polynomial background.

The measurements in the derivatives' extrema reduce the random errors caused by the instability of the abscissa position of the analytical point lying on the steep branch of the peak.

The above cases are realized in typical analytical spectrometry situations. The example is a single component determination in the presence of unknown impurities which disturb the analytical signal. Since there are no data about the composition and the spectrum of the impurities, one can only approximate the interfering signal using a polynomial background. Some samples, such as technical raw materials, intermediate products, and wastewater, contain suspended mechanical particles that cause an additional background. It is common to approximate this background by a straight line, which is absent in the 2nd-order derivative. The 1st-order differentiation transforms the line to the constant signal, which is compensated for by the horizontal baseline.

The second important task is a single component determination in a mixture where interfering components have close spectra. In this case, the columns of the regression matrix for $q = 0$ (Eq. (4.1-26)) become nearly dependent (the collinearity problem [1]). Therefore, the solution of Eq. (4.1-26) suffers from small experimental errors [1].

The differentiation may simplify the solution since the interfering components' spectra have close null-points and regions of weak signal.

We will illustrate these theoretical considerations using the analytical methods developed in 1970s in the central industrial laboratory of chemical trust in Russia. The principles of quantitative derivative spectroscopy, which were developed in these works almost half a century ago, have not lost their relevance. They are widely used in chemical and pharmaceutical laboratories. Also, we recommend this material to be used in a tutorial for courses in analytical spectrometry. The chemical-technological details of these analytical methods are beyond the realms of our interest.

All derivative spectra were obtained by the RC differentiation attachment to the spectrophotometer SPECORD UV VIS (Chapter 2.1). The special home-made device allowed us to set precise wavelength marks on the derivative plots. Now, each industrial spectrophotometer does not require such smart tricks.

Determination of epoxy resin in chloroparaffins

Epoxy resin serves as the stabilizing additive in the manufacturing of chloroparaffins. The amount of resin determines the thermal properties of the paraffins. However, the epoxy resin's precise dozing was impossible due to the high calibration errors of the chemical reactors and the mixing of different resin samples. Also, the analytical control of the resin concentration is crucial since it is an expensive product.

The nonlinear background in the chloroparaffins spectra depends on the qualitative and quantitative composition of the sample (Fig. 4.2-1, a). This background cannot be compensated using the differential spectrophotometric analysis since the non-stabilized chloroparaffin is decomposed. The baseline method (dashed lines) gave erroneous results. In the 1st-order derivative (Fig. 4.2-1, b), the order of the polynomial background decreases. The segment, which is approximated by the linear baseline, becomes significantly narrower than in the spectra (Fig. 4.2-1, a). Therefore, the differentiation eliminates, to a large extent, the impact of the background on the epoxy resin analysis at the point, which is approximately, 34700 cm^{-1}. Table 4.-2-1 shows the accuracy of the analysis.

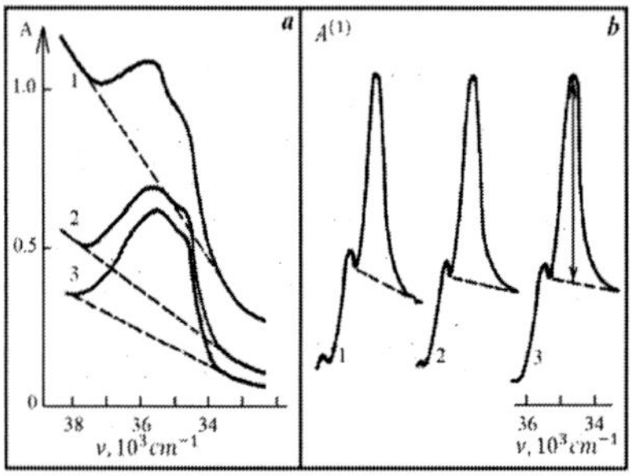

Figure 4.2-1 (adapted from [2]). Chloroparaffins. Spectra (a) and their derivatives (b). Additions of epoxy resin: 1 - 2.00%, 2 - 1.99%.

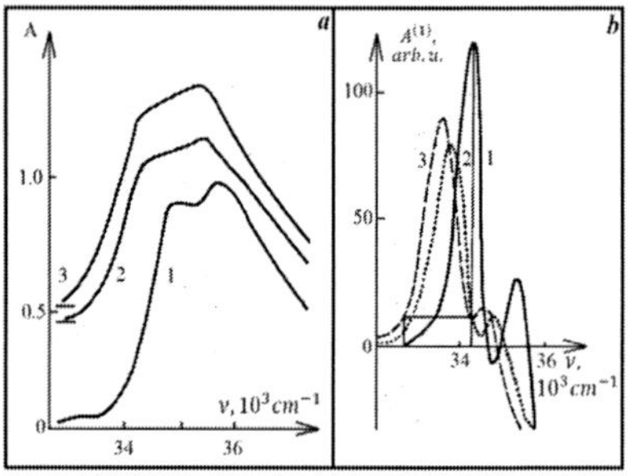

Figure 4.2-2 (adapted from [3]). Dichlorophenols (1-2.5, 2-3.4, 3-2.4). Spectra (a) and their derivatives (b).

Table 4.2-1.
Analysis of Epoxy resin in chloroparaffins

Epoxy resin in chloroparaffins, %		Relative error, %
Chemical analysis	**Derivative method**	
0.51	0.49	-3.9
0.96	0.98	2.1
1.46	1.48	1.4
2.00	1.97	-1.5
2.50	2.52	0.8
3.01	3.00	-0.3

Determination of 2.5-dichlorophenol in a mixture with 2.4- and 3.4-dichclorophenols

The spectrophotometric determination of 2.5-dichlorophenol in a mixture with 2.4- and 3.4-dichclorophenols requires the simultaneous analysis of all mixture components. Since the last two compounds have similar spectra (Fig. 4.2-2, a), the collinearity problem arises (Chapter 4.2-1).

We developed a simple single-point determination of 2.5-dichlorophenol in this mixture using the 1st-order derivative [3]. In the maximum of the derivative of the 2.5-dichlorophenol spectrum (approximately 34300 cm^{-1}), the interfering signal of 2.5-dichlorophenol is compensated for by the horizontal baseline. The baseline begins at the point 32750 cm^{-1}, where the impact of the 2.5-dichlorophenol is negligible. The derivative of 3.4-dichclorophenol, the contents of which less than 20% in the mixture, slightly disturbs the baseline position.

Table 4.2-2 shows that the absolute relative errors in determining the amounts of 2.5-dichlorophenol in the model mixtures are less than 8.0%.

In 1983 we quantified all three dichlorophenols in the mixture using the Walsh transform [4].

Determination of impurities of 2.4-dichlorophenol in butapon

Butapon (butyl ether of 2,4-dichlorophenoxyacetic acid (2,4-D)) [5] is an herbicide. The technical product is 50-60% solution of the butapon in higher alcohols, which contains small impurities: less than 2% 2.4-D and 0.5-2% 2.4-diclorophenol (2.4-DCP). From an ecological point of view, quantitative control of 2.4-DCP pollutants in the herbicide is essential. Chemical analysis of 2.4-DCP by steam distillation of phenol is cumbersome.

Table 4.2-2.
Analysis of 2-5 dichlorophenol in the model mixtures

Content, %		Relative error, %
Accurate	Found	
5.0	4.8	-4.0
10.0	9.2	-8.0
25.0	25.0	0.0
50.0	51.2	2.4
75.0	73.9	-1.5
85.0	83.9	-1.3

Spectra of pure butapon and its mixture with 2.4-DCP (Fig. 4.2-3, curves 1, 3, and 4) show that butapon absorbance becomes negligible in the wavenumber region that starts from 31500cm^{-1} and continues to the right-hand side. However, the spectrum of the industrial product demonstrates a strong background (curve 5). To eliminate the impact of this background, the analytical signal of 2.4-DCP was measured from the horizontal baseline in the 2nd-order derivative (curve 7). The measurement from a baseline in the 1st-order derivative (curve 6) is not applicable due to the disturbance caused by the butapon.

The relative differences between the amounts of 2.4-DCP, obtained by the spectrophotometric and chemical methods, were less than 15-20%.

This accuracy was satisfied with the requirements of the impurities control in herbicides.

Determination of chlorinated phenoxyacetic acids in wastewater

Orto- and para-chlorophenoxyacetic acids (oCPA and pCPA, respectively) are non-core products of herbicide manufacturing. From an ecological point of view, quantitative control of these acids in the wastewater of the herbicide manufacturing plant is essential.

Figure 4.2-4 demonstrates the baseline measurements of the analytical signals of the acids in the 1st-order derivatives. The horizontal baseline goes through the null-point (ν_3) of the pCPA. The line suppresses the signal of the oCPA at point ν_1.

The concentrations of the products are as follows:

$$c_{\text{pCPA}} = 0.70A^{(1)}(\nu_1);\ c_{\text{oCPA}} = 1.56\left[A^{(1)}(\nu_2) - 0.15A^{(1)}(\nu_1)\right]. \qquad (4.2-1)$$

The second term in the brackets takes into account the impact of the pCPA at point ν_2.

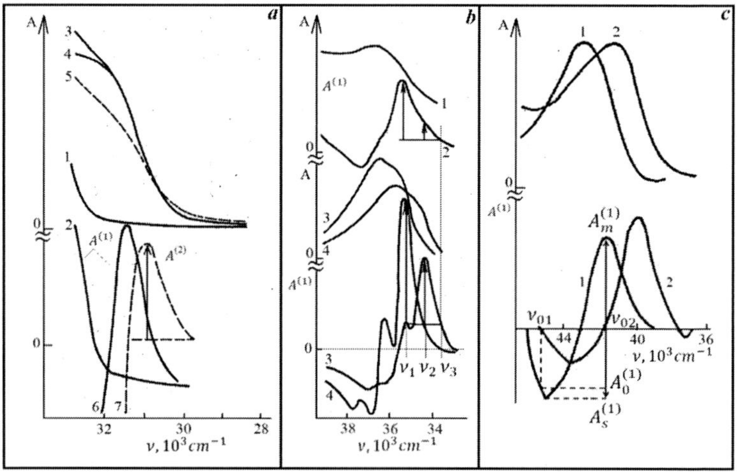

(a) Figure 4.2-3 (adapted from [6]). Butapon spectra (curves 1, 3-5). (2) The 1st-order derivative of (1). 2.4-DCP additions of 2% (3) and 1% (4). Industrial product (5).The 1st-(6) and 2nd–order (7) derivatives of the 2.4-DCP spectra. The curves 3-5 are vertically shifted so that the right tails coincide.

(b) Figure 4.2-4 (adapted from [2]). Wastewater spectrum (1) and its derivative (2). Spectra and their derivatives of oCPA (3) and pCPA (4).

(c) Figure 4.2-5 (adapted from [2]). Spectra and their derivatives of phenol (1) and 22DPP (2).

The accuracy of the determination of the chlorinated phenoxyacetic acids in the model mixtures (Table 4.2-3) has demonstrated the potential features of the method for wastewater control.

Determination of phenol in wastewater

The wastewater from the 2.2-Diphenylpropane (DPP) manufacturing plant contains phenol, DFP (less than 15-20% of all impurities due to the poor solubility in water), and small amounts of non-core products. Chemical analysis of wastewater is cumbersome. The spectrophotometric determination of phenol at the analytical wavelength of 270nm ($37000 cm^{-1}$), which was developed in the 1960s, had a low sensitivity.

Table 4.2-3. Amount of the chlorinated phenoxyacetic acids in the model mixtures, mgL^{-1}

Accurate		Found			
oCPA	pCPA	oCPA	Relative error, %	pCPA	Relative error, %
0.0	100.0	0.0	-	99.8	-0.2
10.0	90.0	9.8	-2.0	90.3	-0.8
20.0	80.0	18.2	-9.0	81.1	1.4
50.0	50.0	48.0	-4.0	50.7	1.4
80.0	20.0	80.0	0.0	20.7	3.5
90.0	10.0	90.5	0.5	9.5	-5.0
100.0	0.0	101.0	1.0	0.3	-

To improve sensitivity, we used the 1st-order derivative of the short-wave phenol band (maximum $42500 cm^{-1}$), whose extinction is six times larger than that of the band at 270nm (Fig. 2.4-5). The null-points' positions in the DPP spectrum (ν_{01} and ν_{02}) and the extrema in the phenol derivative spectra are close to each other. Therefore, due to the small impurities of DFP, the phenol's analytical signal is the length of the straight-line segment between two extrema $A_m^{(1)} - A_s^{(1)}$ instead of the interval $A_m^{(1)} - A_0^{(1)}$. This measurement has a low sensitivity to the small shifts along the abscissa scale.

The relative errors of the phenol determination were estimated using the standard additives of phenol to the wastewater free of this compound (Table 4.2-4). The errors increase significantly for low phenol concentrations due to the more substantial impact of the null-point shifts.

Exercise 4.2-1

List the typical analytical tasks solved by the methods described in this chapter.

Table 4.2-4. Analysis of phenol in waste water

Content, mgL^{-1}		Relative error, %
Added	Found	
0.40	0.31	-22.5
0.80	0.83	3.7
2.00	2.07	3.5
4.00	4.03	0.8
6.00	5.95	-0.8

CHAPTER THREE

THE SELECTIVITY OF QUANTITATIVE DERIVATIVE SPECTROMETRY

The definition of the term "selectivity" and its meaning in analytical chemistry has been the subject of numerous discussions [1]. One must distinguish between analytical methods and instrumental selectivity in, for example, a sensor system.

The IUPAC Recommendations state the following[2]:

> "Selectivity refers to the extent to which the method can be used to determine particular analytes in mixtures or matrices without interferences from other components of similar behavior."

Let us represent the analytical signal of the m-component additive mixture by the system of linear equations (similar to Eq. (4.1-18) but the pure signal of the pth-component is outside the summation symbol):

$$F_i = \sum_{k=1}^{m} F_{ik} = F_{ip} + \sum_{k \neq p} F_{ik} = \varepsilon_{i,p} c_p + \sum_{k \neq p} \varepsilon_{i,k} c_k, \qquad (4.3-1)$$

where $i \in [1, L]$ is the serial number of the analytical point; $\varepsilon_{i,k}$ is the constant coefficient (e.g., the extinction in absorption spectrometry); and c_k is the amount (concentration) of the kth-component. The absorption pathlength is unity.

Eq. (4.3-1) shows that signals of different mixture components in the same point l may overlap since a set of the coefficients $\varepsilon_{l,k}$ includes more than a single non-zero value. The term "sensitivity" is a measure of the overlap.

Danzer's definition of sensitivity in analytical chemistry [3] is as follows: the sensitivity of the determination of an analyte A (index p in Eq. (4.3-1)) at point $t = t_A$ is the derivative

$$S_{AA} = dF_A/dx_A = \varepsilon_{t_A,p}, \qquad (4.3-2)$$

where F_A and x_A are the analytical signal and analyte amount or concentration, respectively. The sensitivity equals the slope of the linear calibration plot [3].

The partial sensitivity S_{AB} reflects the impact of component A on the analytical signal of component B at the analytical point $t = t_B$:

$$S_{AB} = dF_t(A \rightarrow B)/dc_A. \quad (4.3-3)$$

By substituting Eq. (4.3-1) into Eq. (4.3-3), we obtain

$$S_{AB} = \varepsilon_{t_B,p}, \quad (4.3-4)$$

where p and q are indexes of the mixture components A and B, respectively.

The selectivity of the determination of the analyte B in the presence of the interfering component A is the ratio [3]:

$$SE_{AB} = \varepsilon_{t_B,q}/\varepsilon_{t_B,p}. \quad (4.3-5)$$

If $\varepsilon_{t_B,p} \rightarrow 0$ (no interference), then $SE_{AB} \rightarrow \infty$.

The quantification of a compound selectivity relative to all others involves the concept of the Net Analyte Signal (NAS) [4-8]. NAS is the projection of the vector of the analyte signal ($\boldsymbol{F}_A$) orthogonal to the background ($\boldsymbol{F}_{bkg}$), disturbing $\boldsymbol{F}_A$ [9]:

$$(\boldsymbol{F}_A)_\perp^{bkg} = \boldsymbol{F}_A - \boldsymbol{F}_{bkg}\boldsymbol{F}_A^T\boldsymbol{F}_{bkg}/\|\boldsymbol{F}_{bkg}\|^2, \quad (4.3-6)$$

where T is the symbol of transpose; $\| \; \|$ is the symbol for the Euclidian norm.

The interested reader can readily find a rigorous mathematical description of the orthogonalization (Eq. (4.3-6)) in the well-known book [10].

Consider the general case of the linear multicomponent system represented in the matrix form (see Eq. (4.1-26)):

$$\boldsymbol{F}_{mix} = \boldsymbol{E} * \boldsymbol{C} + \boldsymbol{\eta}, \quad (4.3-7)$$

where $\boldsymbol{E}$ and $\boldsymbol{C}$ are the regression and concentration matrices, respectively; and $\boldsymbol{\eta}$ is a vector of measurement errors of $\boldsymbol{F}_{mix}$.

Let $\boldsymbol{E}_{dist}$ be the matrix, whose columns are the components that distort the spectrum of the analyte. Then, according to the orthogonal projection theory [10], we have

$$(\boldsymbol{F}_A)_\perp^{dist} = \{\boldsymbol{I} - \boldsymbol{E}_{dist}(\boldsymbol{E}_{dist}^T\boldsymbol{E}_{dist})^{-1}\boldsymbol{E}_{dist}^T\}\boldsymbol{F}_A, \quad (4.3-8)$$

where $\boldsymbol{I}$ is the identity matrix; the superscript -1 denotes matrix inverse. Opposite to component $(\boldsymbol{F}_A)_\perp^{bkg}$ (Eq. (4.3-6)), which is only orthogonal to the single background vector, the component $(\boldsymbol{F}_A)_\perp^{dist}$ is orthogonal to each column vector of the matrix $\boldsymbol{E}_{dist}$.

The NAS-based figures of merit:

(a) The sensitivity vector (for the unit concentration):

$$\boldsymbol{S}_A = \|(\boldsymbol{F}_A)_\perp\|. \quad (4.3-9)$$

(b) The selectivity:

$$SE_A = \|(\boldsymbol{F}_A)_\perp\| / \|\boldsymbol{F}_A\|. \qquad (4.3-10)$$

The sensitivity vector (Eq. (4.3-9)) is the generalization of the single-point partial sensitivity (Eq. (4.3-4)), which involves only pair interactions of the mixture components.

The selectivity (Eq. (4.3-10)) is the dimensionless measure of the portion of the mixture component analytical signal, which is not lost due to spectral overlap. $SE_A \in [0,1]$. These boundaries correspond to full and zero overlap, respectively.

All of the above definitions are valid for the derivative spectra.

The relative systematic error when determining the analyte B in the presence of the interfering component A in analytical point t is reciprocal to the selectivity (Eq. (4.3-5)):

$$\delta_{BA} = 1/SE_{AB} = \varepsilon_{t,p}/\varepsilon_{t,q}. \qquad (4.3-11)$$

If $\varepsilon_{t,p} \to 0$ (no interference), then $\delta_{BA} \to 0$.

This error was studied by numerical modelling using Gauss doublet and its 1st- and 2nd-order derivatives (see details in Chapter 4.1). Theoretical estimations are given in the research [11].

Numerical experiment

The following numerical experiment illustrates the concepts of sensitivity and selectivity. The model is the Gaussian quartet:

$$Q_G(i) = \sum_{k=1}^{4} F_{0k}\, exp\left\{-\beta\left[\left(i - i_{M,k}\right)/w_k\right]^2\right\}, \qquad (4.3-12)$$

where $i = [1,100]$; $\beta = 4ln2$; $\{F_{0k}, i_{M,k}, w_k\}$ ={1.0, 29, 10; 1.5, 43, 15; 0.8, 57, 12; 1.2, 64, 5} are the triples of the kth peak intensity, maximum position, and width.

The parameters of the 2nd-order SG differentiation filter are $2t = 2$ and $m = 2$.

Figure 4.3-1 represents the quartet, its components, their 2nd-order derivatives, and two orthogonal projections of the first peak on the background $\left((\boldsymbol{F}_1)_\perp^{bkg}\right)$ (Eq. (4.3-6)) and all components of the disturbance matrix $\left((\boldsymbol{F}_k)_\perp^{dist}\right)$ (Eq. (4.3-8)). The background is the total spectrum of the second, third, and fourth components. The columns of the disturbance matrix are all component-spectra.

The analytical points in the mixture spectrum are the peak maxima. In the 2nd-order derivative, the null-points are taken for the first and the third components (Fig. 4.3-1c). In the last case, these points are close to the extrema of the derivatives.

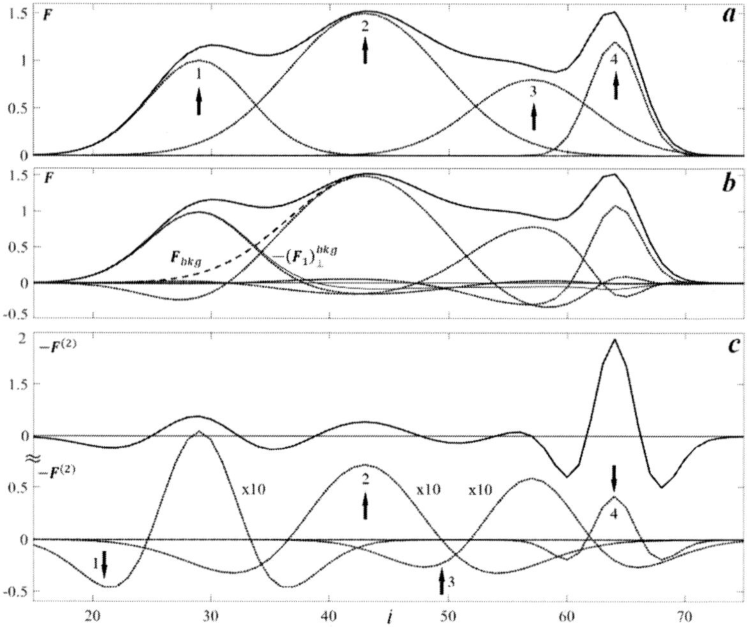

Figure 4.3-1. The Gaussian quartet (a, b, solid curves) and its components: pure and orthogonal to the disturbance matrix (dotted curves in panels a and b, respectively). (b) The 1st component spectrum ($(\boldsymbol{F}_1)_\perp^{bkg}$) orthogonal to the background ($\boldsymbol{F}_{bkg}$) (thin curve). (c) The normalized negative 2nd-order derivative of the quartet and its components (solid and dotted curves, respectively). The arrows point at the analytical points.

Eq. (4.3-2) and Eq. (4.3-4) represent the diagonal and non-diagonal elements of the sensitivity matrix (K-matrix) [3] estimated using the quartet components:

$$\boldsymbol{K} = \begin{vmatrix} 1.000 & 0.134 & 0.000 & 0.000 \\ 0.004 & 1.500 & 0.018 & 0.000 \\ 0.000 & 0.134 & 0.800 & 0.005 \\ 0.000 & 0.007 & 0.311 & 1.200 \end{vmatrix}. \qquad (4.3-13)$$

The analytical points are the peak maxima (which are indicated by the arrows in Fig. 4.3-1, a).

Matrix $\boldsymbol{K}$ includes the quantitative data, which characterize the visual pattern of the overlapped peaks (Fig. 4.3-1, a). The first row shows that only the second component significantly interferes with the accurate determination of the first one. The distortions at the analytical point of the second component are small (the second row). However, the third and the fourth rows demonstrate noticeable perturbations in the analytical signals.

To eliminate the peak overlapping partly, the matrix $\boldsymbol{K_D}$ involves 2nd-order derivatives (Fig. 4.3-1, c). The null-points for the first and the third components are used instead of the peak maxima.

$$\boldsymbol{K_D} = 0.1 * \begin{vmatrix} 0.229 & 0.011 & 0.000 & 0.000 \\ 0.027 & -0.360 & 0.048 & 0.000 \\ 0.000 & -0.030 & 0.124 & 0.000 \\ 0.000 & 0.017 & 0.099 & -2.097 \end{vmatrix}. \qquad (4.3-14)$$

Only the sensitivity of the first component determination increases because of the differentiation. In the case of the second and the third components, the derivative data worsen the sensitivity.

Eq. (4.3-5) defines the elements of the selectivity matrices, which were calculated from Eqs. (4.3-13) and (4.3-14):

$$\boldsymbol{SEL} = \begin{vmatrix} 1 & 7.5 & * & * \\ 344 & 1 & 82 & * \\ * & 6.0 & 1 & 153 \\ * & 183 & 3.9 & 1 \end{vmatrix}, \qquad (4.3-15)$$

$$\boldsymbol{SEL_D} = \begin{vmatrix} 1 & 20.8 & * & * \\ -13.5 & 1 & -7.5 & * \\ 502 & -4{,}2 & 1 & * \\ * & -125 & 21.3 & 1 \end{vmatrix}, \qquad (4.3-16)$$

where the asterisk denotes a huge value.

Exercise 4.3-1.

Readers are invited to prove the above conclusions, which were drawn from $\boldsymbol{K}$-matrices, using selectivity matrices (Eqs. (4.3-15) and (4.3-16)).

The NAS-based figures of merit, including the sensitivity (Eqs. (4.3-9)) and selectivity (Eq. (4.3-10)) of the zero and 2nd-order derivative spectrometric analysis, are

$$S_k = [2.6957 \quad 4.8047 \quad 2.1582 \quad 2.1473], \qquad (4.3-17)$$

$$S_k^{(2)} = [0.1171 \quad 0.0875 \quad 0.0653 \quad 0.3504], \qquad (4.3-18)$$

$$SE_k = [0.9826, 0.9533, 0.8976, 0.9224], \qquad (4.3-19)$$

$$SE_k^{(2)} = [0.9352, 0.8325, 0.8433, 0.9529], \qquad (4.3-20)$$

where $k = 1 \div 4$.

Sensitivities S_k and $S_k^{(2)}$ take the mutual effect of the interfering components on a given component into account.

Eqs. (4.3-17) and (4.3-18) show that the zero and the 2nd-order derivative analyses of the second and fourth components in the presence of the disturbing components yield the best sensitivities, respectively. SE_1 is the best selectivity of the zero-order analysis (Eq. (4.3-19)). $SE_4^{(2)}$ is the maximum value in the derivative method (Eq. (4.3-20)).

Exercise 4.3-2.

Readers are invited to estimate the figures of merit using various sets of quartet parameters. We recommend studying the strong overlapped peaks to compare the zero-order and derivative analysis. The Matlab program is orthogProj.m (Appendix SW7).

The above numerical example has demonstrated that the selectivity may increase due to the improved resolution in the derivative spectra and the null-point measurements. However, maximum selectivity of analysis does not mean that it is the best. This figure of merit does not consider the systematic errors that occur because of the unstable positions of the analytical points and the random distortions of the analytical signals.

We conclude that choosing the differentiation order, analytical points, and the intensity measurement method is a complicated task. The success of its solution depends on the experimenter's experience and intuition. Computer modelling may also help to solve the problem.

CHAPTER FOUR

INFORMATIONAL ASPECTS OF THE QUANTITATIVE DERIVATIVE SPECTROMETRY

Let us compare the quantitative zero-order and derivative spectrometry from an informational-theoretical point of view [1]. To achieve this goal, suppose that the probability density function of the analyte concentration is uniform. In simple terms, the possibilities that any unknown concentration appears in some range $[c_1, c_2]$ are equal. Also, suppose that the random measurement errors of the concentration have a normal distribution with zero mean and standard deviation σ_c. Then, the quantity of the information that was obtained by an analytical procedure, performed using the zero-order spectrum, equals

$$I^{(0)} = ln\left[(c_2 - c_1)/\left(\sigma_c\sqrt{2\pi e}\right)\right] - 0.5B_c^2, \qquad (4.4-1)$$

where $B_c = \delta_c/\sigma_c$; δ_c is the systematic error when determining concentration.

Similar to Eq. (4.4-1), for the nth-order derivative method, we have

$$I^{(n)} = ln\left[(c_2 - c_1)/\left(\sigma_{cd}\sqrt{2\pi e}\right)\right] - 0.5(\delta_{cd}/\sigma_{cd})^2, \qquad (4.4-2)$$

where subscript d stands for derivative.

It follows from Eqs. (4.4-1) and (4.4-2) that the information change is

$$\Delta I = I^{(n)} - I^{(0)} = ln(\sigma_c/\sigma_{cd}) - 0.5(B_{cd}^2 - B_c^2). \qquad (4.4-3)$$

The theoretical estimations of ΔI were based on numerous simplified suggestions relative to real analytical practice [2, 3]. It was shown that the information obtained because of the differentiation is only achieved if the systematic errors, which are significantly larger than random$(B_c \gg 1)$, are substantially suppressed in the derivative analytical signal.

To make this statement more transparent and understandable, we estimated ΔI in the following numerical experiment. The goal was to demonstrate that, despite the poor selectivity and sensitivity of the 2nd-order derivative-based procedure, the differentiation may extract useful information from the quantitative analysis of the mixture with strongly overlapped component spectra.

The mixture spectrum is (Fig. 4.4-1a)

$$\boldsymbol{F}_{mix} = \boldsymbol{c} * \boldsymbol{E} + \boldsymbol{F}_{syst} + \boldsymbol{\eta}, \qquad (4.4-4)$$

where $\boldsymbol{c} = [0.7, 0.1, 0.8, 0.1]$;

$\boldsymbol{E} = \begin{vmatrix} \boldsymbol{E}_1(i) \\ \boldsymbol{E}_2(i) \\ \boldsymbol{E}_3(i) \\ \boldsymbol{E}_4(i) \end{vmatrix}$; $\boldsymbol{E}_k(i)$ is the vector, whose elements are the Gaussians, similar to the components of Eq. (4.3-12), but the parameters were changed:

$$E_k(i) = E_{0k} exp\left\{-\beta\left[\left(i - i_{M,k}\right)/w_k\right]^2\right\}, \qquad (4.4-5)$$

where $i = [1{,}100]$; $\beta = 4ln2$; $\{E_{0k}, i_{M,k}, w_k\}$ ={10, 32, 14; 1.5, 44, 11; 1.2, 62, 16} are the kth peak intensity, maximum position, and width; $F_{syst}(i) = 0.02exp\{-\beta[(i-62)/50]^2\}$ simulated the unidentified component, introducing the systematic errors; $\boldsymbol{\eta}$ is the normally distributed random vector with zero mean and the standard deviation $\sigma_\eta = 0.01$.

Eq. (4.4-4) has the following form at the four analytical points [32, 44, 56, 62] (the arrows in Fig. 4.4-1):

$$\begin{vmatrix} F_{mix}(32) \\ F_{mix}(44) \\ F_{mix}(56) \\ F_{mix}(62) \end{vmatrix} = \boldsymbol{c} * \boldsymbol{E}_{anal} + \begin{vmatrix} F_{syst}(32) \\ F_{syst}(44) \\ F_{syst}(56) \\ F_{syst}(62) \end{vmatrix} + \begin{vmatrix} \eta_1 \\ \eta_2 \\ \eta_3 \\ \eta_4 \end{vmatrix}, \qquad (4.4-6)$$

where $\boldsymbol{E}_{anal} = \begin{vmatrix} E_1(32) & E_1(44) & E_1(56) & E_1(62) \\ E_2(32) & E_2(44) & E_2(56) & E_2(62) \\ E_3(32) & E_3(44) & E_3(56) & E_3(62) \\ E_4(32) & E_4(44) & E_4(56) & E_4(62) \end{vmatrix}$.

In the 2nd-order derivative mode:

$$\boldsymbol{F}_{mix}^{(2)} = \boldsymbol{c} * \boldsymbol{E}^{(2)} + \boldsymbol{F}_{syst}^{(2)} + \boldsymbol{\eta}^{(2)}. \qquad (4.4-7)$$

The parameters of the SG derivative filter: $2t = 2$ and $m = 17$.

The solution to Eq. (4.4-6):

$$\hat{\boldsymbol{c}}_x = (\boldsymbol{F}_{mix} + \boldsymbol{F}_{syst})\boldsymbol{E}_{anal}^{-1}, \qquad (4.4-8)$$

where the superscript -1 denotes the inverse of the matrix $\boldsymbol{E}_{anal}$.

The solution to Eq. (4.4-7) is similar to Eq. (4.4-6). These equations may also be solved using Cramer's rules based on the determinants [4].

The random and the systematic errors in determining the mixture concentrations (Table 4.4-1) were obtained from 1000 repetitions each time with regenerated noise.

The random errors were estimated from the well-known equation [18]:

$$\boldsymbol{\sigma}_{est} = \sigma_\eta\{\|\boldsymbol{E}_{anal}^{-1}(32)\|, \|\boldsymbol{E}_{anal}^{-1}(44)\|, \|\boldsymbol{E}_{anal}^{-1}(56)\|, \|\boldsymbol{E}_{anal}^{-1}(62)\|\}, \quad (4.4-9)$$

where || || denotes the Euclidian norm.

Table 4.4-1 shows that the errors ($\boldsymbol{\sigma}_{est}$) estimated theoretically for the zero-order analysis match the random errors obtained numerically. The discrepancies between these errors, which were observed for the derivative method, are due to the correlations between the analytical points introduced by the SG filter.

The total error is

$$e = \sqrt{\delta^2 + \sigma^2}. \quad (4.4-10)$$

The vector of the information gain obtained because of the derivative-based analysis is (Eq. (4.4-3))

$$\Delta\boldsymbol{I} = [\,0.6288, 0.5506, 0.5933, 0.6921]. \quad (4.4-11)$$

The positive components of $\Delta\boldsymbol{I}$ reflect the decreasing of the total errors e_{cd} compared to e_c (Table 4.4-1).

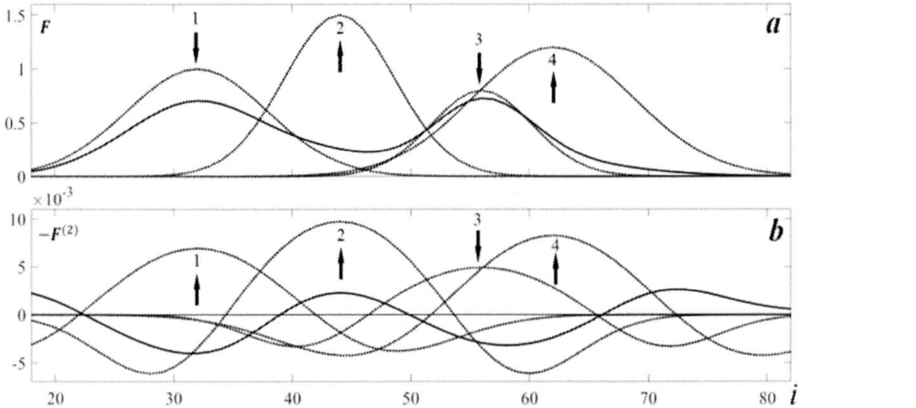

Figure 4.4-1. (a) The spectrum, its Gaussian components (solid and dotted curves, respectively), and (b) negative 2nd-order derivatives. The arrows indicate the analytical points.

The following figures of merit were obtained for the four analytical points located at the peak maxima (pointed by the arrows in Fig. 4.4-1) (For more details, see Chapter 4.3).

The sensitivity matrices (K-matrices) (the subscript **D** denotes the derivative analysis):

$$\boldsymbol{K} = \begin{vmatrix} 1.0000 & 0.0553 & 0.0000 & 0.0001 \\ 0.1304 & 1.5000 & 0.0148 & 0.0359 \\ 0.0003 & 0.0553 & 0.8000 & 0.8126 \\ 0.0000 & 0.0009 & 0.2949 & 1.2000 \end{vmatrix}. \qquad (4.4-12)$$

$$\boldsymbol{K_D} = 0.01 * \begin{vmatrix} -0.69 & 0.35 & 0.05 & 0.06 \\ 0.18 & -0.97 & 0.19 & 0.42 \\ 0.16 & 0.35 & -0.49 & -0.48 \\ 0.02 & 0.55 & -0.29 & -0.83 \end{vmatrix}. \qquad (4.4-13)$$

Table 4.4-1. Errors of multicomponent analysis

Parameter	Mixture component			
	1	2	3	4
$\boldsymbol{\delta_c}$	0.0070	0.0086	0.0096	0.0135
$\boldsymbol{\sigma_c}$*	0.0098	0.0067	0.0208	0.0120
	0.0101	0.0068	0.0201	0.0118
$\boldsymbol{e_c}$	0.0121	0.0109	0.0229	0.0181
$\boldsymbol{\delta_{cd}}$	0.0019	0.0051	0.0046	0.0064
$\boldsymbol{\sigma_{cd}}$*	0.0066	0.0056	0.0119	0.0084
	0.0080	0.0076	0.0138	0.0099
$\boldsymbol{e_{cd}}$	0.0069	0.0076	0.0128	0.0106

* The second row represents the errors obtained by Eq. (4.4-8).

The selectivity matrices:

$$\boldsymbol{SEL} = \begin{vmatrix} 1 & 18.07 & * & * \\ 11.50 & 1 & 101.61 & 41.77 \\ * & 14.45 & 1 & 0.98 \\ * & * & 4.07 & 1 \end{vmatrix}, \qquad (4.4-14)$$

where the asterisks denote huge values;

$$\boldsymbol{SEL_D} = \begin{vmatrix} 1 & -2.00 & -13.07 & -12.21 \\ -5.25 & 1 & -5.07 & -2.29 \\ -3.08 & -1.42 & 1 & 1.02 \\ -33.45 & -1.50 & 2.89 & 1 \end{vmatrix}. \qquad (4.4-15)$$

The sensitivity and selectivity data (Eqs. (4.4-12)-(4.4-15)) contain no information which allows us to give preference to any of the two analytical methods.

The Net Analyte Signal (NAS) - based figures of merit: the sensitivity (Eqs. (4.3-9)) and selectivity (Eq. (4.3-10)) of the zero the 2nd-order derivative spectrometric analyses ($k = 1 \div 4$) are represented by the following equations. The subscript "errComp" denotes the orthogonalization to the matrix, which consists of the mixture components ($\boldsymbol{E}$) and the unidentified spectrum $\boldsymbol{F}_{syst}$ (Eq. (4.4-4)). Additional orthogonalization step to $\boldsymbol{F}_{syst}$ decreases the sensitivity and the selectivity (compare pairs with and without the subscript "errComp").

$$S_k = [3.1133, 4.0802, 1.5144, 2.9032]. \qquad (4.4-16)$$

$$S_{k,errComp} = [2.9118, 3.3860, 1.5125, 2.0184]. \qquad (4.4-17)$$

$$S_k^{(2)} = 0.1 * [\,0.186,\ 0.239\,,\ 0.146,\ 0.209\,]. \qquad (4.4-18)$$

$$S_{k,errComp}^{(2)} = 0.1 * [\,0.182,\ 0.224,\ 0.146,\ 0.169\,]. \qquad (4.4-19)$$

$$SE_k = [0.9591,\ 0.9453,\ 0.6900,\ 0.6972\,]. \qquad (4.4-20)$$

$$SE_{k,errComp} = [\,0.8970,\ 0.7845,\ 0.6891,\ 0.4847\,]. \qquad (4.4-21)$$

$$SE_k^{(2)} = [0.7524,\ 0.6243,\ 0.7365,\ 0.6694\,]. \qquad (4.4-22)$$

$$SE_{k,errComp}^{(2)} = [\,0.7355, 0.5857, 0.7364, 0.5408\,]. \qquad (4.4-23)$$

Eqs. (4.4-20)-(4.4-23) show that the 2nd-order differentiation decreases the selectivity of the mixture analysis. Corresponding figures of merit were obtained using full-range orthogonalization. Theoretically, the NAS - based multicomponent analysis of the mixture, whose exact composition is known, must have the same metrological characteristics as the standard method [18]. However, the differentiation is a non-optimal transform, but orthogonalization is the optimal one. Therefore, the full-range derivative data decrease the selectivity of the mixture analysis and, consequently, its accuracy and precision.

In the above numerical example, the analysis was only performed in four analytical points. The derivative method allowed us to decrease the errors when determining mixture concentrations and to increase the amount of the information extracted from the analysis.

However, shifts of the first and the second analytical points $i_{M,1} = 32 \rightarrow 24$, $i_{M,2} = 44 \rightarrow 47$, which have been performed to reduce systematic errors, eliminate the advantages of the derivative spectrometry (Table 4.4-2) since the random components dramatically increase. The information gain becomes negative:

$$\Delta \boldsymbol{I} = -[\,2.4271, 2.0115, 1.8904, 0.7099]. \quad (4.4-24)$$

Exercise 4.4-1.

Readers are invited to repeat the above numerical experiments by changing the model's parameters: the differentiation order, the width (m), and the polynomial power ($2t$) of the SG filter (Appendix SW8). What conclusions can you draw?

Table 4.4-2. Errors of the multicomponent analysis in the shifted first and second analytical points

Parameter	Mixture component			
	1	2	3	4
δ_c	0.0093	0.0103	0.0083	0.0145
σ_c*	0.0242	0.0081	0.0194	0.0112
	0.0247	0.0083	0.0202	0.0118
e_c	0.0259	0.0131	0.0211	0.0183
δ_{cd}	-0.0032	0.0031	0.0025	0.0057
σ_{cd}*	0.2950	0.1355	0.1410	0.0520
	0.3710	0.1721	0.1744	0.0680
e_{cd}	0.2950	0.1355	0.1410	0.0523

* The second row represents the errors obtained by Eq. (4.4-9).

In conclusion, we have underlined the importance of the information measure (Eq. (4.4-3)) for estimating the efficiency of the signal processing method used in quantitative analysis. The advantage of the information gain over selectivity is that it considers random and systematic errors together.

PART V:

DIFFERENTIATION OF ANALYTICAL SIGNALS: INDUSTRIAL AND SCIENTIFIC APPLICATIONS

INTRODUCTION

The industrial and scientific applications of derivative-based analytical methods have been the subject of numerous studies reviewed in books [1-4] and articles (e.g., [5-12]). It is not easy to find a field of analytical chemistry where these methods would not find a practical use. Chapter 4.2 illustrated the use of derivative spectrophotometry for technological control in a Russian industrial laboratory. This final part is the first to describe the whole picture for the analytical signal processing employing mathematical differentiation.

We sincerely apologize to all those researchers whose outstanding works are not cited because the book does not have enough free space to include a complete bibliography.

In the following chapters, we demonstrate potential opportunities for derivative methods to solve typical instrumental problems in analytical practice:

1. Detection of the characteristic points of analytical signals. Identification of the peaks and determination of their parameters in poorly-resolved contours.
2. Resolution and contrast enhancement.
3. Background suppression.

Due to this preprocessing, qualitative and quantitative analyses are simplified. For example, it becomes possible to estimate the analyte amount at the analytical points free from the influence of interfering mixture components.

Despite the significant commonality among the derivative spectroscopic methods, used in different optical ranges and fields (e.g., due to the similar peak shapes), they are significantly different. They depend on the specific instrumental features; properties of the random and systematic errors and their impact on experimental data; and, finally, on the analytical problems solved by the derivative spectrometry.

These issues are the subject of the next chapter.

CHAPTER ONE

THE DERIVATIVE TECHNIQUE IN DIFFERENT TYPES OF SPECTROSCOPY: TECHNICAL FEATURES AND ADVANTAGES

Atomic spectroscopy

Optical atomic spectroscopy is a powerful technique for determining the elemental composition of a sample by analyzing its emission or absorption electromagnetic spectrum. Today, analysts widely use the following methods: Flame Atomic Emission and Absorption Spectroscopy, Graphite Furnace Atomic Absorption Spectroscopy, Laser-Induced Breakdown Spectroscopy, and Inductively Coupled Plasma Optical Emission Spectroscopy [1, 2].

Atomic emission spectroscopy

Figure 5.1-1 shows simplified diagrams of the atomic emission spectrometers.

In the flame designs (panel a), the nebulized sample, which had been premixed with the fuel and oxidant, is introduced in the burner. The atomization and excitation occur in the flame. In another device (panel b), a xenon arc is the continuum excitation source. To prevent the rapid cooling of the atomized analyte species about the graphite rod atomizer, the atomizer is enclosed in flame.

In Inductively Coupled Plasma Optical Emission Spectroscopy (ICP-OES) (panel c), the nebulized sample is entrained in the argon flow. The radiofrequency coil creates an oscillating magnetic field, which stimulates the inelastic collisions between the neutral argon atoms and the charged particles. So, a high-temperature plasma of about 7000 K is generated. To overcome the disadvantages of the pneumatic nebulization in ICP-OES, the graphite furnace is used as an electrothermal atomizer (panel d). This device has outstanding characteristics compared with conventional nebulization, e.g., low transport losses [1]. Also, it decreases non-spectroscopic or spectroscopic interferences from the matrix.

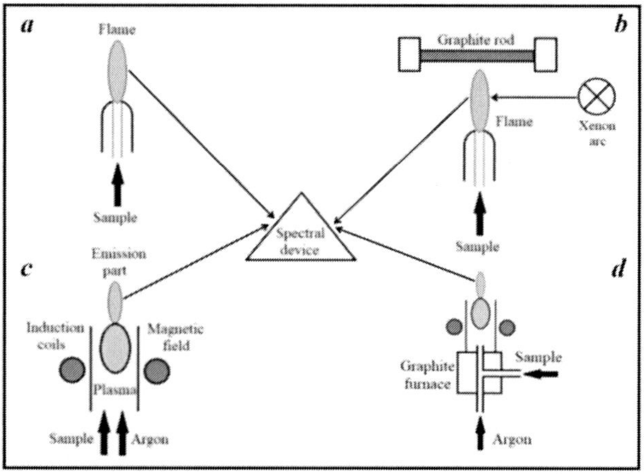

Figure 5.1-1. Diagrams of the atomic emission spectrometers: (a, b) flame devices; (b) electrothermal vaporization and continuum-source excitation (adapted from T. F. Wynn's Thesis, Florida University, 1980); (c) ICP-OE spectrometer; and (d) ICP-OE spectrometer with electrothermal vaporization.

The disadvantage of the non-flame atomizer (Fig. 5.1-1, b) is a strong background emission. Also, the reproducibility of the furnace emission declines over time. Since the background radiation has a broad continuous spectrum, which is linear at narrow intervals, it disappears in the 2nd - harmonic (derivative) of the frequency modulated signal. The use of wavelength modulation (WM) improved detection limits in some cases by several orders of magnitude and increased the applicability of ICP-OES for the sample analysis [2].

A gas flow may contain small amounts of water vapor and other hydrogen compounds. At a high furnace temperature, the spectrum of these impurities, which consist of C_2, CH, and other bands, is observed as a fine background structure. Null-point analysis eliminated the impact of this structure [2].

Atomic analysis of some elements requires high atomization and excitation temperatures which significantly increase the detector flicker noise. The modulation devices reduced this noise; therefore, the detection limits decreased [2].

The modulation suppressed the continuous background caused by the alkali metal ions recombination appeared when the temperature flame (nitric oxide-acetylene) was high [3]. Usually, before each measurement, the combustion mode must be controlled. Modulation eliminated this procedure [3]. A high-resolving monochromator was not needed.

Eliminating the continuous background and the CaOH bands decreased the detection limits of the alkali and alkaline earth elements of 1-2 orders of magnitude in complex systems [4]. However, in pure aqueous solutions, the results were limited to those obtained by other methods.

The advantage of the modulation technique was a significant decrease of the sample volume (less than $1mL^{-1}$), since only the maximum of the 2nd -order derivative was measured [4]. There was no need to measure the total analytical line for background correction. However, the inertia of the electronic unit limited a further decrease of the sample volume.

The 2nd-order WM minimized spectral interferences in matrices in Flow-Injection-ICP-AES [5].

In optical emission spectrometry, the emitted radiation from the analyte zone of the ICP is viewed in two basic configurations (radial and axial), which have different performance characteristics. The analytical performance of a long torch for axially viewed horizontal ICP-OES (or ICP-AES, where "A" denotes "Atomic") was examined using a high-resolution echelle spectrometer with the 2nd-order harmonic WM [6]. Compared to conventional torch measurement, net emission intensities of 24 elements (having spectral lines between 210 and 770 nm) increased by 20~40%, and background intensities of water blanks decreased by 30-50%.

The author of the study [7] found that the resonance line at 670.784 nm provided the greatest isotopic splitting between ^{6}Li and ^{7}Li. The 1st-order numerical derivative of the resonance line quantified the ^{6}Li and ^{7}Li contents.

Atomic fluorescence spectroscopy

Atomic fluorescence spectrometry (AFS) is a high sensitivity analytical method. The instrumental base of AFS is similar to that of the atomic emission spectrometers. However, AFS requires very efficient atomization since the matrix molecules can quench fluorescence.

The WM reduced the flicker noise of the analyte emission, and that, which was caused by the interfering radiation that was incident on the detector, and the scattering by evaporating particles [1, 2].

Laser excitation sources significantly increased the sensitivity of AFS. Theoretically, the detection limits may be impressively small. However, the atomizer produces a strong background (the black body radiation and molecular emission by thermal excitation). The use of the powerful lasers for decreasing the detection limits causes light scattering from atomic and molecular species. The laser radiation frequency modulation minimized the impact of the scattering and laser fluctuation on the measurements [3].

The study on pulsed-laser-excited AFS demonstrated that the "wavelength modulation corrected, effectively and quantitatively, for flame background, blackbody emission from a graphite furnace, and scatter of laser radiation off aluminum chloride matrix particles in both the furnace and the flame" [4].

Laser-induced Breakdown Spectroscopy

Laser-Induced Breakdown Spectroscopy (LIBS) is an elemental analysis based on laser-generated plasma [1]. Powerful pulses from a laser atomizer, when focused on a sample, produce a high-temperature micro-plasma of a small amount of material. The plasma formation only begins when the laser achieves a certain threshold for optical breakdown (a spark causes the medium to electrically conduct). The excited atoms in the plasma emit a light, whose spectrum provides the "fingerprints" of the elements in the sample.

We found only three studies based on derivative LIBS data. The partial least square method was applied to the 1st-order derivative of the LIB-spectra [2, 3]. The differentiation SG filters performed the background correction [4].

Atomic absorption spectrophotometry

A simplified diagram of the atomic absorption spectrometer (Fig 5.1-2) explains its operating principles. In order to analyze a sample for its atomic constituents, it is atomized (transformed to the cloud of free metallic ions) by flame (b), or graphite furnace devices (c). The light emitted by the radiation source is absorbed by the cloud. Analytical information is extracted from the absorption spectrum.

Like the hollow cathode lamp, the linear spectrum radiation source contains a cathode made of the element to be determined. In multi-elemental analysis, a lamp that is appropriate to the element being measured, must be selected.

The source of the continuum radiation is a high-pressure xenon short-arc lamp.

The interested reader may find all technical details in the professional literature [1, 2].

The continuous spectra radiation source in Atomic Absorption Spectrophotometry (AAS) has numerous advantages over the discrete

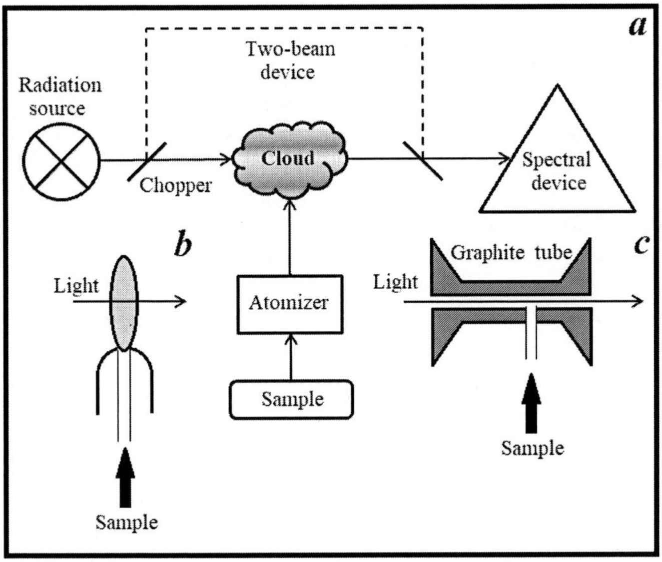

Figure 5.1-2. Diagram of the atomic absorption spectrometer (a). Flame (b) and graphite furnace (c) atomizers.

spectra source [2]:

1. Ease of user controls (it is unnecessary to change the source).
2. Low cost.
3. It is possible to measure the intense lines, which are the most suitable for analysis in the atomic spectra of metals.
4. Simultaneous multi-element analysis.

However, the sensitivity decreases approximately by 1-2 orders of the magnitude because the analytical signal becomes weak despite the large absorption pathlength.

The detector's photocurrent noise includes a strong flicker component that occurs because of the instability of the background absorption and emission of the flame. This component significantly increases for the intense radiation. The pioneering studies [3-5] have experimentally demonstrated that the WM effectively suppresses the low-frequency flicker noise, the Rayleigh scattering on small particles (non-vaporizing in the flame), and the molecular absorption end emission occurring within the burned gas.

The calibration curves were linear over more than 3 orders of magnitude in concentration for Ag, Ca, Cd, Cr, Cu, Fe, Mg, and Ni [4].

The simultaneous multi-element AAS analysis used an air acetylene flame as an atomization unit. Three radiation sources (hollow cathode lamps) were modulated in the frequency range 300-700 Hz. [6]. The 2nd-order harmonic of the modulated AAS signal was free from background interferences. The elements, Mn, Cd, Pb, Cu, Ni, Cr, and Zn, were determined simultaneously (three at a time) in sewage sludge, estuarine sediment, and phosphate rock with good feasibility.

A group of Chinese scientists contributed to developing of the derivative AAS based on the electronic unit [7-15]. This unit included an operational amplifier differentiator circuit connected with the RC-smoothing circuit (Chapter 2.1).

The signal models and their derivatives in a cold vapor AAS (CVAAS) and hydride generation AAS were represented. Detailed analytical data were tabulated [15]. The CVAAS requires no vaporization step because the sample is a volatile heavy metal, such as mercury, which is a vapor at room temperature. A method, which was developed to determine total mercury by CVAAS with derivative signal processing, was applied to the determination of total mercury content in cosmetic samples with a recovery range of 92–102% [7]. Differentiation suppressed a strong background signal.

Sun et al. [10] developed the derivative Atom-Trapping Flame AAS (ATF-AAS) [10]. ATF-AAS "involves the generation of volatile species, usually hydrides, trapping these species on the surface of an atom trap held at an optimized temperature and, finally, re- volatilizing the analyte species by rapid heating of the trap and transporting them in a carrier gas to a heated quartz tube, as commonly used with hydride generation AAS systems" [16].

The derivative ATF-AAS was applied to the determination of lead in water and liqueur samples with a recovery range of 94-108% and a relative standard deviation of 3.5-5.6%. The detection limit and sensitivity were 2 and 3 orders of magnitude higher than those of conventional FAAS [10].

Direct determination of trace cadmium using the derivative ATF-AAS improved the detection limit and sensitivity by 1 and 2 orders of magnitude, respectively, compared to conventional FAAS [8]. The method was applied to the determinations of cadmium in water samples. The recovery range and the relative standard deviation were 91 ~ 111% and 4.7 ~ 5.6%, respectively.

Differentiation improved the sensitivity of chromium (III) determination by more than 14 times compared to conventional flow-injection FAAS [9].

Hydride-generation AAS (HGAAS) involves converting the vaporized metal samples, such as As, Sb, and Se into volatile hydrides [17]. The derivative HGAAS was applied to the determination of traces of lead in water [11]. The detection limit and sensitivity of the proposed method were 26 and 8.8 times better than conventional HGAAS. The recovery range was 92.5-103%.

The derivative HGAAS was applied to the determination of tellurium in urine [12]. The characteristic concentration and the detection limit for tellurium were, respectively, 52 and 26 times better than conventional HGAAS. The determination of tellurium in urine samples from a small population of healthy individuals showed a recovery range of 89-98%.

The study of atomic spectral lines using nanocells, with widths equal to the laser radiation wavelength, allowed the observation of super-fine structure which, for example, appeared in the magnetic field [18]. The 2nd-order derivative processing of the absorption spectra of alkali-metal atomic vapor nanocells recovered narrow homogeneous spectral features from a broad inhomogeneous overlapped profile. The authors studied the following issues: the measurements of hyperfine splitting and atomic transition probabilities, atom-surface interactions, the determination of isotopic abundance, and the magnetic-field-induced modification of atomic transition frequency and probability [19].

Molecular spectroscopy

Infrared absorption spectrophotometry

Derivative IR spectroscopy (DIR) was at its peak in the 1970-1980s [1]. Numerous instrumental electronic and numerical differentiation methods for resolution enhancement were developed and successfully used for analytical applications. The similarity of the visual patterns observed in the spectra and their better-resolved even-order derivatives was the primary factor of DIR popularity for identifying the peak maxima. This peak parameter is one of the essential characteristics of spectra. Also, thermal IR detectors are the source of the white noise in spectra, and their denoising is relatively straightforward using analog and digital smoothing (Part II). The wide distribution of IR spectrometers in scientific and industrial physical, chemical, biochemical, and pharmaceutical laboratories encouraged the use of DIR in analytical applications.

Quantitative DIR was not very popular, since the conventional IR spectrum, which usually consists of many bands, often allowed the selection of analytical peaks free of strong interferences. The linear equations used in the multicomponent analysis took these interferences into account. The previous chapter pointed out some drawbacks of derivative spectroscopy. However, new smart preprocessing algorithms, which partly solved these problems, did not displace the derivative technique [2]. The emergence of computerized spectral devices, which were supplied with powerful mathematical software, transformed DIR into a routine analytical method.

Tutorial [3] represents an example of the baseline correction using the 2nd-order SG derivative of NIR soil data.

The derivative method proved useful for determining the initial peak parameters (guesses) in the least square fitting of *a priori* given spectral model to the experimental data. For example, the IR spectra of the glass surface of hydrolyzed silica optical fibers were decomposed into elementary peaks [4]. The guesses for the model's parameters were estimated using the 2nd- and the 4th-order derivatives.

Two studies [5, 6] demonstrated the medical applications of derivative NIR spectroscopy. 1st-order derivative FTIR spectroscopy allowed quantitatively analyzing atorvastatin, rosuvastatin, and simvastatin in their binary mixtures with ezetimibe [7].

The 1st-order derivative FT-IR spectra were used for the direct determination of ethanol in alcoholic beverages, from beers to spirit samples [8], and quantification of binary mixtures of the anti-hyperlipidemic statin drugs and ezetimibe [9].

The 2nd-order derivative FT-IR spectra enhanced the separation of overlapped peaks in the analysis of the spatial distribution of main tissue components of articular cartilage (collagen and proteoglycans) [10].

Analysis of the 2nd-order derivative FT-IR discriminated between the cooked-up samples of perilla oils with soybean oils and/or corn oils added at concentrations of ≥ 5 vol % [11].

UF-VIS absorption spectrophotometry

Contrary to IR spectroscopy, the absorption bands of significant organic and inorganic compounds in the UF-VIS region are wide and poorly structured. Differentiation of these bands narrows the spectrum, thus revealing its fine structure elements. It is possible to identify the chemical composition of the sample, detect spectral patterns of intra- and intermolecular interactions, and other information.

Quantitative analytical derivative spectrometry has demonstrated its advantages over conventional methods due to background suppression, thereby extending the limits of the applicability of the baseline correction procedure and the appearance of the zero absorption intervals of the interfering components.

From an instrumental perspective, the differentiation of UF-VIS spectra is a relatively simple technical task due to the high signal-to-noise ratio. However, the flicker, as well as the white, noise must be reduced. Usually, signal processing is performed using numerical differentiation. Modulation devices are most common in the spectral gas analysis.

Modulation Far UF spectrometry used a high-intense source for the stable polarized radiation-synchrotron [1]. A magnetic field accelerates a particle beam in this device and travels around a fixed, closed-loop path. The synchrotron emission has a continuous spectrum with no superimposed structure. The high intensity is needed for measuring tiny changes of the radiation (about 10^{10} photons). However, synchrotrons are expensive and require high-vacuum equipment.

Higher-order derivatives of UV Vis absorption spectra of amorphous thin films of barium strontium titanate, which had been deposited by RF magnetron sputtering on SiO2 at different substrate temperatures, were obtained [2]. The critical points, which were related to the optical interband transitions at various regions in the Brillouin zone, were identified.

A similarity index algorithm, in conjunction with the 4th-order absorbance spectra, provided discrimination among microscopic algae (unicellular species-photosynthetic eukaryotic organisms) [3].

The 1st-4th-order derivatives UF-VIS absorption spectra allowed turbidity correction in chemical oxygen demand measurements [4].

In the last few decades, the derivative spectrophotometric analysis of food additives [5] and drugs in pharmaceutical formulations has become a routine procedure. There have been hundreds of articles about this published in journals of analytical chemistry and pharmaceuticals (see reviews [6-8] and some later works [9-65]).

The temperature derivative spectroscopy was used to measure the decay kinetics of oxy-ferrous complexes in cytochromes P450 [66].

The partial least square regression on the 1st-order derivative spectra of dyes gave better optimization of the calibration matrices with respect to the original data [67].

Laser Modulation Spectroscopy

One of the first reviews, dedicated to the analytical applications of lasers [1], pointed out their remarkable properties: (1) high spectral and spatial power density; (2) tunable output with variable bandwidth in a vast frequency region; and (3) continuous or pulse output.

The first property may result in decreasing the detection limit in fluorescence analysis, which sensitivity depends on the source power.

Also, the excited states' saturation becomes possible (e.g., laser-excited atomic fluorescence spectroscopy-AFS). The tunable output uses the laser as a radiation excitation of several transitions. Very short pulses can be employed in the measurements of the excited-state lifetimes.

Laser Modulation Spectroscopy (LMS) has opened up new possibilities for the physical and chemical research due to the modulation technique (Chapter 2.10) (reviews [2-5]). LMS is the most suitable for determining small molecules where the spectra consisted of relatively simple intense peaks [6]. Careful selection of the analytical peaks and their narrowing by reducing the pressure eliminated interfering signals. The strong background and IR laser fluctuations were significantly suppressed because of the use of the 2nd-order harmonics of the modulated light. If the modulation frequency is enough high (hundredths Hz), then the low-frequency (flicker) noise is efficiently reduced. In the non-scanning mode (single-point measurement), the time interval of the signal collection and averaging may be made as large as possible (if there is no drift). Therefore, white noise is significantly suppressed.

Table 5.1-1 illustrates some applications of the modulation-based laser measurements.

Table 5.1-1. Modulation-based laser measurements

A long-wavelength vertical-cavity surface-emitting laser with an extended tuning range was used to detect benzene vapor at atmospheric pressure. The 2nd-order WM harmonic's response was measured as a function of the modulation index.	[7]
Stable carbon isotopes of CO_2 were measured using WMS with a distributed feedback laser diode in the 2-µm wavelength range. The limit of detection was 16±1 parts per billion by volume at the most substantial absorption peak. The 3-min response and high precision of this measurement allowed for precise continuous measurements of stable carbon isotopes in ambient CO_2.	[8]
A calibration-free WMS was used for real-time, in situ detection of nitric oxide in particulate-laden combustion-exhaust gases up to temperatures of 700 K. The 2nd - harmonic, normalized to the first one, demonstrated better noise immunity for non-absorption transmission, than wavelength-scanned direct absorption. A 0.3 ppm×m detection limit was estimated using the R15.5 transition near 1927 cm^{-1} with 1 s averaging.	[9]
Propane' concentration was measured by 3.37 µm GaInAsSb/AlGaInAsSb DFB laser. WM enhanced a sharp feature in the broad propane spectrum by around 3370.4 nm. A minimum detectable concentration of 30 ppb×m (a response time of 0.5 s.) was the sensitivity improvement by order of magnitude.	[10]
Measurement of O_2 pressure with an uncertainty of ±1.3% of the full scale (760 Torr) was performed.	[11]
NO concentrations in the few ppb range in diluted exhaust-gas bag samples, which were collected in the vehicle certification process, were detected.	[12]
Optimal sensitive detection of optical absorption by species with structured spectra at elevated pressures was demonstrated.	[13]

Table 5.1-1 (continued). Modulation-based laser measurements

Atmospheric nitrous oxide (N_2O), methane (CH_4), and water vapor (H_2O) were simultaneously detected by the continuous wave quantum cascade laser-based absorption sensor system. The 2nd-order WM harmonic achieved minimum detection limits of 1.7 ppb for N_2O, 8.5 ppb for CH_4, and 11 ppm for H_2O with a 2 s integration time for individual gas detection.	[14]
Concentrations of CH_4 and CO in ambient air were measured simultaneously by an extended-wavelength (2.33 μm) multi-mode diode laser. The use of the 2nd-order WM harmonic improved the detection sensitivity of the system. The system's detection limits were about 81 ppbv and 31 ppbv for CH_4 and CO, respectively.	[15]
The performance of highly sensitive trace gas detection was improved using scanning baseline suppression and increasing resistance to external factors (temperature and humidity). The best measurement accuracy of the system in trace water vapor detection was one part per million by volume for an optical path length of 10 cm.	[16]
Ethylene in combustion exhaust from a portable power generator was detected using a distributed feedback interband cascade laser (3.3 μm). The 2nd-harmonic WM achieved detection limit equals to 96 parts per billion by volume at a signal-to-noise ratio of 2 under 3 kPa of pressure.	[17]
The overlapping spectra of the gas mixture (4% CO and 6% CO_2) were separated using the least square method. The model was a weighting sum of the 1st-, 2nd-, and 3rd-order derivatives of the Lorentzian function. The linear correlation coefficients of the relationship between CO and CO_2 concentrations and amplitudes of the separated second harmonics were more than 0.997.	[18]
Acetylene concentrations were detected in the range of 1-400 parts per million by volume using the absorption NIR line at 1530.36 nm. System sensitivity, detection precision, and limit were markedly improved because of the self-calibration method, which has better detecting performance than the conventional WM spectroscopy.	[19]
Carbon monoxide (CO) sensing in high-pressure, fuel-rich combustion gasses was associated with the internal conditions of hydrocarbon-fueled liquid bipropellant rockets. A scanned WM technique was utilized to infer species concentration from CO absorption and mitigate the influence of non-absorption transmission losses and noise associated with the combustor environment.	[20]
NO, and NH_3 in the urban area were simultaneously monitoring. The sub-ppb detection limit was achieved for all gasses with an average time of about 100 s.	[21]
A CO sensor, based on the 2nd-order WM spectroscopy, was applied to SF_6 decomposition analysis in an electric power system. A nonlinear least-squares fitting allowed using all 2f spectrum. A measurement precision of approximately 40 ppb was achieved with a data update rate of 0.6 s.	[22]
A cesium vapor was analyzed by band-resolved WM spectroscopy using large modulation indices. This study has insights for the measurement of cooperative emission effects in a bulk atomic ensemble.	[23]
A calibration-free method, WM - direct absorption spectroscopy (WM-DAS), combines the advantages of measuring absolute absorbance profile from calibration-free DAS with enhanced noise rejection and high sensitivity of WMS. WM-DAS is a perspective method for the high-precision measurement of spectral line parameters and gas information with weak absorptions.	[24]

Table 5.1-1 (continued). Modulation-based laser measurements

FM spectra of CH_4 transitions around 3070 cm^{-1} were measured. The concentration-time profiles of HCl at 2925.90 cm^{-1} allowed studying the reaction Cl + CH_4, following the 193 nm excimer laser photolysis of oxalyl chloride. OH radicals were generated by UV photolysis of H_2O_2, and its transients were recorded. The detection limit is about a factor of 4 above the shot-noise limit.	[25]
The first harmonic-phase-angle WMS (that is immune to the laser intensity and the demodulation phase) was used for measuring the absorption line of CO_2 around 6362.5 cm^{-1}. This method has great potential for WMS analysis in strong turbulence or high-pressure environments.	[26]
Hydroxyl radical (OH) is a critical intermediate reactive species during combustion processes relevant to power production, transportation, and manufacturing. An industrial WM laser absorption spectroscopy OH sensor was capable of in situ, quantitative detection of OH down to mole fraction values of 10−5 over a 75-cm pathlength. The concentration of OH radical above a catalytic combustor was measured under different operating conditions.	[27]
A breath acetone intensity-normalized 2nd-order harmonic WM-sensor was paired with a multi-pass optical gas cell to attain signals related to differential absorbance of acetone, water vapor, and methane. A spectral fitting routine corrected the acetone signal for interfering absorption. A detection limit (2σ) of approximately 0.11 ppm acetone was achieved. Acetone was detected with very high levels of exhaled methane (up to 40 ppm). The sensor was insensitive to methanol and ethanol in normal human breath. It was used to track ketosis during several subjects' regimented ketogenic diet.	[28]
A multi-species detection method was developed using a single quartz crystal tuning fork (QCTF), based on a photoelectric detector and dual-frequency WM. The light intensity measurements utilized the piezoelectric effect and resonant effect of the QCTF (instead of semiconductor detectors and lock-in amplifier). The FFT-decomposition of overlapping peaks was performed.	[29]
Acetylene was detected in a novel multi-pass cell by the WMS. The 2nd-order harmonic was normalized to the distance between its two minima. The proposed sensing system is suitable for trace gas sensing in a weight-limited unmanned aerial vehicle and an exhalation diagnosis for smoking test.	[30]
Low (1 kHz) and high (100 MHz) modulated frequency WMS, and two-tone FMS (390±5 MHz) were compared by estimating the minimum detectable absorption measured by AlGaAs diode laser tuned on a third-overtone methane transition at 886 nm.	[31]
The dependence of photoacoustic spectra on experimental parameters was investigated. WM provided the superior signal-to-noise ratio compared to the AM and eliminated background drifts and fluctuations.	[32]
The effects of pressure and modulation broadening for simultaneous AM and high-order WM spectra were studied.	[33]
The higher-order harmonic WMS increased the sensitivity while monitoring concentration fluctuations of gaseous species. There were regions where the commonly used the 2nd-order derivative showed a negligible variation of signal magnitude with concentration fluctuations, whereas a higher harmonic demonstrated a much improved signal. The measurements of WMS of lines in the oxygen A band were presented.	[34]

Table 5.1-1 (continued). Modulation-based laser measurements

The dependence of the amplitude of phase-sensitive harmonic WM signals the optical depth, temperature, pressure of the sample gas, and the modulation amplitude was investigated theoretically using the Lorentzian and Gaussian lines. The 2nd-order harmonic, which was normalized by the direct current signal, restored the unsaturated signal.	[35]
FM-enhanced remote sensing was considered to be a general method for monitoring atmospheric trace gases at ambient levels.	[36]
Quantitative analysis algorithms using the 2nd-order harmonic spectra for arbitrary combinations of line width and modulation depth were validated by measurements on CO lines.	[37]
Atomic absorption in an argon discharge was studied by WMS with a frequency-doubled KNbO3 diode laser. Minimum 3σ-detectable absorbances of 7.7×10^{-5} and 1.9×10^{-4} were estimated for 2f- and 4f-harmonic detection with a time constant of 0.1 s. The concentrations of argon were in the range of 3×10^{8} to 1.2×10^{11} cm^{-3}.	[38]
Thermal tuning and modulation of a 1556-nm distributed feedback fibre laser by the resistive heating of a thin silver film chemically deposited on the fiber were tested by recording part of the ν1+ν3 combination band spectrum of $13C_2H_2$ by scanning of the fiber laser.	[39]
Cavity-enhanced absorption spectroscopy and WMS were combined to measure the integrated output of unlocked cavities. The absorption of excited carbon atoms in microwave-induced plasma was measured with detection limits equivalent to optical depths below 10^{-6}.	[40]
Multiplexed fiber-coupled diode lasers are used to probe second-harmonic line shapes of NIR water absorption features to infer temperatures in gases containing water vapor, such as combustion flows. An optimal selection of the modulation indices simplified data interpretation over extended temperature ranges and minimized the need for calibration when performing 2f ratio thermometry.	[41]
The coupling of the RF signal to the quantum cascade laser through the cryostat was studied using low-pressure N_2O and CH_4 gases. Enhancement of factor six was measured with respect to the direct absorption method.	[42]
Adaptive SVD was used to eliminate the optical interference fringes due to unwanted etalons, which introduce systematic baseline changes in diode laser spectroscopic trace gas measurements. The acetylene detection limit was 20 ppb (1σ) over the one-week measurement.	[43]
A new mid-IR laser spectrometer was applied to high-precision measurements of isotopic ratios of nitrous oxide molecules at ppm concentrations.	[44]
Ammonia was detected by the WM cavity-enhanced IR tunable diode laser absorption spectrometer. The spectral resolution and sensitivity were sufficient to measure ammonia isotopomers ($^{14}NH_3$, $^{15}NH_3$) in planetary atmospheres.	[45]
An instrument, which was based on an improved off-axis alignment of integrated cavity output spectroscopy (OA-ICOS) in conjunction with a WM technique, was developed using a DFB NIR diode laser. The minimum detectable absorption was approx. 3.6 ppmv using the second harmonic. OA-ICOS-WM technique in NIR improved the detection sensitivity by a factor of 14 compared to that obtained with OA-ICOS. Measurements for the CO_2 mixing ratios in ambient air have been performed.	[46]

Table 5.1-1 (continued). Modulation-based laser measurements

For a constant laser modulation current, the WM amplitude was determined by the analysis of the distance between the two zero crossings of the measured 2nd-order harmonic spectrum of gas absorption. A frequency analyzer (instead of a LIA) provided a higher bandwidth and allowed a very precise detection of the zero crossings due to the logarithmic output.	[47]
A convolution Fourier-domain-based model for the WMS harmonic and harmonic signals with arbitrary transmission function was given. A recursion formula and a mean value property for n^{th} harmonic spectra have been found.	[48]
Temperature measurements in water vapor, based upon WM and two-line thermometry, were performed by the tunable diode laser absorption spectroscopy utilizing a standard tunable distributed feedback diode laser working in the telecom C-band.	[49]
A new sensor based on the tunable WM diode-laser absorption of CO2 near 2.7 μm was used for the sensitive and accurate measurement of the temperature behind the reflected shock waves in a shock-tube. The sensor was also used in reactive shock-tube experiments of n-heptane oxidation. Seeding of relatively inert CO_2 in the initial fuel-oxidizer mixture enabled measurements of the pre-ignition temperature profiles.	[50]
Difference frequency, which was generated within a 5% MgO doped PPLN crystal by coupling approx.735 nm radiation from a tunable external cavity diode laser with high powered 532 nm radiation from both Nd:YVO3 and Nd:YAG lasers, was combined with the WMS and cavity-enhanced absorption spectroscopy. Rotationally resolved transitions in the combination bands of NH_3 and CO_2 in the 1.9 μm region were investigated.	[51]
Proportional, shot and thermal noise effects and their reduction in the WMS and the direct absorption diode laser AS were studied.	[52]
Multi-mode absorption spectroscopy (MUMAS) has been combined with the WMS, and cavity-enhanced absorption spectroscopy (CEAS), to record multiple molecular transitions using a single laser and a single detector. Enhancement of the signal-to-noise ratio demonstrated an increased detection sensitivity for the MUMAS-based gas-sensing.	[53]
Absorption features of the oxygen A-band transitions in the optically thick regime, using the high order harmonics of WMS, were studied. The absorption saturation resulted in suppression of the line-center lobes of the harmonic signals depending on the optical pathlength and the modulation index.	[54]
A WMS absorption sensor, utilizing a quantum cascade laser, was used for CO detection at 4.6 μm. The detection limit at a room-temperature was 0.03 ppm per meter of absorption path length at a 1-kHz detection bandwidth (an order-of-magnitude increased sensitivity compared to scanned-wavelength direct absorption).	[55]
A fiber-amplified Distributed Bragg Reflector diode laser was used to pump a continuous wave singly resonant Optical Parametric Oscillator combined with WMS. The system's sensitivity was 0.8 ppbv for ethane for the absorption peak at 2996.9 cm^{-1} recorded in 1.3 sec.	[56]
WMS was performed in the oxygen A-band at different harmonics for different modulation indices and optical pathlengths. The Lorentzian and Voigt models were used to study the relationship between the higher harmonics' structure and the lineshape. The optical path length saturation in WMS served as a diagnostic to the lineshape.	[57]

Table 5.1-1 (continued). Modulation-based laser measurements

Acetylene at atmospheric pressure and the room temperature was detected using quartz-enhanced photoacoustic spectroscopy based on the wavelength modulated fiber-coupled distributed feedback diode laser. A minimum detectable limit (1σ) of 2 ppmv was achieved.	[58]
The sensitivity enhancement was attained by combining WM and the integrated cavity output spectroscopy (ICOS). WM appreciably improved the SNR of an ultrasensitive ICOS system, if the cavity transmission is so low that the detector noise is not negligible. The performance of ICOS and WM-ICOS was compared in a high sensitivity ambient-air methane detection experiment.	[59]
Two multi-harmonic detection methods for WMS systems-the simultaneous curve fitting of harmonic spectra and the reconstruction of the transmission from harmonic coefficients-were considered. Multi-harmonic detection was better than the single-harmonic method.	[60]
The use of integrating spheres as multipass gas cells combined with the 2nd-order harmonic WMS was studied. The gas lineshape was distorted at high concentrations because of the exponential pathlength distribution of the sphere, which introduced nonlinearity.	[61]
A real-time, in situ, water vapor sensor, based on the 2nd-order harmonic-WM tunable diode laser, continuously monitored water vapor in the synthesis gas of an engineering-scale high-pressure coal gasifier. The 1f-normalized, WMS-2f signal was insensitive to non-absorption transmission losses (e.g., beam steering and light scattering by the particulate in the synthesis gas). The sensor demonstrated a detection limit of approx. 800 ppm (25 Hz bandwidth) at conditions with more than 99.99 % non-absorption transmission losses.	[62]
The water vapor concentration (25-1048 ppmv) was measured with the sensitivity 52 ppbv for a 10-cm optical path length using the dual-beam WMS. The impact of the laser power variation and the residual AM was eliminated by photocurrent normalization through a balanced ratiometric detector.	[63]
Apodized 2f/1f WMS calibration-free method was validated experimentally by the trace detection of carbon monoxide using an air-broadened transition line in the first overtone band in the NIR region around 2.33 μm.	[64]
The line width under high absorption was measured using the ratios of the 2nd- and 4th-order harmonics at the line center. The transitions of CO_2 and H_2O were selected to measure the line width, gas pressure, and concentration.	[65]
A two-color tunable diode laser sensor for measurements of temperature and H_2O in an ethylene-fueled model scramjet combustor was developed. A calibration-free scanned-WMS spectral-fitting method was used to infer the integrated absorbance of each transition without *a priori* knowledge of the absorption line's shape.	[66]
A tunable WM diode laser sensor for measurements of the temperature and H_2O in high-pressure and -temperature gases was developed. The sensor was validated in a shock tube at temperatures and pressures ranging from 1,000 to 2,700 K and 8 to 50 bar.	[67]
A photoacoustic spectroscopy system based on the distributed feedback WM laser diode allowed the highly sensitive detection of ammonia.	[68]
The flame flatness of a flat-flame burner was validated using a resolution-doubled one-dimensional WMS tomography. The distributions of temperature and H_2O mole fraction in an axisymmetric premixed flame were simultaneously reconstructed, thus validating the flame's flatness.	[69]

Table 5.1-1 (continued). Modulation-based laser measurements

The FM efficiency was more than four times larger than that of conventional NIR distributed feedback lasers; it decreased monotonically with increasing the modulation frequency.	[70]
A H_2O-temperature sensor, based on scanned calibration-free WMS, was applied to real-time in situ measurements and temperature time histories (0.25-s time resolution) in the hot gases 2–11 mm above biomass pellets during atmospheric combustion in the reactor. Temperatures between 1,200-1,600 K and H_2O concentrations up to 40 % were detected above the biofuels.	[71]
An apodized 2f/1f WM-based optical polymeric fiber sensor was used for the sensitive detection of CO_2 gas in the NIR region (1.57 μm). A minimum detectable absorption of approx. 0.9×10^{-4} was achieved.	[72]
Laser heterodyne technology was combined with linear FM technology to measure the thickness of plate glass. The maximum relative measurement error was 0.01 %.	[73]
The 1st- and 2nd-order harmonics were used to measure the absolute absorbance, and then obtain gas pressure and concentration.	[74]
A calibration-free wavelength-scanned current-WM quantum cascade laser (QCL) was used for gas concentration detections. The CH_4 concentration was estimated by fitting a simulation spectrum, based on spectral line parameters, to the background-subtracted 1f-normalized 2f signal.	[75]
A frequency-modulated-continuous-wave laser detection system, based on the four-quadrant photodetector and, using the cross-power-spectral-density algorithm demonstrated a better signal-to-noise ratio for the target echo to smoke interference.	[76]
Review. Laser spectroscopy for breath analysis.	[77]
A non-absorbing interference occurs in shock tubes because of significant beam-steering noise and imperfect optical alignment, which was rejected using WMS.	[78]
The WM system, which is based on the fast modulating the injection current of the Fabry–Perot-type green diode laser and which is equipped with a Littrow grating to increase the laser-mode power density, significantly simplified the data processing for the extraction of small mercury absorption signals from a large and complex light background.	[79]
New wavelength-modulation spectroscopy with 1f-phase detection (WMS-θ1f) was analyzed, and the optimal operating regimes were established. A WMS-θ1f sensor for measuring temperature and H_2O mole fraction in the exhaust of a CH_4/air flat-flame burner was developed.	[80]
A robust breath acetone sensor was based on the detection of the 2nd-order harmonic WM laser absorption of multiple species near 8.2 μm. The signals were related to differential absorbance of acetone, water vapor, and methane. The sensor was tested in exhaled breath samples with varying amounts of interfering species (e.g., methane up to 40 ppm). The sensor was used to track ketosis during a regimented ketogenic diet of several subjects, which demonstrated regular non-intrusive monitoring of breath acetone over a clinically-relevant range (0.4–74 ppm) and sensitivity to dietary changes.	[81]
A customized long fiber ring etalon improved the accuracy and time response rate of the WM measurements. The dynamic wavelengths and phases of a distributed feedback diode laser were measured. The data for the application of quantitative and high-repetition WMS technique in combustion diagnostics were provided.	[82]

Raman spectroscopy

Differentiation suppressed the fluorescence background of the laser radiation and the sample luminescence observed in Raman spectra (RS) (RS also stands for Raman spectroscopy) [1, 2].

The intensity of the radiation incident on the detector is

$$I_D = I_R(\nu_E, \nu_R) + I_L(\nu_E, \nu_R), \qquad (5.1-1)$$

where $I_R(\nu_E, \nu_R)$ and $I_L(\nu_E, \nu_R)$ are the intensities of the Raman scattering and luminescence, respectively; ν_E, ν_R are the frequencies of the excited and scattered light, respectively; $I_L(\nu_E, \nu_R) = L(\nu_R)E(\nu_E)$; $L(\nu_R)$ and $E(\nu_E)$ are the luminescence and scattering components.

The following equations were obtained from Eq. (5.1-1) in the absence of the resonance effects [1]:

$$\partial I_R/\partial \nu_E = -\,\partial I_R/\partial \nu_R + 4I_R/\nu_R. \qquad (5.1-2)$$

$$\partial I_L/\partial \nu_E = L(\nu_R)\, \partial E/\partial \nu_E. \qquad (5.1-3)$$

In the local minimum of $E(\nu_E)$, $\partial E/\partial \nu_E \approx 0$, and the term $\partial I_L/\partial \nu_E \ll \partial I_R/\partial \nu_E$. The term $4I_R/\nu_R$ (Eq. (5.1-2)) was estimated numerically.

The derivative of I_D, which is relative to the excitation frequency, is equal to the negative derivative of the Raman scattering over the frequency of the scattered light:

$$\partial I_D/\partial \nu_E = -\,\partial I_R/\partial \nu_R. \qquad (5.1-4)$$

Coherent Anti-Stokes RS (CARS) involves the nonlinear mixing of three different wavelengths of a pump, Stokes, and probe beams. Anti-Stokes line intensities are orders of magnitude stronger than those of spontaneous Raman emission. The latter is resonantly enhanced when the frequency difference between the pump and the Stokes beams coincides with Raman resonance frequency. Solid-state (pump) and tunable dye lasers are used in the CARS.

A measured coherent optical signal at the anti-Stokes frequency is a cubic nonlinear susceptibility, which consisted of a non-resonant (background) $\left(\chi_{nr}^{(3)}\right)$ and resonant $\left(\chi_r^{(3)}\right)$ components [3]:

$$\chi^{(3)}(\nu_1 - \nu_2) = \chi_{nr}^{(3)} + \chi_r^{(3)}, \qquad (5.1-5)$$

where $\nu_1 - \nu_2$ is proportional to the combinational resonant frequency; ν_1 and ν_2 are the frequencies of the solid state and tunable dye lasers.

Active Raman spectroscopy analysis is impossible for a weak resonant (e.g., in spectra of impurities) $\chi_{nr}^{(3)} \gg \chi_r^{(3)}$. Since $d\chi_{nr}^{(3)}/d\nu_2 = 0$, the frequency modulation of the tunable laser suppresses $\chi_{nr}^{(3)}$, and the background appears because of the luminescence and spurious optical effects.

The study of the surface layer of substances by stimulated Raman spectroscopy produced a strong thermal background, which may be up to 10^4 times larger than the analyte signal [4, 5]. The luminescence of the laser-excited hydroxyl groups, lying on the surface layer was a strong background source, making it challenging to measure the RS of the samples absorbed or lying on the oxide surface [6]. In both cases, the frequency modulation technique reduced the background by a factor of 10^4-10^5, thereby permitting researchers to measure the vibrational spectra of the monolayers [5, 6]. The system achieved ultrahigh sensitivity, which was only limited by the shot noise [5].

The study [7] suggested taking the difference of two spectra, obtained with slightly shifted excitation frequencies, to eliminate fluorescence. This method, which is called Shifted Excitation Raman Difference Spectroscopy (SERDS), is similar to the two-wavelength difference method (Chapter 2.9). Since the shift of the excitation frequency is small, the measured signal is close to the 1st-order derivative over the Raman shift.

SERDS has some advantages over the WM technique [7]: (1) changing the wavelength of the excitation laser did not require special equipment; (2) the signal-to-noise ratio was limited by the shot noise of the fluorescence background due to multichannel detection.

The meat, fat, connective tissue, and bone from pork and beef were objects of the biological studies using SERDS [8]. Diode lasers emitted an excitation light at 783nm, 671nm, and 488nm. The spectral shifts were about 10 cm^{-1}.

SERDS was improved using an auxiliary tunable diode laser to excite the Raman spectra, and the WM technique [9-24].

Suppose that the Raman spectrum $S_j(\nu_i)$ is measured N times ($j = 1, 2, \dots, N$) in the discrete wavelength intervals ν_i. In WM mode [75],

$$S_j(\nu_i, \Delta\nu_j) = S_F(\nu_i + \Delta\nu_j) + S_R(\nu_i + \Delta\nu_j) = S_F(\nu_i) + S_R(\nu_i + \Delta\nu_j), (5.1-6)$$

where subscripts F and R stand for the fluorescence and Raman emission; $\Delta\nu_j$ is the shift of the Raman excitation laser. According to Kasha's rule [12], for a small $\Delta\nu_j$, the fluorescence component does not depend on the shift, while the Raman spectrum does.

The Raman signal was estimated mathematically from the matrix, whose elements were the spectra $S_j(\nu_i, \Delta\nu_j)$ measured for various $\Delta\nu_j$. The Principal Component Analysis (PCA) gave a remarkable noise-free 1st-order derivative better than SERDS. To simplify the mathematical processing, a recurrent reconstruction algorithm was developed [13]. However, it required the existence of a spectral region free of any Raman bands.

Table 5.1-2 summarizes some applications of the SERDS.

Luminescence spectroscopy

Luminescence spectroscopy is the collective name of molecular fluorescence, phosphorescence, and chemiluminescence spectroscopy. The intensity of the emitted radiation relates directly to the source power (e.g., laser). High-powered lasers, emitted at the proper wavelength of radiation to excite a sample, are good radiation sources for high sensitivity measurements. However, the interference may appear because of background fluorescence from the solvents, light scattered by turbid solutions, Rayleigh and/or Raman scatter, and instrumental problems (e.g., emission from the optical components, stray light passing through the optics). These interferences are sources of systematic errors in analysis.

Table 5.1-2. Applications of SERDS

Analysis of biological objects.	[14]
Identification of ethanol in rum using pairs of high-power laser diodes with a fixed-wavelength separation.	[15]
The two-wavelengths excitation. The PCA classification of meat species (beef, pork, chicken, and turkey) separated them into four distinct groups for both wavelengths. The variables were the myoglobin content and gradual differences of protein Raman band intensities and positions.	[16]
The auto-fluorescence of the sample and the fiber's fluorescence background were suppressed in a reasonable acquisition time of the 30s using the pure signal's PCA reconstruction. The samples were bovine bone tissues and adipose tissue derived from chicken, pork, beef, and lamb.	[17]
Raman microscope-based human sells analysis. Experimental optimization of the acquisition parameters.	[18]
Non-invasive monitoring of the cultivation of phototrophic microorganisms producing complex molecules of pharmaceutical relevance in a bioreactor. The use of Partial Linear and Nonlinear Support Vector Regressions.	[19]
Pharmaceutical applications.	[20]
The quantitative analysis of trace amounts of methanol in red wines.	[21]
Classification of the Raman spectra of healthy and bladder cancer cells.	[22]
Comparison of reconstruction algorithms. Analysis of tissues of pigs and from the oral cavity of humans, and a model solution of dye dissolved in ethanol.	[23]
Two-fold improvement in signal-to-background-noise-ratio was achieved because of the rapid SERDS operation in the kilohertz range.	[24]

The first study on derivative luminescence spectrometry [1] showed that RC differentiation significantly reduced the systematic errors but at the expense of a slightly weaker signal-to-noise ratio. Many analytical examples are in the book chapter [2]. Broad overlapped peaks are severe limitations of room-temperature phosphorescence and fluorescence multicomponent quantitative analyses of organic compounds, adsorbed on solid surfaces. The 1st- and 2nd-order derivatives allowed improving the resolution [3, 4].

Synchronous fluorescence spectroscopy (SFS) [5] combined two conventional modes of fluorescence measurements: (1) scanning the excitation radiation while the detected emission wavelength is a constant; (2) detecting of the emission radiation in a wide range at the constant wavelength excitation. So, SF spectra are obtained by scanning both the excitation and emission monochromators simultaneously. SFS demonstrated spectral simplification, bandwidth narrowing, and reduced scattering interference [6]. The derivative SFS (DSFS) gave remarkable results in the multicomponent analysis [6] (see, also Table 5.1-3).

Table 5.1-3. Applications of the Derivative FS

Quantitative residue determination of pesticides in crops. The derivative amplitudes were measured graphically using the interference-free character of the isodifferential points in the calibration curves.	[7]
Eleven polycyclic aromatic hydrocarbons (PAH) were detected in the technical mixture of a solution of eighteen PAHs in hexane by the 2nd-order DSFS using eight different wavelength shifts.	[8]
Quinine determination in tonic waters.	[9]
The identification of polyaromatic hydrocarbons in petroleum fractions using the 2nd-order DSFS.	[10]
The identification of individual dimethyl naphthalenes that have very closely lying electronic transitions using the 2nd-order DSFS.	[11]
The studies of the peculiarities of the 2nd-order derivative fluorescence spectra (SDFS) of tryptophan in proteins. The turbidity observed in whole membrane extracts was eliminated by using DSFS.	[12]
Studies of the peculiarities of the DSFS of indole compounds.	[13]
Simultaneous determination of naproxen and diflunisal in pharmaceuticals and human serum samples using the 1st-order DSFS.	[14]
Simultaneous determination of chlorzoxazone and ibuprofen in drugs and human plasma using the null-point measurements in the 2nd-order DSFS.	[15]
Analysis of duloxetine hydrochloride and vardenafil by the DSFS.	[16]
Simultaneous determination of tadalafil and dapoxetine HCl in binary mixtures using null-point measurements in the 1st-order SG DSFS.	[17]
The 1st-order derivative emission spectrofluorometric determination of the supramolecular complex of valsartan and sacubitril in tablets.	[18]
The stability of dapoxetine hydrochloride was studied in the presence of its acidic degradation products and drugs Vardenafil and Tadalafil using DSFS.	[19]
Simultaneous determination of binary mixtures of ibuprofen (IBF) and phenylephrine hydrochloride (PHE) in pure powder, synthetic mixture, and tablets using null-point 1st- and 2nd-order derivative spectrofluorimetry. The direct analysis of IBU and PHE in spiked human plasma was satisfactory.	[20]

Table 5.1-3 (continued). Applications of the Derivative FS

Determination of dissolved phenanthrene and its metabolites in aqueous solution by the DSFS.	[21]
Determination of trimebutine and its degradation product (eudesmic acid) using the 1st-order derivative fluorimetry.	[22]
Detection of cefadrine and cefadroxil in water samples by the DSFS.	[23]
Simultaneous determination of simeprevir and ledipasvir by the DSFS.	[24]
Determination of amlodipine and celecoxib in pharmaceuticals and human plasma by the DSFS.	[25]

Photoacoustic spectroscopy

In photoacoustic spectroscopy (PAS) (Fig. 5.1-3), the light pulses generate the acoustic waves in a cell where the sample is held. Two microphones pick up the acoustic waves and transform them into electrical signals. In the conventional PAS device, the light pulses have the same wavelength. Figure 5.1-3 shows the modulation spectrometer, which combines the periodic interruption of the radiation incident on the sample by the chopper with the wavelength shift to produce the derivative output signal [1] (for technical details see Chapter 2.9).

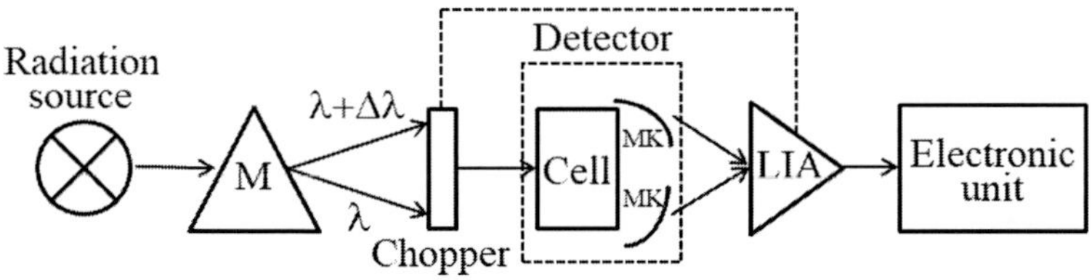

Figure 5.1-3. Diagram of the photoacoustic modulation spectrometer. M, monochromator; MK, microphone; LIA, Lock-in amplifier (Appendix J).

The amplitude of the PAS signals depends on the sample's ability to absorb the light [2]. Since PAS is unaffected by scattering, it can be used to analyze fine powders, crystalline, and biological samples. However, there are losses due to reflection at the sample surface during sample excitation, and the thermal wave reflected in the sample. The modulated light excitation of a sample induces several additional responses that can be used to analyze the sample.

To generate a noticeable photoacoustic effect, the radiation must be intensive. Therefore, a wide monochromator bandwidth is needed. However, the resolution decreases. There is no such problem with the laser source. Despite the zero-background nature of PAS, in both cases the cell window and broadband absorption from gas constituents other than the target species produce background. WM allowed reducing resolution and suppressing the background [3]. The principal difference between the WM spectroscopy and WM-PAS is the type of the energy measured by detector: transmitted and absorbed by the sample, in the first and second cases, respectively. The theoretical model of WM-PAS showed that the 1st-order harmonic is offset-free contrary to the WMS [4]. This property is due to the zero-background nature of PAS.

The approximate 1st- and 2nd-order derivatives of the photoacoustic spectrum of $Pr(Gly)_3Cl_3 \cdot 3\ H_2O$ and $PrCl_3 \cdot 6$ H2O were estimated using continuous wavelet transform [5].

The use of the 2nd-order harmonic of modulated NIR diode lasers increased the ammonia detection's sensitivity by more than ten times [6].

The sensitivity of gas detection was increased by a factor of 35 due to the high-frequency WM [7].

The derivative PA spectra, which was obtained by the optical differentiation, allowed the detection of unnoticeable peaks in the PA spectrum of carbon [8].

A multipoint gas sensing WM-based photoacoustic system detected the water vapor at 1368.597 nm with a minimum detection limit: 479-630 ppb depending on the vapor concentration [9].

The PAS-based carbon monoxide (CO) gas sensor achieved a minimum CO detection limit of 9.8 ppm using the 2nd-order harmonics of the WM signal to reduce the background [10].

Remote Sensing Data and Applications

The introduction to the course “Remote Sensing Data and Applications” of Harvard University states [1] that

“The growing concern about human impact on the environment has led to the development of new observation and analysis tools to tackle and monitor types, magnitudes, and rates of environmental changes. Timely observations by Earth observation (EO) satellite systems and improved mapping and analysis tools are enabling a better understanding of the ecological interactions that underlie our Earth systems, which is critical for developing sustainable solutions.”

Remote hyperspectral sensors on-board various satellite and aircraft platforms allow the collection of a huge amount of hyperspectral data together with a geographical information system. The extraction of required information from these spatial databases brings new challenges for analysts. Multi-spectral scanning systems with high spatial resolution record the reflectance of the Earth’s surface in the range 400-2500 nm [2].

Table 5.1-4 includes some applications of remote sensing (RSE) and remote hyperspectral sensing (RHS) for environment control.

Table 5.1-4. Applications of Remote Sensing Spectroscopy Technique

A reduction of the soil background reflectance in the RSE of vegetation and the resolution of the complex spectra of several target species within individual pixels in RSE were demonstrated. The derivative spectral indices for monitoring chlorosis in vegetation showed their advantage over the near-infrared / red reflectance ratio data.	[3]
RES data were used to estimate the suspended sediment concentration (SSC) in water. The correlation between the derivative spectral reflectance and SSC in seawater was substantially more significant than that of the spectral reflectance data and SSC.	[4]
Reflectance bands spectra (252) represented the dominant botanical and substrate classes within Prentiss Bay, Horseshoe Bay, and Lake Huron which were transformed into relative percent reflectance. These resampled spectra emulated the band configurations of the airborne, hyperspectral imagery. 2nd-order derivative analysis of the transformed spectra allowed the selection of 8 visible–NIR wavelength bands that were the most botanically explanative for the differentiation of coastal wetland vegetation. A reduction of the band number without significant information loss made it possible to utilize small pixels to differentiate the botanical communities.	[5]
In was shown that the derivative vegetation indices were preferred over the simple reflectance-based ones. The chlorophyll-related derivative index D 725/D 702 was suitable for most satellite-based sensors. The index is not sensitive to soil reflectance and can be used to test open crops. The presence of blank reflectance is also unnecessary.	[6]
Derivative-based analysis of high-resolution, spectrally continuous RSE data was performed using several smoothing and derivative computation algorithms.	[7]
A model system, which consisted of carbonate sediments with variable concentrations of microbial pigments, was analyzed by RHS derivative reflectance spectroscopy. The major sediment pigments, chlorophyll A and fucoxanthin were identified and showed a quantitative correlation with sediment pigment concentrations.	[8]
Ratios of the magnitude of the derivatives 725 to that at 702 nm enabled the identification of plant stress caused by gas leakage. The technique was able to identify stress responses to long-term leaks in all the crops tested but only to short-term leaks in grass.	[9]
Hyperspectral data were processed by high-order SG derivatives. Hyperspectral signatures for cotton, sicklepod, and bare soil were obtained.	[10]
Phytoplankton pigment assemblages were identified using the 2nd-order derivative RHS reflectance in simulation-based experiments. Resolution enhancement was demonstrated.	[11]
The retrievals of three particular phytoplankton functional types (diatoms, coccolithophores, and cyanobacteria) were performed using absorption and reflection spectra derivatives.	[12]

CHAPTER TWO

DERIVATIVE RECORDING OF ANALYTICAL SIGNALS EXCLUDING OPTICAL SPECTROSCOPY

This final chapter includes mainly bibliographic data concerting some applications of the derivative method in instrumental analysis, excluding optical spectroscopy. Although these data may be outside of the book's main focus, it will be useful for readers to familiarize themselves with these applications to get an overall picture of derivative analytical techniques. The first review on this issue [1] was published in 1988.

Electrochemical analytical methods

Electrochemical analytical methods are divided into: potentiometry, including the constant-current chronopotentiometry; coulometry; and voltammetry, including polarography [2].

Potentiometry methods measure the solution's potential between the indicator and reference electrodes (an electrochemical cell) under static conditions. The relationship between the analyte concentration and the potential is studied. Potentiometric titration is similar to direct titration, but no indicator is needed.

Coulometry involves the measurements of the amount of substance deposited on an electrode due to an electrochemical reaction.

In voltammetry, a potential (constant, varying, or a combination) is applied to an electrode's surface. The resulting current is measured.

The derivative method was first put into practice in electrochemical analysis [3 and references therein]. The 1st-order derivative was obtained by a simple RC circuit (Chapter 2.1).

A fractional derivative of the voltammogram is called "neopolarogram" [4 and reference therein].

Signal shapes and their derivatives are given in Appendix M.

Table 5.2-1 summarizes the application of the derivative technique in electrochemical analysis.

Table 5.2-1. Derivative electrochemical analytical methods

The study of analytical characteristics of 1st-, 2nd-, and 3rd-order derivative measurements. The 1st-order derivative increased the sensitivity more than an order of magnitude and suppressed the impact of the interfering electroactive species	[5]
Derivative voltammetry, in conjunction with anodic stripping analysis at the hanging mercury drop electrode, demonstrated a significant enhancement in sensitivity. This method was less sensitive to the interferences.	[6]
The derivative measurements when studying rapid reactions demonstrated their advantages. The rate constant was estimated in a single sweep experiment using the peak value of the 1st-order derivative current.	[7]
The modulated polarographic and voltammetric techniques in natural water chemistry demonstrated the high sensitivity to very low concentrations of electroactive components and their chemical form.	[8]
The 1st- and 2nd-order derivative curves of cellulose acetate membrane electrophoretograms of human serum protein revealed multicomponent presence in each globulin fraction.	[9]
An automated slope analyzer for measuring the 1st- and 2nd order time derivatives of the probe response was developed.	[10]
Pseudo-derivative polarograms (PDP) produced better signal-to-noise ratios, and, therefore, lower detection limits than differential pulse polarography. The low sensitivity of the PDP to the influences of homogeneous chemical kinetics allowed this method to be used for metal speciation studies.	[11]
Derivative linear sweep (DLSVA) and derivative cyclic voltabsorptometry (DCVA) are the optical analogs of linear sweep and cyclic voltabsorptometry. In these methods, the absorbance of the electrode reaction product is differentiated relative to the linear scan potential.	[12]
Derivative hydrodynamic-modulation voltammetry was developed to improve the resolution in highly sensitive mixture analyses using solid electrodes. The voltammetric current-potential waves were transformed into peak-shaped curves, which are more convenient for further data processing. Single and multiple peak systems were evaluated at the micromolar concentration level using stopped-rotation and stopped-flow voltammetry. The test compounds were dopamine, ascorbic acid, NADH, homovanillic acid, and chlorpromazine.	[13]
The peak-shaped voltammograms arising from the derivative and differential normal pulse voltammetry were compared for reversible and heterogeneous charge transfer control systems for varying kinetic and experimental parameters.	[14]
A microprocessor-based voltammetric and polarographic analyzer was developed for peak differentiation. The analyzer was used for the determination of $UO2^{+2}$ ions by the adsorbed cathodic stripping voltammetry and in vivo analysis of dopamine traces on a carbon fiber microelectrode.	[15]
The derivative of the absorbance of the reaction product (at a particular wavelength) relative to the time was obtained during a slow potential scan at an optically transparent thin layer electrode. The differentiation was useful for the background-free optical signal when dealing with the systems characterized by high faradaic/capacitive background and/or multiple products.	[16]

Table 5.2-1 (continued). Derivative electrochemical analytical methods

The fractional 0.5-order differentiation performed the extraction of the single peaks from the complex linear sweep and cyclic voltammograms; replacing of all undesired peaks by baselines obtained using the least-squares curve-fitting; and restoring a single-peak voltammogram by the fractional 0.5-order integration.	[17]
The theory for the DLSVA and DCVA of the long-path-length spectroelectrochemical (SEC) cell for the single reversible electrode reaction was suggested. The results showed that the DLSVA and DCVA of the long-path-length SEC cell are more sensitive (by several decades or more) than those of the optically-transparent electrode.	[18]
The derivative current-potential curves were used for the studies of heterogeneous electron-transfer reactions. Various peak asymmetry parameters were evaluated. The heterogeneous kinetic parameters were determined by fitting the theoretical derivative equation to the experimental data.	[19]
The estimated semi-integer-order derivative and semi-integral curves of theoretical cyclic voltammograms were found to agree with semi-differentiation and semi-integral electroanalysis theory.	[20]
Analytical expressions deduced for derivative voltammetry and differential pulse voltammetry, as well as corresponding to EC, CE, and catalytic mechanisms at spherical electrodes, established criteria to discriminate between last three mechanisms. (The abbreviations E and C represent an electron transfer and a chemical reaction, respectively).	[21]
Overlapping peaks in capillary electrophoresis of eleven benzoic acids were resolved by differentiation of the spectrophotometric detector's signal. Compared with standard detection, the derivative mode demonstrated slightly lower reproducibility, comparable sensitivity, better separation, and baseline suppression.	[22]
The electrochemical reactions under conditions of linear semi-infinite diffusion with an irreversible electron transfer process were simulated.	[23]
Theoretical expressions for the 1st-, 2nd-, and 3rd-order derivatives of voltammetric curves were analytically derived and analyzed for reversible, quasi-reversible, and irreversible processes under spherical diffusion.	[24]
The instantaneous corrosion rate and corrosion mechanism were studied using the electrochemical frequency modulation (EFM). In EFM, the corrosion parameters were measured using the Fourier harmonics and their combinations. This method demonstrated high sensitivity and accuracy.	[25]
The 2nd-order derivative linear sweep voltammetry was used for the ultrasensitive determination of 4-nitrophenol based on acetylene black paste and graphene hybrid electrode. Numerous applications of this method were cited for the enhancement of sensitivity and specificity in the quantitative analysis.	[26]
The 1st- and 2nd-order derivatives of the cyclic voltammograms were used to determine the standard electrochemical potentials from the curves that exhibit chemical irreversibility. The voltammogram's reversibility was estimated.	[27]

Chromatography

Chromatography is a separation method. The separation is based on differential partitioning between the mobile (which carries a sample) and stationary (through which the sample passes) phases. The method's principal problem is selecting the appropriate phases and experimental parameters that allow full separation of the mixture component. A lot of effort and experience is required to solve this problem. Therefore, in the case of partial mixture separation, a mathematical resolution enhancement may be useful, e.g., differentiation of chromatograms over time.

In the pioneer study [1], the 1st-order time derivative of the effluent transmission curve from the column chromatogram was obtained by the RC circuit. This article was published eight years later than the first reports on derivative spectroscopy. Despite the primitive technique, the conclusions drawn by Tracey [1] are still relevant:

> "The position of peaks can be more accurately determined [in derivatives] over a very wide concentration range. This increased accuracy may be particularly useful in determining small differences in elution volume between a standard preparation of known behavior and a series of unknowns […] In the purification of complex mixtures it gives more immediate evidence of multiplicity of components."

Table 5.2-2 summarizes, in chronological order, some analytical applications of the Derivative Chromatography made in 1967-2019.

Table 5.2-2. Derivative Chromatography

Type and Reference	Description
Gas [2]	The time concentration derivative signal was measured by two thermal conductivity cells separated by a dead space auxiliary column.
Ion exchange [3]	The measurement of the difference in the electrical conductances at the inlet and outlet of a dead space column produced the elution curve derivative. The derivative method was tested in an experiment using cupric ion eluted with ethylene diamine hydrochloride.
Liquid, [4]	Analysis of a mixture of aromatic hydrocarbons.
Liquid, [5]	The null-point analysis of a mixture of phenylalanine and tryptophan using the 1st-order SG derivative.
Molecular sieve, [6]	Electronic differentiation resolved the shoulders and inflections of a single broad peak, which characterizes these chromatograms.
Gas, [7]	Analysis of trace components (hydrocarbons), eluting on the tail of a broad solvent peak, using the 1st- and the 2nd-order derivatives.
HP liquid, [8]	Analysis of a mixture of porphyrin esters (in nanogram quantities) in the culture medium of the chick embryo liver cell. The desired sensitivity was obtained by using a second derivative-fluorometric detection system.

Table 5.2-2 (continued). Derivative Chromatography

Type and Reference	Description
Thin-layer, [9]	The trace analysis and detailed studies of unresolved compounds were significantly simplified in pharmaceutical research due to the coupling photodensitometry with a derivative recording device.
Gas, [10]	Analysis of hydrocarbon mixtures using RC differentiation.
HP thin-layer, [11]	The use of a spectrophotodensitometer signal differentiated by an electronic device allowed the analysis of small amounts of a component masked by major peaks. The study of some conservants used in pharmaceutical formulations was performed.
HP thin-layer,[12]	Similar to the previous section, but using the 4th-order derivatives. Study of a commercial colorant and a syrup formulation.
HP thin-layer, [13]	The resolution of overlapping model systems, noscapine and palavering, and the red pigments, R112 and R3, was improved using a 2nd-order derivative.
[14] All	The determination of chromatographic peak purity: (1) at the specific wavelength, where the major compound has a zero derivative; (2) by screening derivative curves for possible overlapping impurities (mapping technique).
[15] All	The 2nd-order derivatives of some Gaussian doublets were studied to establish the integration limits of chromatographic peaks.
All [16]	The 2nd-order derivatives of some model doublets were studied to establish the limiting properties of quantitative derivative analysis.
All [17]	Peaks were detected using a combination of the derivatives of different orders.
HP Liquid [18]	Mixtures of piperonyl butoxide, neopynamine, and fenitrothion characterized by overlapped peaks in liquid chromatography were quantified by derivative spectra obtained by a diode-array spectrophotometer around the maxima signal of the chromatographic peak.
All [19]	The multivariate PCR model's application gave better accuracy in quantifying partially overlapped UV peaks of enantiomers than the 1st- and 2nd-order differentiation.
Gas-MS [20]	The 1st-3rd-order derivative gas chromatography interfaced to the isotope ratio mass spectrometry was used to analyze the mixtures of 2,3,3′,4,4′,5,5′-heptachloro-1′-methyl-1,2′-bipyrrole (Q1) and 2,2′,4,5,5′-pentachlorobiphenyl (PCB 101). Differentiation enabled the identification of the interference of Q1 with PCB 101 even when both peaks fully co-eluted. The derivative method was used to study the peak purity of 2, 2′,3,4,4′-pentabromodiphenyl ether (BDE 85) in technical pentabromo diphenyl ether (DE-71).
Gas-MS [21]	The 1st-order derivatives of chromatograms of mixtures of benzo(b)fluoranthene and benzo(k)fluoranthene (for the m/z 252 ion) resolved the overlapped peaks of these two compounds with the identical mass spectra. This method made it possible to use the 250 and 248 ions to confirm the identity of these analytes.
LC-MS [22]	Noisy chromatographic peaks for the analysis of impurities in pharmaceutical products were detected using the 2nd-order digital differentiator filter automatically adapted to the data.

Table 5.2-2 (continued). Derivative Chromatography

Type and Reference	Description
Gas, Liquid,[23]	Peaks were identified using the 1st- and 2nd-order derivatives of the one-dimensional chromatograms. The retention times were estimated.
All [24]	The boundaries of chromatographic peaks were identified using the 2nd-order derivative.
All [25]	Automatic processing in clinical and toxicology laboratories.
HP Liquid [26]	The 1st-order derivative of UV spectra allowed the quantification of the plasticizers [bis (2-ethylhexyl) adipate and di-isononylphtalate], which co-eluated with other components [di(2-ethylhexyl) phthalate and dioctyl terephthalate, respectively].
HP Liquid [27]	The resolution enhancement used combinations of derivatives for symmetrical and asymmetrical peak models.

X-ray spectroscopy

X-ray spectroscopy (XS) methods detect photons that have wavelengths in the X-ray region. The main branches of the XS are X-ray Absorption (XAS), Emission (XES), and Diffuse Scattering (XDS) Spectroscopy. The wavelengths of the X-rays, which were emitted or absorbed during electronic transitions between atomic energy levels, are characteristics of the elements. The XAS is used to determine the local structure around the absorbing atom (bond distance, coordination number, and chemical identity of the elements). The X-ray absorption near-edge structure (XANES) provides information about the oxidation state of an excited atom and its coordination symmetry. Extended X-Ray Absorption Fine Structure (EXAFS) is used to determine the chemical state of practically important species, which occur in very low abundance or concentrations.

The fine lines of the XES give data on the density of filled states. The XES allows the oxidation and total spin state of the emitting atom to be studied. In the XDS, photons that be elastically scattered, contain information about the structural dynamics in materials.

Differentiation and modulation techniques are used in the XS for resolution enhancement and background suppression.

Photoelectron spectroscopy

X-ray photoelectron spectrum (XPS) is a dependence of the electron binding energy on the number of electrons that escape from the tiny surface layer (up to 10 nm) of the sample, which is irradiated with an X-ray beam. The XPS consists of bell-shaped peaks that are sometimes poor-resolved (e.g., due to the small chemical shifts). Therefore, recording derivatives improves resolution.

In some cases, peaks of Auger Electron Spectra (AES) are observed in the XPS but not the XPS peaks in the AES. AES methods are more sensitive to the chemical environment than those of the XPS. In the AE spectrometer, an electron beam, focused onto a sample, stimulates the electron emission. The AE data are obtained in the derivative mode.
However, quantitative derivative analysis must consider that the intensities in derivatives depend on the peak shape, which may differ in various matrices [1].

The book [2] includes numerous applications of the derivative photoelectron spectroscopy.

Table 5.2-3. X-ray spectroscopy

The digitized DC induced in a detector was measured, and the data placed on paper tape was numerically differentiated. Computer differentiated spectra were close to those obtained by the modulation technique.	[1]
A new method of the quantitative analysis of demodulated EXAFS spectra resolved small changes in the sample's structure, with greatly enhanced precision. This method revealed the formation of ruthenium oxide species upon partial oxidation of a Ru metal particle, which cannot be detected in a standard EXAFS experiment.	[2]
The X-ray absorption fine structure at the copper K-edge in a mixture containing cuprous oxide and cupric oxide was studied using four methods (including differentiation).	[3,4]
The 2nd-order derivative (obtained by a spline method) of an X-Ray diffraction pattern was used as fingerprints of the Tibetan medicine snow lotus herb.	[5]
The sensitivity of extended X-ray absorption fine structure spectroscopy for minute structural changes was enhanced by combining the modulated excitation approach and phase-sensitive analysis.	[6]
Resolution enhancement and background suppression using the 2nd-order derivative.	[7]

Miscellaneous

Electron Paramagnetic Resonance

The 1st-order differentiation of the absorption spectrum is the most common way to record Electron Paramagnetic Resonance (EPR) (or Electron Spin Resonance spectra [1, 2]). The modulation technique allows for reducing noise. The 1st-order derivative of the continuous-wave EPR spectrum of 1, 1-diphenyl-2-picrylhydrazyl powder was obtained by the frequency (1.6 MHz) modulation [3]. Measuring the 1st-order derivative relative to the microwave frequency helps to investigate the electron-spin-related phenomena in ecological and biological objects [1].

Nuclear Magnetic Resonance

The 1st-order derivative of the Nuclear Magnetic Resonance (NMR) spectra is usually obtained by the magnetic field modulation with a linear variation of the strength of the static magnetic field. The phase-sensitive detection improves the signal-to-noise ratio [4].

The continuous autodyne spectrometer, supplied by the modulation unit, generated the 1st-order derivatives of the broad peaks of the proton magnetic spectra of moisture and methane-rich coals [5]. The necessary information was extracted from the derivatives.

The 1st-order derivatives of the NMR spectra of ^{7}Li and ^{6}Li nuclei in a LiTaO3 single crystal sample of congruent composition allowed the referencing of weak sidelines of ^{7}Li nuclei to cation sublattice defects [6].

Table 5.2-4. Photoelectron and Auger spectroscopy

The 2nd-order derivative of the photoelectron energy distribution from cesiated Cu revealed structure that was not detectable in the energy distribution itself.	[3]
Differentiation revealed significant features of gold valence band spectra.	[4]
The combination of the even-order derivatives of poorly resolved curves was used for the nonlinear least-squares curve fitting.	[5]
The peak detection with peak/background ratio was down to 0.01.	[6]
The quantitative surface analysis of metals (Ag, Cd, In, and Sn), indium-tin alloys and films (indium, tin oxides, and indium-tin oxide) was performed using AES.	[1]
The study of O (1s) and C (1s) derivative spectra obtained from carbon fiber.	[7]
The C (electron shell transitions KLL) spectra from natural diamond, graphite, and single-crystal β-SiC were investigated using the 1st-order derivative X-Ray Excited AES (XAES). Enhanced fine structure XAES provided a fingerprint of the carbon bonding state.	[8]
The peaks were identified using a combined method, which included 2nd-order derivatives.	[9]
The sp^2/sp^3 ratio was estimated from the 1st-order derivative of the Auger C (KLL) spectrum of carbon films (from diamond to graphite).	[10]

Mossbauer spectroscopy

Mossbauer spectroscopy is based on the Mossbauer Effect. This effect consists of the emission and absorption of nuclear gamma rays, produced by a source of γ radiation, in solids.

The source moves back and forth to generate a Doppler effect, that is, a wave frequency is changed if the wave detector is moving relative to the source. The absorption Mossbauer spectrum is a function of transmittance versus velocity.

A modulation procedure similar to that used in EPR spectrometers was applied to the ^{57}Fe-γ rays absorption spectrum of metallic iron [7].

The precision analysis of hyperfine splittings of the very high resolution 93.3 KeV Mossbauer resonance in ^{67}Zn was carried out by a frequency modulation Mossbauer spectrometer [8].

Mossbauer radiation was frequency modulated by passing through a vibrating resonance medium, which was induced by pulsed laser excitation. The ratio of the 4th- and 2nd-order Fourier harmonics of the modulated radiation was found to be very sensitive to the amplitude of nuclear vibrations [9].

Gamma-ray spectroscopy

Peaks of the gamma-ray spectra were identified using 1st- and the 2nd-order derivatives [10, 11].

The resolution of the γ-ray spectra obtained by the scintillation (NaI (TI)) detectors was improved using the weighted sum of the signal and its 2nd- and 4th-order derivatives [12].

Differential scanning calorimetry

Differential scanning calorimetry (DSC) involves measuring the heat flow of the sample versus temperature. In the power-compensating DSC, sample and reference cells are heated with separate heaters. Usually, the temperature of both cells is linearly varied as a function of time. The separate sensors measure the cell temperatures, and the differential power signal is recorded.

DSC has some drawbacks [13]: (1) complex transitions are not properly analyzed since only the sum of all the thermal events in the sample is measured; (2) baseline drift; and (3) a compromise between sensitivity and resolution is needed.

The modulation DSC (MDSC) technique eliminated these drawbacks. In MDSC, a sinusoidal modulation was overlaid on the conventional linear temperature regime [13-15].

The 1st- and 2nd-order derivatives of the thermograms of lipid samples were used to define extrema and the start points of thermal events [16].

Mass Spectrometry

Overlapped peaks are detected by examining the 2nd-order derivative of the raw native electrospray-ionization mass spectra from large protein complexes [17].

The summing of the spectrum with its weighting 2nd-order derivative enhanced the resolution of overlapping peaks using an adaptive algorithm [18]. The relationship between the sharpening ratio and weighting factor was established.

APPENDIX A

ANALYTICAL SIGNAL PROCESSING USING FOURIER TRANSFORM AND SPLINES

A1. Continuous Fourier transform

Given an integrable function $F(\lambda)$:

$$\int_{-\infty}^{\infty} |F(\lambda)|\, d\lambda < \infty. \qquad (A1-1)$$

Function $F(\lambda)$ is represented by the Fourier integral (transform-FT) [1]:

$$F(\lambda) = (1/2\pi)\int_{-\infty}^{\infty} \tilde{F}(\omega)\exp(j\omega\lambda)\, d\omega, \qquad (A1-2)$$

where $\tilde{F}(\omega)$ is FT of $F(\lambda)$:

$$\tilde{F}(\omega) = \int_{-\infty}^{\infty} F(\lambda)\exp(-j\omega\lambda)\, d\lambda, \qquad (A1-3)$$

ω is the angular frequency (argument in the frequency domain) with an inverse dimension of λ (argument in the time domain); $j = \sqrt{-1}$.

Interested readers can find a simple derivation of the FTs of the Gaussian and Lorentzian functions [2].

The integral convolution in the time domain (e.g., Eq. (1.1-3))

$$F_m(\lambda) = \int_{-\infty}^{\infty} I(\lambda + \lambda')F_T(\lambda')d\lambda' \qquad (A1-4)$$

is the product in the Fourier domain:

$$\tilde{F}_m\omega = \tilde{I}(\omega)\tilde{F}_T(\omega). \qquad (A1-5)$$

Suppose that the analytical signal (AS) is a stationary random process. In simple terms, stationarity refers to time invariance of some statistics of a random process, e.g., mean or autocorrelation. In other words, these characteristics do not depend on the shifts in the time domain [3]. Consider only a wide-sense-stationary random process [3]. The autocorrelation function of this process is a mean

$$K(\Delta\lambda) = \overline{F(\lambda)F(\lambda + \Delta\lambda)}. \qquad (A1-6)$$

The dispersion is

$$\sigma^2 = K(0). \qquad (A1-7)$$

According to the Wiener–Khinchin theorem, the autocorrelation function is the power spectrum of the process:

$$W(\omega) = \int_{-\infty}^{\infty} K(\Delta\lambda)\exp(-j\omega\Delta\lambda)\, d\Delta\lambda. \qquad (A1-8)$$

According to Eq. (A1-2),

$$K(\Delta\lambda) = (1/2\pi)\int_{-\infty}^{\infty} W(\omega)\exp(j\omega\Delta\lambda)\, d\omega. \qquad (A1-9)$$

Suppose that AS passes through a linear device, which frequency characteristic (generally, a function of the complex argument $j\omega$), is the ratio of the output and input signals:

$$G(j\omega) = S_{output}(j\omega)/S_{input}(j\omega). \qquad (A1-10)$$

Then, the power spectrum of the output signal is

$$W_{out}(\omega) = |G(j\omega)|^2 W_{input}(\omega). \qquad (A1-11)$$

By substituting Eqs. (A1-9) and (A1-11) into Eq. (A1-7), we have

$$\sigma_{output}^2 = (1/2\pi)\int_{-\infty}^{\infty} |G(j\omega)|^2 W_{input}(\omega)\, d\omega. \qquad (A1-12)$$

The frequency characteristic of the ideal n^{th}-order differentiator is

$$G^{(n)}(j\omega) = (j\omega)^n. \qquad (A1-13)$$

According to Eq. (A1-13), the differentiator increases the higher Fourier frequencies and decreases the lower ones. This effect is the most significant for the high-order derivatives.

The FT of the AS, which contains normal noise, concentrates useful information in the low-frequency harmonics, while the high-frequency components mainly represent the noise. Therefore, the high-order derivatives are strongly disturbed by the noise. We readily explain this conclusion by substituting Eq. (A1-13) into Eq. (A1-12):

$$\sigma_{output}^2(n) = (1/2\pi)\int_{-\infty}^{\infty} \omega^{2n} W_{input}(\omega)\, d\omega. \qquad (A1-14)$$

If $W_{input}(\omega)_{\omega\to\infty} \sim \omega^{-q}$ $(q \le 2n)$, then the integral may diverge: $\sigma_{output}^2(n) \to \infty$. For the convergence, the stabilizing factor is introduced into the integrand [4].

Another way is clipping in the Fourier spectrum using a maximum Fourier frequency (ω_{max}) (Appendix A2).

A2. Discrete Fourier Transform

The chapter "Harmonic analysis" in a practical guide, written by C. Lanczos [5], is an accessible introduction to the numerical processing in the Fourier domain, which is performed using the Fourier series.

Suppose that $f(x)$ is a one-dimensional continuous periodic function with a period $T = 2\pi$. If $f(x)$ satisfies Dirichlet conditions (discontinuity points of the first kind only, and the finite number of local extrema at the interval $[a, a+T]$), then $f(x)$ is represented at the range $[-\pi, \pi]$ by the Fourier series:

$$f(x) = \frac{a_0}{2} + \sum_{k=1}^{\infty} a_k coskx + b_k sinkx, \qquad (A2-1)$$

where $a_k = (1/\pi)\int_{-\pi}^{\pi} f(x)coskxdx\,;\; b_k = (1/\pi)\int_{-\pi}^{\pi} f(x)sinkxdx.$

Eq. (A2-1) is valid in all points where $f(x)$ is continuous. At discontinuity point x_0, $f(x)$ is equal to the half-sum of its left and right limits at x_0:

$$f(x) = (f(x_0 - 0) + f(x_0 + 0))/2. \qquad (A2-2)$$

Consider the finite-dimensional analog of Eq. (A2-1), which is named the Discrete FT (DFT).

Suppose, that the one-dimensional analytical signal $\boldsymbol{F}$ is defined at the $L+1$ equidistant points of the abscissa axis (e.g., wavelength) in the range $[x_0, x_L]$; the sampling interval $h = (x_L - x_0)/L$. Then,

$$F_0 = F(x_0), F_1 = F(x_1), \dots, F_L = F(x_L). \qquad (A2-3)$$

It is common to say that the function $\boldsymbol{F}$ is defined in the time domain.

From a mathematical point of view, the spectrum can be formally extended to the region of negative arguments $(x_{-1}, \dots, x_{-L})$ in an even or odd way. Below, we will consider the latter case:

$$F(-x) = -F(x). \qquad (A2-4)$$

Also, suppose that F is the periodic function with the period $[-x_L, x_L]$ (Fig. A2-1). Let us approximate this function by a series of complex exponents ("harmonics"):

$$\hat{F}(x) = \sum_{k=-m}^{m} c_k \exp(jk\pi x/L), \qquad (A2-5)$$

where $m \le L$; $j = \sqrt{-1}$;

$$c_k = (1/2L)\sum_{\alpha=-L}^{L} F(x_\alpha)\exp(-jk\pi\alpha/L). \qquad (A2-6)$$

The first and the last terms of the sum in Eq. (A2-6) have the weights 0.5. The coefficients c_k, which are known as *complex Fourier harmonics*, represent the Fourier spectrum (transform) of the function $\boldsymbol{F}$. In other words, $\boldsymbol{F}$ is defined in the frequency domain.

If $m = L$, then the values obtained by Eq. (A2-5) are $\hat{F}(x_\alpha) = F(x_\alpha)$. Eq. (A2-6) is the sum of the real and the imaginary parts according to Euler's formula:

$$c_k = a_k - jb_k, \tag{A2-7}$$
where $a_k = (1/2L)\sum_{\alpha=-L}^{L} F(x_\alpha)\cos(k\pi\alpha/L)$;
$b_k = (1/2L)\sum_{\alpha=-L}^{L} F(x_\alpha)\, sin(k\pi\alpha/L)$.

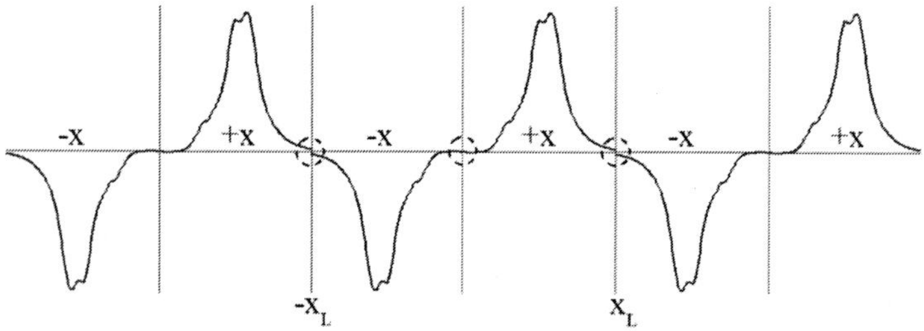

Figure A2-1. Representation of the non-periodic function as the periodic function.

If $F(x)$ is the odd function (Eq. (A2-4)), then
$$\hat{F}(x) = \sum_{k=1}^{m-1} b_k \sin(k\pi x/L), \tag{A2-8}$$
where $b_k = (1/L)\sum_{\alpha=1}^{L-1} F(x_\alpha)\, sin(k\pi\alpha/L)$.
The coefficients c_k and their real and imaginary parts are readily obtained by the MATLAB function, fft.m, using the fast algorithm named *"The Fast Fourier Transform"*.

The power density Fourier spectrum (or for short, the power spectrum) is
$$P(k) = (a_k^2 + b_k^2)/L^2. \tag{A2-9}$$
FT has some essential drawbacks: signal periodicity assumption and the Gibbs phenomenon [1]. The last fact is due to the jump discontinuity of function ***F*** in its "junction points" (Eq. (A2-2) (circles in Fig. A2-1). This discontinuity causes the weak convergence of the Fourier series. Therefore, the restored signal obtained by the inverse FT (IFT) contains oscillations. The same distortions are inherent in the Fourier-smoothed data for the truncated Fourier series (the high-frequency harmonics are rejected, $m \ll L$ in Eq. (A2-5)). To improve convergence, it is necessary that $F_0 = F_L$. To meet this condition, the linear trend (a baseline, BL) (Eq. (A2-10)) is subtracted from function F.
$$BL = a + bx, \tag{A2-10}$$

where $a = F_0$; $b = (F_L - F_0)/x_L$. This method is most suitable for high-order derivative spectroscopy since the derivatives of the order $n \geq 2$ suppress the trend.

While studying the non-stationary processes, new smart transform algorithms were suggested to overcome the drawbacks of FT [6].

In the main text, we will study the filtering properties of the digital filters using DFT.

Panels a and b (Fig. A2-2) show the absolute DFT spectra of the derivatives of the Gaussian and Lorentzian peaks, respectively. Generally speaking, graphical data demonstrates that the spectra broaden while increasing the order of the derivatives. The impact of factor ω^n on the DFT-derivative spectra explains this result. The DFT of the Lorentzian is significantly broader than that of the Gaussian. Approximately eight FT harmonics are needed to represent the Gaussian data (Fig. A2-2a). This number must be more than three times larger for the Lorentzian curves (Fig. A2 - 2b).

A3. Splines

The $(2t + 1)^{th}$-order degree polynomial spline of degree $2t$ consists of the $N + 1$ piecewise continuous polynomials jointed at the N points are known as knots or nodes. The spline continuity at the nodes must be up to the $(2t - 1)^{th}$-order derivatives. For example, consider the interpolation cubic spline [7].

Suppose that a function $F(\lambda)$ is defined at N knots λ_i given in the interval $[\lambda_a, \lambda_b]$:

$$\lambda_a = \lambda_1 < \lambda_2 <\ldots< \lambda_{N-1} < \lambda_N = \lambda_b. \qquad (A3-1)$$

$F_N(\lambda)$ is named the cubic spline (Fig. A3-1) with knots (Eq. (A3-1)) if the following conditions are valid:

1. In each interval $[\lambda_i, \lambda_{i+1}]$

$$F_{Ni}(\lambda) = \sum_{k=0}^{3} a_{ik}(\lambda - \lambda_i)^k, \qquad (A3-2)$$

where a_{ik}=const.

2. The continuity in the interval $[\lambda_a, \lambda_b]$ means that the functions $F_{N,i-1}$ and $F_{N,i}$ and their derivatives must be equal at each adjacent point λ_i (Fig. A3-1): that is,

$$F_{N,i-1}^{(n)}(\lambda_i) = F_{N,i}^{(n)}(\lambda_i), \qquad (A3-3)$$

where $n = 0, 1$, and 2.

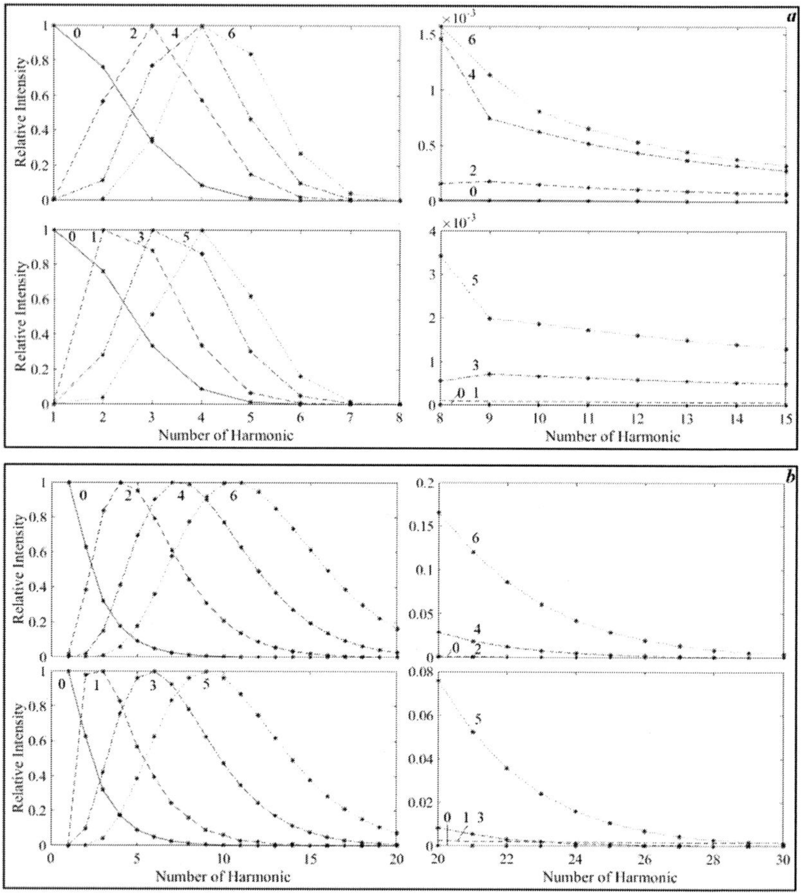

Figure A2-2. The relative intensities of the absolute DFT of the Gaussian (a) and Lorentzian (b) peaks, and corresponding n^{th}-order derivatives. The numbers near the curves are the n values.

The 1st- and 2nd-order derivatives of Eq. (A3-2) are

$$F_N'(\lambda) = a_{i1} + 2a_{i2}(\lambda - \lambda_i) + 3a_{i3}(\lambda - \lambda_i)^2; \qquad (A3-4)$$

$$F_N''(\lambda) = 2a_{i2} + 6a_{i3}(\lambda - \lambda_i). \qquad (A3-5)$$

The spline at the endpoints (λ_a and λ_b) is also calculated under particular conditions.

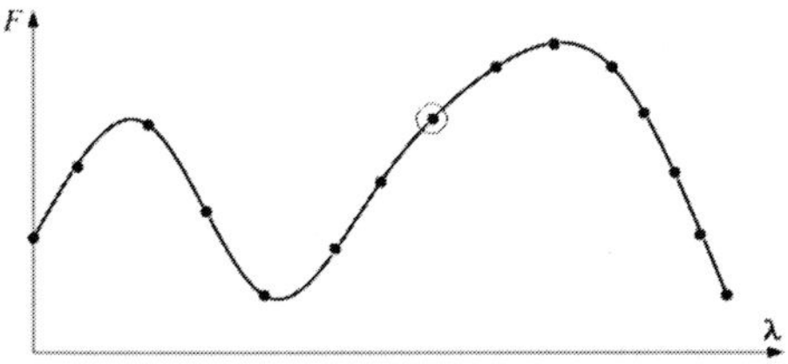

Figure A3-1. Cubic spline. (adapted from [7]).

Tutorial [7] gives a detailed mathematical description of the spline approximation.

The B-spline basis (B is short for basis spline) defines a set of knots and their coordinates. A random B-spline has an arbitrary number of knots and their random coordinates. A cardinal B-spline has a constant separation between knots. A linear combination of B-splines constricts all possible spline functions.

In the general case, selecting the spline parameters and the number of knots is a cumbersome mathematical problem [8], which is out of the remit of this book. Instead, we will focus on some remarks concerning spectroscopy problems [9].

To decrease the spline dimension and increase the signal-to-noise ratio of the smoothed data, the number of knots must be minimal. One of the knot's selection criteria requires the maximum smoothness of the spline. Mathematically this statement is equivalent to the minimizing of the maximum value over all of the pieces of the spline:

$$max_i \left\{\Delta\lambda_i{}^k \left\| F_N^{(k)}(\lambda_{i,i+1}) \right\| \right\}, \qquad (A3-6)$$

where $\Delta\lambda_i = \lambda_{i+1} - \lambda_i$; $\lambda_{i,i+1} = [\lambda_{i+1}, \lambda_i)$. Since $F_N^{(k)} = 0$ for the spline of the order k, the kth-order derivative must be independently calculated, e.g., using the SG method.

Another minimax solution follows from the following function:

$$[1/(N+1)] \sum_{\lambda=\lambda_a}^{\lambda_b} \left| F_N^{(k)}(\lambda_{i,i+1}) \right|^{1/k}. \qquad (A3-7)$$

Gans and Gill [9] also suggested the empirical function similar to Eq. (A3-7).

Since the interpolation spline is disturbed by the measurement noise, the regularized smoothing spline is often useful in practice.

According to the regularization principle, the smoothing spline $F_{N,\alpha}(\lambda)$ minimizes the functional [4]

$$\psi(F_{N,\alpha}) = \sum_{i=1}^{N} p_i^{-1} \left[F(\lambda_i) - F_{N,\alpha}(\lambda_i)\right]^2 + \alpha \sum_{i=1}^{N} \left| F''_{N,\alpha}(\lambda_i) \right|^2 . (A3-8)$$

The first term of $\psi(F_{N,\alpha})$ is the squared Euclidean norm of the discrepancy between measured and smoothed values. The weighting factors p_i are proportional to the noise dispersion. For normal noise, the first member is optimal. The second term of $\psi(F_{N,\alpha})$ defines the degree of the smoothing. We will discuss the estimation of the regularization parameter α in Chapter 2.6 and Appendix G.

Antonov [10] transformed the uniformly spaced spectra's coordinates, which were defined in the wavelength domain, to the uniform wavenumber scale by the spline interpolation. Furthermore, he applied the 4th-order differentiation SG filter. The spline interpolation restored the original wavelength scale.

References

1. Gray, R. M., Goodman, J. W. *Fourier Transforms. An Introduction for Engineers.* New York: Springer. 1995.
2. Shachaf, L., E120: Fourier Transform of simple functions. https://physics.bgu.ac.il/ COURSES/QuantumMechCohen/ExercisesPool /EXERCISES/ex_1200_sol_Y11.pdf
3. Lindgren, G. *Stationary Stochastic Processes: Theory and Applications*. Chapman and Hall/CRC. 2012.
4. Tikhonov, A. N., Arsenin. V. A. *Solution of Ill-posed Problems.* Washington: Winston & Sons. 1977.
5. Lanczos, C. *Applied analysis*. New York: Dover publications. 1956.
6. Nigmatullin, R. R., Gubaidullin, I. A. (2018). NAFASS: Fluctuation spectroscopy and the Prony spectrum for description of multi-frequency signals in complex systems. Communications in Nonlinear Science and Numerical Simulation, 56, 252-269.
7. Pollock, D. S. G., (1999). Smoothing with Cubic Splines. https://www. researchgate.net/publication/241171192_Smoothing_with_Cubic_Splines
8. Wang, Y. *Smoothing Splines: Methods and Applications.* Chapman and Hall/CRC. 2011,
9. Gans, P, Gill, J. B. (1984). Smoothing and differentiation of spectroscopic curves using spline functions. Applied Spectroscopy, 38, 370-376.
10. Antonov, L. (1997). Fourth derivative spectroscopy - a critical view. Analytica Chimica Acta, 349, 295-301.

APPENDIX B

DIFFERENTIATION OF THE PMG AND DOBOSZ FUNCTIONS

B1. The PMG function

$F_{PMG}(y) = exp\ (-y^2/(1+\tau y)^2),$ $(B1-1)$

where τ is the asymmetry parameter.

The form that is more suitable for differentiation:

$F_{PMG}(x) = \exp(ARG),$ $(B1-2)$

where $ARG = -z^2$; $z = y/(1+\tau y) = (1-B)/\tau$; $B = 1/(1+\tau y)$.

The 1st-order derivative:

$F_{PMG}^{(1)} = F_{PMG} ARG^{(1)}.$ $(B1-3)$

For further differentiation ($n \geq 2$), we use the general Leibniz rule:

$F_{PMG}^{(n)} = \sum_{k=0}^{n-1} C_{n-1}^{k}\, F_{PMG}^{(n-k-1)} ARG^{(k+1)},$ $(B1-4)$

where $C_n^k = n!/[(n-k)!\,k!]$.

Function	Equation	Number
$F_{PMG}^{(2)}$	$F_{PMG}^{(1)} ARG^{(1)} + F_{PMG} ARG^{(2)}$	$B1-5$
$F_{PMG}^{(3)}$	$F_{PMG}^{(2)} ARG^{(1)} + 2F_{PMG}^{(1)} ARG^{(2)} + F_{PMG} ARG^{(3)}$	$B1-6$
$F_{PMG}^{(4)}$	$F_{PMG}^{(3)} ARG^{(1)} + 3F_{PMG}^{(2)} ARG^{(2)} + 3F_{PMG}^{(1)} ARG^{(3)} + F_{PMG} ARG^{(4)}$	$B1-7$
$ARG^{(1)}$	$-2zz^{(1)}$	$B1-8$
$ARG^{(2)}$	$-2\left[\left(z^{(1)}\right)^2 + zz^{(2)}\right]$	$B1-9$
$ARG^{(3)}$	$-2[3z^{(1)}z^{(2)} + zz^{(3)}]$	$B1-10$
$ARG^{(4)}$	$-2\left[3\left(z^{(2)}\right)^2 + 4z^{(1)}z^{(3)} + zz^{(4)}\right]$	$B1-11$

The nth-order derivatives of z and B are:

$$z^{(n)} = -B^{(n)}/\tau; \qquad (B1-12)$$

$$B^{(n)} = -(-\tau)^n n!/(1+\tau y)^{n+1}. \qquad (B1-13)$$

We use these equations in the MATLAB function pmgDerivative.m (Appendix SW9).

B2. The Dobosz function

The Dobosz function in the form more suitable for differentiation:

$$F_D = exp(-\tau(1-tan^{-1}y))/(1+y^2) = R_y^{(0)}K_y^{(1)}, \qquad (B2-1)$$

where $R_y^{(0)} = Cexp(K_y)$; $C = exp(-\tau)\,/\,\tau$; $K_y = \tau tan^{-1}y$; $K_y^{(1)} = \tau/(1+y^2) = \tau F_L^{(0)}$; $F_L^{(0)} = 1/(1+y^2)$ is the Lorentzian function.

Similar to Appendix B1, we use the general Leibniz rule for differentiation.

Function	Equation	Number
$F_D^{(1)}$	$R_y^{(1)}K_y^{(1)} + R_y^{(0)}K_y^{(2)}$	$B2-2$
$F_D^{(2)}$	$R_y^{(2)}K_y^{(1)} + 2R_y^{(1)}K_y^{(2)} + R_y^{(0)}K_y^{(3)}$	$B2-3$
$F_D^{(3)}$	$R_y^{(3)}K_y^{(1)} + 3R_y^{(2)}K_y^{(2)} + 3R_y^{(1)}K_y^{(3)} + R_y^{(0)}K_y^{(4)}$	$B2-4$
$F_D^{(4)}$	$R_y^{(4)}K_y^{(1)} + 4R_y^{(3)}K_y^{(2)} + 6R_y^{(2)}K_y^{(3)} + 4R_y^{(1)}K_y^{(4)} + R_y^{(0)}K_y^{(5)}$	$B2-5$
$R_y^{(1)}$	$R_y^{(0)}K_y^{(1)}$	$B2-6$
$R_y^{(2)}$	$R_y^{(1)}K_y^{(1)} + R_y^{(0)}K_y^{(2)}$	$B2-7$
$R_y^{(3)}$	$R_y^{(2)}K_y^{(1)} + 2R_y^{(1)}K_y^{(2)} + R_y^{(0)}K_y^{(3)}$	$B2-8$
$R_y^{(4)}$	$R_y^{(3)}K_y^{(1)} + 3R_y^{(2)}K_y^{(2)} + 3R_y^{(1)}K_y^{(3)} + R_y^{(0)}K_y^{(4)}$	$B2-9$

Using derivatives of the Lorentzian (Chapter 1.2), we obtain

$$K_y^{(n)} = \tau F_L^{(n-1)}. \qquad (B2-10)$$

We use these equations in the MATLAB function difDobosz.m (Appendix SW9).

APPENDIX C

NOISE

It is generally accepted to classify noises according to their frequency domain properties, using the power spectral density (PSD) [1]. The PSD of a signal $x(t)$ is

$$S_{xx}(\omega) = \lim_{T\to\infty} \boldsymbol{E}\{|\tilde{x}(\omega)|^2\}, \qquad (C-1)$$

where $\boldsymbol{E}$ stands for the mathematical expectation (mean); ω is the angular (Fourier) frequency; $\tilde{x}(\omega) = \left(1/\sqrt{T}\right)\int_0^T x(t)\exp(-j\omega t)\,dt$ is the amplitude spectral density (the integral is the truncated Fourier transform of $x(t)$ in the interval $[0, T]$); $j = \sqrt{-1}$.

It is assumed in practical applications that the white noise's PSD is constant in some frequency range. White normal (Gaussian) noise has a normal distribution in the time domain with a zero mean.

The PSD of the colored noise depends on the Fourier frequency [1, 2]:

$$S_{xx}(\omega) \sim |\omega|^{\alpha}, \qquad (C-2)$$

where $\alpha = -1, -2$, and 1 for pink, red, and blue noise, respectively.

If $\alpha \leq -1$, then the noise is non-stationary [1, 2].

Figure C-1 illustrates Eq. (C-2). The MATLAB operator (abs(fft(noise)) transformes the colored noise, which is generated by software [2], into the frequency domain.

References

1. Stoica P., Moses, R. *Spectral analysis of signals*. Prentice Hall, New Jersey. 2005.

2. Zhivomirov, H. (2018). A method for colored noise generation. Romanian Journal of Acoustics and Vibration, 15, 14-19.

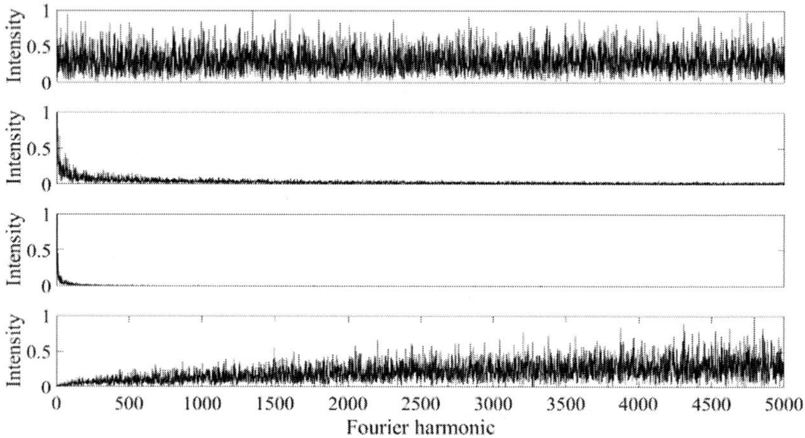

Figure C-1. The absolute Fourier spectra of white, pink, red, and blue noise, from the top to the bottom panels, respectively.

APPENDIX D

ESTIMATION OF THE INFORMATION MEASURE OF THE RESOLUTION

Gaussian doublet

Consider the symmetrical doublet (Eq. (1.4-1)). The information measure to distinguish between the doublet components (Eq. (1.4-5)) is

$$\mu = (2/\pi)\int_{-\infty}^{+\infty}\left|\tilde{F}_1\right|^2[1-\cos(p\delta/2)]^2/G(\omega)\,d\omega. \qquad (D1-1)$$

For white noise, $G(\omega) = \pi\sigma^2/\omega_0$. Using the FT of the Gaussian peak $\tilde{F}_1 = C_G F_0 w exp(-p^2/4\beta_G^2)$, we have

$$\mu = (2p_0\xi^2/\pi^2)C_G^2\int_{-\infty}^{+\infty}\Psi\,[1-\cos(p\delta/2)]^2 dp, \qquad (D1-2)$$

where $p_0 = w\omega_0$; $C_G = \sqrt{\pi}/\beta_G$; $\Psi = exp(-p^2/2\beta_G^2)$; $\beta_G = 2\sqrt{ln2}$; $\xi = F_0/\sigma$ is the signal-to-noise ratio; F_0 is the peak amplitude.

The integrand of Eq. (D1-2) is a trinomial:

$$\Psi[1-\cos(p\delta/2)]^2 = \Psi[1-2\cos(p\delta/2)+cos^2(p\delta/2)] =$$
$$0.5\Psi[3-4\cos(p\delta/2)+\cos(p\delta)]. \qquad (D1-3)$$

By substituting β_G into Eq. (D1-3) and using the integration formulas [1], we obtain the integrals of the trinomial:

$$1.5\int_{-\infty}^{\infty}\Psi dp = 3\sqrt{2\pi ln2}. \qquad (D1-4)$$

$$-2\int_{-\infty}^{\infty}\Psi\cos(p\delta/2)\,dp = -4\sqrt{2\pi ln2}\exp(-\delta^2 ln2/2). \qquad (D1-5)$$

$$0.5\int_{-\infty}^{\infty}\Psi\cos(p\delta)\,dp = 2\sqrt{2\pi ln2}\exp(-\delta^2 2ln2). \qquad (D1-6)$$

Finally,

$$\mu = (2p_0\xi^2/\pi^2)C_G^2\Phi =$$
$$p_0\xi^2 T_G[3-4\,exp(-\delta^2 ln2/2)+exp(-\delta^2 2ln2)], \qquad (D1-7)$$

where $T_G = 1/\sqrt{2\pi ln2}$; $\Phi = 3\sqrt{2\pi ln2} - 4\sqrt{2\pi ln2}\,exp(-\delta^2 ln2/2) + \sqrt{2\pi ln2}\exp(-\delta^2 2ln2)$.

For the strongly overlapped doublet ($\delta \ll 1$), using the quadratic approximation of the exponent [2]: $exp(x) \approx 1 - x + x^2/2$, from Eq. (D1-7), we have

$$\mu \cong T_G p_0\xi^2[-2(\delta^2 ln2/2)^2 + 0.5(\delta^2 2ln2)^2] = Q_G p_0\xi^2\delta^4, \qquad (D1-8)$$

where $Q_G = 1.5(ln2)^2/\sqrt{2\pi ln2}$=0.3453.

General solution for the asymmetrical doublet

The integrand of Eq. (1.4-5) is

$$|(\tilde{s}_1 - \tilde{s}_2)|^2 = (CwQ(p))^2\Phi^2 = 2(CwQ(p))^2[A - Bcos(p\delta/2) + Ccos(p\delta)], \quad (D1-9)$$

where $\Phi = |(1 + C) - exp(-jp\delta/2) - Cexp(jp\delta/2)|$; $A = 1 + C + C^2, B = (1 + C)^2, C = Rr$.

We perform integration similar to Eqs. (D1-4)-(D1-6):

$$\mu = p_0\xi^2 T_G[A - B\exp(-\delta^2 ln2/2) + C\exp(-\delta^2 2ln2)]. \quad (D1-10)$$

Lorentzian doublet

For the white noise $G(\omega) = \pi\sigma^2/\omega_0$ and $\tilde{F}_1 = C_L F_0 w exp(-|p|/2)$, similar to the previous section (but in the new integration limits), from Eq. (1.4-1) we have

$$\mu = 0.5p_0\xi^2 \int_0^{+\infty} exp(-p)\,[3 - 4\cos(p\delta/2) + \cos(p\delta)]dp. \quad (D1-11)$$

Using the integration formulas [1], we obtain

$$3\int_0^{\infty} exp(-p)\,dp = 3. \quad (D1-12)$$

$$-4\int_0^{\infty} exp(-p)\cos(p\delta/2)\,dp = -4/(1 + \delta^2/4). \quad (D1-13)$$

$$\int_0^{\infty} exp(-p)\cos(p\delta)\,dp = 1/(1 + \delta^2). \quad (D1-14)$$

Finally,

$$\mu = 0.5p_0\xi^2[3 - 4/(1 + \delta^2/4) + 1/(1 + \delta^2)]. \quad (D1-15)$$

For the strongly overlapped doublet ($\delta < 0.5$), using the quadratic approximation [2]: $1/(1 + x^2) \approx 1 - x^2 + x^4$, we have from Eq. (D1-15) we have

$$\mu \cong 0.375p_0\xi^4. \quad (D1-16)$$

General solution for the asymmetrical doublet

Similar to the general solution for Gaussian doublet, we have

$$\mu = 0.5p_0\xi^2 \int_0^{+\infty} exp(-p)\,[A - B\cos(p\delta/2) + C\cos(p\delta)]dp. \quad (D1-17)$$

Similar to Eqs. (D1-12)-(D1-14), after integration we obtain

$$\mu = 0.5p_0\xi^2[A - B/(1 + \delta^2/4) + C/(1 + \delta^2)]. \quad (D1-18)$$

References

1. Gradshteyn, I. S., Ryznik, I. M. *Table of Integrals, Series, and*

Products. 7th ed., Academic Press. 2007.
2. Courant, R. *Differential and Integral Calculus*. Vol. 1. Ishi Press. 2013.

APPENDIX E

RC CIRCUITS

E1. Solution of Eq. (2.1-4)

Let

$U'_{out} + (1/\tau)U_{out} = U'_{in},$ $(E1-1)$

where $U'_{in} = -2(t - t_0)exp[-(t - t_0)^2]$.

Let

$U_{out} = u(t)v(t),$ $(E1-2)$

then from Eq. (E1-1) we have

$v(u' + u/\tau) + uv' = U'_{in}.$ $(E1-3)$

Suppose that $u' + u/\tau = 0$,

then $u = exp(-t/\tau)$. $(E1-4)$

By substituting Eq. (E1-4) into (E1-3), we have

$v' = exp(t/\tau)\, U'_{in}.$ $(E1-5)$

Finally, from Eqs. (E1-2), (E1-4), (E1-1), and integral of Eq. (E1-5), we obtain

$U_{out} = -2\,exp(-t/\tau)\int_0^t (t - t_0)\,exp(-(t - t_0)^2 + t/\tau)\,dt.$ $(E1-6)$

Eq. (E1-6) was integrated numerically.

E2. Frequency characteristic of a linear electronic device

In the linear electronic devices (e.g., RC circuits), there is a linear relationship between the output (U_{out}) and input (U_{in}) signals (Fig. 2.1-1). The mathematical expression of the device frequency characteristic is

$G(p) = U_{out}(p)/U_{in}(p),$ $(E2-1)$

where $p = j\omega$; $j = \sqrt{-1}$; ω is the angular (Fourier) frequency.

Let us calculate the frequency characteristics of the differentiator and integrator RC-circuits (Fig. 2.1-1, a and c, respectively), using complex impedances (Laplace-domain circuit analysis).

In the first case, according to Ohm's law, we have

$$G^{(1)}(p) = \frac{\check{I}\check{R}}{\check{I}(\check{R} + \check{C})} = \frac{\tau_{RC}p}{1 + \tau_{RC}p} = A^{(1)}\, exp(j\varphi), \qquad (E2-2)$$

where $\check{I}, \check{R},$ and $\check{C} = 1/pC$ are the complex impedances of the current, active, and the capacitor resistance, respectively; τ_{RC} is the time constant; $A^{(1)} = \tau^2\omega^2/(1 + \tau^2\omega^2)$, and $\varphi = atan(1/\tau\omega)$ are the amplitude and the phase angle of the exponential form of a complex number, respectively.

For the integration circuit (Fig. 2.1-1 b),

$$G^{(0)}(p) = \frac{\check{I}\check{C}}{\check{I}(\check{R} + \check{C})} = \frac{1}{1 + \tau_{RC}p} = A^{(0)}\, exp(j\varphi), \qquad (E2-3)$$

where $A^{(0)} = 1/(1 + \tau^2\omega^2)$; $\varphi = atan(\tau\omega)$.

Eqs. (E2-2) and (E2-3) show that the RC circuits disturb the input signal phase due to the exponential terms. These terms introduce inertial distortions, which depend on the angular frequency. If the frequency ω increases, then the amplitude $A^{(1)}$ (Eq. (E2-2)) increases approaching one, but $A^{(0)}$ (Eq. (E2-3)) decreases to zero. So, the differentiation and integration give more weight to the higher and lower Fourier harmonics, respectively.

APPENDIX F

COMBINATION OF THE TRIANGLE MULTI-PASS MOVING AVERAGE FILTER WITH THE 2ND-ORDER DIFFERENTIATOR

The multi-pass filter, which was obtained by the self-convolution of the N points moving average filter, repeated $r = N$ times, has a triangle form [1]. For example, for $r = N = 5$:

1	1	1	1	1				
	1	1	1	1	1			
+		1	1	1	1	1		
			1	1	1	1	1	
				1	1	1	1	1
1	2	3	4	5	4	3	2	1

Generally, for $r = N$, the width of the multi-pass filter is $2r - 1$. The normalization factor is

$$M_{mpmv} = \sum_{\alpha=1}^{2r-1} W_\alpha = \sum_{\alpha=1}^{r} W_\alpha + \sum_{\alpha=1}^{r-1} W_\alpha =$$

$$\{r(1+r) + (r-1)[1+(r-1)]\}/2 = r^2. \qquad (F-1)$$

The noise coefficient:

$$K_{mpmv} = \left(1/M_{mpmv}^2\right) \sum_{\alpha=1}^{2r-1} W_\alpha^2 =$$

$$(1/r^4)\{(1+2^2+\cdots+r^2) + (1+2^2+\cdots+(r-1)^2)\} =$$

$$(1/r^4)(1/6)\{r(r+1)(2r+1) + (r-1)r(2r-1)\} =$$

$$(1/3)(2r^2+1)/r^3 \approx 0.67/r. \qquad (F-2)$$

Consider the 2nd-order differentiator based on the central differences method (Eq. (2.2-2)) applied to the data smoothed by the triangle filter. For example, $r = N = 5$:

1	2	3	4	5	4	3	2	1		
+	-2	-4	-6	-8	-10	-8	-6	-4	-2	
		1	2	3	4	5	4	3	2	1
1	0	0	0	0	-2	0	0	0	0	1

The normalization factor:

$$M_{combdif} = M_{mpmv} = r^2. \qquad (F-3)$$

The noise coefficient:

$$K_{combTrdif} = (1 + 2^2 + 1)/M_{combdif}^2 = 6/r^4. \qquad (F-4)$$

References

1. Smith, S. W. *The Scientist and Engineer's Guide to Digital Signal Processing*. 2nd ed. San Diego: California Technical Publishing. 1997-1999.

APPENDIX G

TIKHONOV REGULARIZATION

Suppose that we want to approximate the nth-order derivative (u) of some function (f), measured experimentally, by a smooth function $g^{(n)}$:

$$u(t) \approx g^{(n)}(t), \qquad (G-1)$$

where $t = \{t_1, t_2, .., t_N\}$.

E.g., in the SG method, g is a polynomial. However, the polynomial parameters are not unique.

There are many possible solutions to the approximation (Eq. (G-1)), which belongs to the ill-posed problems.

Tikhonov regularization involves the well-known fact that differentiation is the inverse operation of computing an integral. According to [1], the derivative u is estimated from the integral equation (convolution):

$$\int_0^t K(t-\tau)\, u(\tau) d\tau = f(t), \qquad (G-2)$$

where $K(t-\tau) = (t-\tau)^{n-1}/(n-1)!$ is the kernel.

Consider a simple example when $f(t) = t^3$ is a noise-free function.

$n = 3$. $K(t-\tau) = (t-\tau)^2/2$. $u(\tau) = f^{(3)} = 6$.

$3\int_0^t (t-\tau)^2\, d\tau = 3[t^3 - t^3 + t^3/3] = f(t)$.

If $f(t)$ is contaminated by noise, the approximated function $\hat{f}(t) = \int_0^t K(t-\tau)\, \hat{u}(\tau) d\tau$ is obtained by substituting the solution $\hat{u}(\tau)$ into Eq. (G-2).

The regularized solution of Eq. (G-2) in the Fourier domain has an elegant form [1]:

$$\tilde{u}_\alpha(\tau) = \frac{1}{2\pi} \int_{-\infty}^{\infty} \frac{\tilde{K}(-\omega)\tilde{f}(\omega)\exp(-j\omega t)}{\left|\tilde{K}(\omega)\right|^2 + \alpha M(\omega)} d\omega, \qquad (G-3)$$

where tilde is the FT symbol; $\tilde{K}(\omega) = (j\omega)^n; j = \sqrt{-1}$; ω is the angular frequency; α is the regularization parameter; $M(\omega) = \sum_{k=0}^{p} q_k \omega^{2k}$ is the stabilizer, generally, $q_k > 0$.

The solution (Eq. (G-3)) minimizes the functional [1]:

$$\Psi_\alpha = \int_{-\infty}^{\infty} \left(\hat{f}(t) - f(t)\right)^2 dt + \alpha \int_{-\infty}^{\infty} M(\omega) |\tilde{u}_\alpha(\omega)|^2 \, d\omega. \qquad (G4-4)$$

The authors of the articles [2, 3] obtained derivative spectra in different regions using Tikhonov regularization in the time domain. They estimated the regularization parameter by the cross-validation method. It is not easy to give significant preference to the regularization in comparison with other methods. However, some particular cases described in the literature reveal the advantages of regularized differentiation. The remarkable review [4], which is dedicated to the regularization problem in numerical differentiation, will be useful for interested readers.

References

1. Tikhonov, A. E., Arsenin, V. Y. *Solutions of Ill-posed Problems.* John Wiley & Sons, New York, 1977.
2. Yeow, Y. L., Leong Y. K. (2005). A general computation method for converting normal spectra into derivative spectra. Applied Spectroscopy, 59, 584 - 592.
3. Yeow, Y. L., Azali, S., Ow, S., Y., et al. (2005). Evaluating the third and fourth derivatives of spectral data. Talanta, 68, 156-164.
4. Tinguan, X. Regularization of numerical differentiation: methods and applications. http://sourcedb.igg.cas.cn/cn/zjrck/200907/ W020100801406250190094.pdf

APPENDIX H

ESTIMATION OF THE OPTIMAL FILTER FOR A LINEAR SIGNAL

Given matrix $\boldsymbol{\Phi}_f = \begin{vmatrix} f_1 & f_2 & \cdots & f_N \\ f_2 & f_3 & \cdots & f_{N+1} \\ \vdots & \vdots & \vdots & \vdots \\ f_{L_N} & f_{L_N+1} & \cdots & f_L \end{vmatrix}$, where $f_i = f_0 + hi,$ $L_N = L - N + 1.$

For simplicity, consider the case $h = 1$.

According to the definition suggested by J. Domsta [1],

$\boldsymbol{M} = \boldsymbol{\Phi}_f^T \boldsymbol{\Phi}_f + z\boldsymbol{I},$ where $z = L_N \sigma^2$; $\boldsymbol{I}$ is the identity matrix.

$\boldsymbol{\Phi}_f^T \boldsymbol{\Phi}_f = (a\boldsymbol{U} + b\boldsymbol{V})(a\boldsymbol{U} + b\boldsymbol{V})^T + c^2 \boldsymbol{U}\boldsymbol{U}^T,$

where $a = \epsilon_0 \sqrt{NL_N}; \epsilon_0 = k_0 + j_0 - 1;\ k_0 = (1 + N)/2;$ $j_0 = (1 + L_N)/2;\ b = h\sqrt{K_2 L_N};\ c = h\sqrt{NJ_2};$

$K_2 = \boldsymbol{v}_k^T \boldsymbol{v}_k, \boldsymbol{v}_k = \begin{vmatrix} 1 \\ 2 \\ \vdots \\ N \end{vmatrix} - k_0 \mathbf{1}_N;\ \mathbf{1}_N = \begin{vmatrix} 1 \\ 1 \\ \vdots \\ 1 \end{vmatrix};\ c = h\sqrt{J_2 N}\ ;\ J_2 = \boldsymbol{v}_j^T \boldsymbol{v}_j, \boldsymbol{v}_j =$ $\begin{vmatrix} 1 \\ 2 \\ \vdots \\ L_N \end{vmatrix} - k_0 \mathbf{1}_{L_N};\ \boldsymbol{U} = \mathbf{1}_N/\sqrt{N};\ \boldsymbol{V} = \boldsymbol{v}_k/\sqrt{K_2}.$

Inverse matrix [1]:

$\boldsymbol{M}^{-1} = (1/z)(\boldsymbol{I} - \boldsymbol{U}\boldsymbol{U}^T - \boldsymbol{V}\boldsymbol{V}^T) + \boldsymbol{O}_1 + \boldsymbol{O}_2 - \boldsymbol{O}_3,$

where $\boldsymbol{O}_1 = R_z(1,1)\boldsymbol{U}\boldsymbol{U}^T; \boldsymbol{O}_2 = R_z(1,2)(\boldsymbol{U}\boldsymbol{V}^T + \boldsymbol{V}\boldsymbol{U}^T) + R_z(2,2)\boldsymbol{I};$

$\boldsymbol{O}_3 = R_z(2,2)(\boldsymbol{I} - \boldsymbol{V}\boldsymbol{V}^T);$

$\boldsymbol{R}_z = \boldsymbol{Q}_z/det\boldsymbol{Q}_z, \boldsymbol{Q}_z = \begin{bmatrix} b^2 + z & -ab \\ -ab & a^2 + b^2 + z \end{bmatrix}.$

Although the factor $1/z$ is enormous, the terms $\boldsymbol{O}_i$ cannot be neglected.

According to (Eq. 2.5-7),

$$\boldsymbol{w}^T = (1/z)(\boldsymbol{I} - \boldsymbol{U}\boldsymbol{U}^T - \boldsymbol{V}\boldsymbol{V}^T + \boldsymbol{O}_1 + \boldsymbol{O}_2 - \boldsymbol{O}_3)\boldsymbol{T}, \qquad (H-1)$$

where $\boldsymbol{T} = \boldsymbol{\Phi}_F^T \boldsymbol{F}_{red}^T$.

In what follows, we will prove that all components of the vector $\boldsymbol{w}^T$ are constants.

Being normalized to the unit sum, the components of $\boldsymbol{w}^T$ give the coefficients of the moving average filter. Therefore, the values of these constants are not important.

The ith component of the vector $\boldsymbol{T}$:

$T_i = \sum_{k=1}^{L_N}[k + (i-1)][k + (N-1)/2] = T_1 + (i-1)\Delta T$,

where $T_1 = (L_N/2)(L_N + 1)[(2L_N + 1)/3 + (N-1)/2], \Delta T =$

$(L_N/2)(L_N + N)$.

Consider impact of the terms $(\boldsymbol{I} - \boldsymbol{UU}^T - \boldsymbol{VV}^T + \boldsymbol{O}_1 + \boldsymbol{O}_2 - \boldsymbol{O}_3)$ (Eq. (H-1)) on the vector $\boldsymbol{w}^T$. Since all components of the matrix $\boldsymbol{UU}^T$ are constants $(1/N)$, all components of the vector $\boldsymbol{UU}^T\boldsymbol{T}$ are also constants. This result is valid for the term $\boldsymbol{O}_1\boldsymbol{T}$.

The ith component of the vector $\boldsymbol{Y} = (\boldsymbol{I} - \boldsymbol{VV}^T)\boldsymbol{T}$: $Y_i =$

$T_i - \{[i - (N+1)/2]/K_2\}\sum_{j=1}^{N}[j - (N+1)/2]\,[T_1 + (j-1)\Delta T]=$

$T_i - [i - (N+1)/2]\Delta T(N/12)(N^2 - 1)/K_2 = T_1 - [(N+3)/2]\Delta T$

does depend on i. This result is valid for the vector $\boldsymbol{O}_3\boldsymbol{T}$.

The vector $\boldsymbol{Z} = \boldsymbol{O}_2\boldsymbol{T} =$

$R_z(1{,}2)(\boldsymbol{UV}^T + \boldsymbol{VU}^T + \boldsymbol{I})\boldsymbol{T} + [R_z(2{,}2) - R_z(1{,}2)]\boldsymbol{IT} = \boldsymbol{Z}_1 + \boldsymbol{Z}_2$,

where $\boldsymbol{Z}_1 = R_z(1{,}2)(\boldsymbol{UV}^T + \boldsymbol{VU}^T + \boldsymbol{I})\boldsymbol{T}$; $\boldsymbol{Z}_2 = [R_z(2{,}2) - R_z(1{,}2)]\boldsymbol{IT}$.

Since all rows of both matrices

$$\boldsymbol{UV}^T = \left(1/\sqrt{NK_2}\right)\left\{\begin{vmatrix} 1 & 2 & \cdots & N \\ 1 & 2 & \cdots & N \\ \vdots & \vdots & \vdots & \vdots \\ 1 & 2 & \cdots & N \end{vmatrix} - k_0 \begin{vmatrix} 1 & 1 & \cdots & 1 \\ 1 & 1 & \cdots & 1 \\ \vdots & \vdots & \vdots & \vdots \\ 1 & 1 & \cdots & 1 \end{vmatrix}\right\},$$

and $\boldsymbol{VU}^T = (\boldsymbol{UV}^T)^T$ are identical, vector $\boldsymbol{Z}_1$ also contains identical components. Since T_i does not depend on i, vector $\boldsymbol{Z}_2$ also contains identical components. So, we conclude that the moving average filter is optimal for the linear signal.

Reference

1. https://www.researchgate.net/post/How_can_one_analytically_inverse_the_sum_of _the_singular_and_diagonal_matrices/1 [accessed July 1, 2020]

APPENDIX I

TABLES TO CHAPTER 2.6

Table 2.6-3. Parameters $E_s^{(0)}$ and $E_i^{(0)}$ (Eq. (2.6-14)) for the Gaussian peak

p	ξ_0				
	10	**20**	**50**	**100**	**200**
2	152.493	166.318	185.021	200.309	216.504
	0.2727	0.1412	0.0624	0.0334	0.0180
3	144.071	153.788	166.514	176.297	186.434
	0.1853	0.0932	0.0385	0.0199	0.0103
4	141.197	149.366	159.688	167.348	174.849
	0.1511	0.0737	0.0292	0.0147	0.0076
5	140.033	147.451	156.626	163.213	169.573
	0.1344	0.0641	0.0246	0.0122	0.0061
6	139.518	146.510	155.014	161.088	166.957
	0.1253	0.0590	0.0222	0.0107	0.0052

Table 2.6-4. Similar to Table 2.6-3 but for the Lorentzian peak

p	ξ_0				
	10	**20**	**50**	**100**	**200**
2	170.313	188.364	212.748	231.915	251.955
	0.3527	0.1816	0.0769	0.0405	0.0215
3	164.908	179.765	198.953	213.286	227.618
	0.2784	0.1376	0.0549	0.0277	0.0141
4	163.553	177.305	194.660	207.398	219.715
	0.2526	0.1223	0.0475	0.0233	0.0116
5	163.250	176.487	193.058	205.079	216.586
	0.2410	0.1155	0.0441	0.0214	0.0105
6	163.280	176.265	192.377	203.912	215.182
	0.2351	0.1120	0.0425	0.0206	0.0099

Table 2.6-5. Parameters $K_s^{(0)}$ and $K_i^{(0)}$ (Eq. (2.6-16)) for the Gaussian peak

H	ξ_0				
	10	**20**	**50**	**100**	**200**
0.005	3.6724	3.7560	3.8380	3.8870	3.9237
	2.4327	2.8956	3.5536	4.0709	4.6087
0.01	3.6222	3.7176	3.8105	3.8637	3.9062
	2.2111	2.6590	3.2974	3.8064	4.3366
0.02	3.5657	3.6728	3.7776	3.8386	3.8863
	1.9989	2.4307	3.0564	3.5511	4.0691
0.05	3.4813	3.6055	3.7295	3.8005	3.8562
	1.7136	2.1382	2.7377	3.2179	3.7238

Table 2.6-6. Similar to Table 2.6-5 but for the Lorentzian peak

H	ξ_0				
	10	**20**	**50**	**100**	**200**
0.005	4.6198	4.8174	5.0363	5.1757	5.2957
	1.6065	1.9074	2.3391	2.6916	3.0656
0.01	4.5087	4.7226	4.9585	5.1094	5.2391
	1.4620	1.7535	2.1717	2.5092	2.8719
0.02	4.3879	4.6198	4.8748	5.0364	5.1753
	1.3227	1.6069	2.0111	2.3387	2.6910
0.05	4.2119	4.4713	4.7550	4.9327	5.0868
	1.1416	1.4148	1.8014	2.1179	2.4540

Table 2.6-7. Parameters $L_s^{(0)}$ and $L_i^{(0)}$ (Eq. (2.6-17)) for the Gaussian peak

H	p				
	2	**3**	**4**	**5**	**6**
0.005	0.9088	0.9712	1.0458	1.1365	1.2443
	7.7888	11.1764	14.6629	18.2004	21.7553
0.01	0.9149	0.9855	1.0683	1.1722	1.2927
	7.4478	10.7804	14.2050	17.6605	21.1267
0.02	0.9224	1.0038	1.0989	1.2137	1.3500
	7.1007	10.3634	13.7121	17.0843	20.4453
0.05	0.9394	1.0346	1.1492	1.2852	1.4376
	6.6051	9.7752	12.9967	16.2323	19.4662

Table 2.6-8. Similar to Table 2.6-7 but for the Lorentzian peak

H	*p*				
	2	**3**	**4**	**5**	**6**
0.005	0.9676	1.1429	1.3626	1.6060	1.8634
	8.7362	12.8297	16.9528	21.0907	25.2408
0.01	0.9871	1.1810	1.4180	1.6798	1.9556
	8.3203	12.2787	16.2558	20.2379	24.2230
0.02	1.0099	1.2248	1.4816	1.7586	2.0560
	7.8852	11.6912	15.5091	19.3346	23.1388
0.05	1.0483	1.2931	1.5789	1.8894	2.2081
	7.2632	10.8494	14.4283	17.9959	21.5740

Table 2.6-9. Parameters $M_c^{(0)}, M_s^{(0)}$ and $M_i^{(0)}$ (Eq. (2.6-18)) for the Gaussian peak

p	ξ_0				
	10	**20**	**50**	**100**	**200**
	x10^3	**x10^4**	**x10^4**	**x10^5**	**x10^6**
2	2.5410	1.0267	6.4798	2.5615	1.0285
	0.1478	0.02916	0.07232	0.01428	0.002852
	0.01078	0.001141	0.001224	0.0001285	0.00001347
3	2.8280	1.11354	6.7192	2.6819	1.0657
	0.1663	0.03192	0.07627	0.01505	0.002977
	0.01442	0.001508	0.001593	0.0001658	0.00001722
4	3.0332	1.1853	6.9841	2.8652	1.1011
	0.1840	0.03480	0.08096	0.01614	0.003100
	0.01818	0.001889	0.001978	0.0002049	0.000021118
5	3.3735	1.2818	7.5820	3.0477	1.1462
	0.2074	0.038354	0.08836	0.01725	0.003262
	0.02205	0.002283	0.002379	0.0002452	0.00002517
6	3.8948	1.4195	8.26155	3.1349	1.2068
	0.2373	0.04279	0.09703	0.01821	0.003471
	0.02602	0.002685	0.002789	0.0002863	0.00002935

Table 2.6-10. Similar to Table 2.6-9 but for the Lorentzian peak

p	ξ_0				
	10	20	50	100	200
	x10^3	x10^4	x10^5	x10^5	x10^6
2	2.8773	1.09678	0.6659	2.626	1.0374
	0.1698	0.03195	0.007601	0.01483	0.002911
	0.01199	0.001264	0.0001348	0.0001411	0.00001473
3	3.5072	1.3174	0.7644	2.934	1.1391
	0.2117	0.03901	0.008914	0.01694	0.003255
	0.01672	0.001749	0.0001846	0.0001915	0.00001983
4	4.2585	1.5417	0.8810	3.4233	1.2818
	0.2600	0.04670	0.01049	0.01994	0.003724
	0.02162	0.002253	0.0002367	0.0002449	0.00002524
5	4.9706	1.8529	1.06370	3.9768	1.5175
	0.3086	0.05602	0.01254	0.02325	0.004368
	0.02660	0.002770	0.0002903	0.0002996	0.00003084
6	5.9899	2.1662	1.2170	4.4842	1.6810
	0.3672	0.06564	0.01449	0.02659	0.004942
	0.03168	0.003294	0.0003446	0.0003552	0.00003651

Table 2.6-13. Parameters $E_s^{(2)}$ and $E_i^{(2)}$ (Eq. (2.6-23)) for the Gaussian peak

p	ξ_0				
	10	20	50	100	200
2	736.546	1177. 525	2697. 114	5181.672	10157.931
	5. 1668	2. 8661	2.1509	1.8331	1.6657
3	513.400	770. 663	1712.4	3235.063	6278.474
	3. 3557	1.7304	1.24	1.0281	0.9163
4	309. 087	427.916	912.315	1683.5449	3221.3509
	1. 8989	0. 8894	0.6011	0.4784	0.4149
5	208.011	270.1091	557.418	1009.6982	1909.926
	1.2312	0. 5388	0.3481	0.2685	0. 2784
6	138.499	168.951	337.125	598.8394	1118. 898
	0.8004	0. 3272	0.2022	0.1510	0. 1254

Table 2.6-14. Similar to Table 2.6-13 but for the Lorentzian peak

p	ξ_0				
	10	**20**	**50**	**100**	**200**
2	1053. 877	1896.415	4596.787	9105.817	18173.978
	10.0929	7.0222	6.0948	5.7071	5.5185
3	793.127	1361.208	3236.783	6351.825	12611.754
	6.7723	4.4458	3.7526	3.4617	3.3186
4	520.139	838.274	1941.162	3759.009	74086.606
	3.9356	2.3873	1.9381	1.7506	1.6577
5	367.823	564.984	1281.278	2454.815	4809.231
	2.5896	1.4771	1.1623	1.0322	0.9676
6	255.348	373.588	829.023	1570.710	3057.799
	1.6964	0.9085	0.6918	0.6034	0.5597

Table 2.6-15. Parameters $K_s^{(2)}$ and $K_i^{(2)}$ (Eq. (2.6-24)) for the Gaussian peak

H	ξ_0				
	10	**20**	**50**	**100**	**200**
0.0025	3.9979	4.0942	4.2006	4.2690	4.3316
	0.8579	1.1480	1.5535	1.8725	2.1947
0.005	3.9436	4.0464	4.1604	4.2363	4.3014
	0.7175	1.0055	1.4031	1.7090	2.0316
0.01	3.8864	3.9972	4.1219	4.1999	4.2697
	0.5741	0.8596	1.2463	1.5561	1.8705
0.02	3.8240	3.9435	4.0762	4.1616	4.2358
	0.4331	0.7178	1.1012	1.4017	1.7118

Table 2.6-16. Similar to Table 2.6-15 but for the Lorentzian peak

H	ξ_0				
	10	**20**	**50**	**100**	**200**
0.0025	5.0272	5.2110	5.4341	5.5786	5.7133
	0.1661	0.3689	0.6188	0.8303	1.0414
0.005	4.9262	5.1216	5.3493	5.5078	5.6507
	0.0679	0.2686	0.5337	0.7255	0.9285
0.01	4.8167	5.0270	5.2805	5.4341	5.5786
	-0.0183	0.1674	0.3952	0.6173	0.8303
0.02	4.6994	4.9248	5.1863	5.3550	5.5094
	-0.1038	0.0741	0.3140	0.5175	0.7211

Table 2.6-17. Parameters $L_s^{(2)}$ and $L_i^{(2)}$ (Eq. (2.6-25)) for the Gaussian peak

H	*p*				
	2	3	4	5	6
0.0025	0.6664	0.7845	0.8908	0.9987	1.1144
	7.2724	11.1238	14.8693	18.5979	22.3065
0.005	0.6759	0.7999	0.9132	1.0317	1.1568
	7.0030	10.7888	14.4678	18.1135	21.7478
0.01	0.6883	0.8182	0.9413	1.0680	1.2023
	6.7181	10.4373	14.0391	17.6135	21.1626
0.02	0.7029	0.8388	0.9722	1.1095	1.2543
	6.4196	10.0703	13.5871	17.0717	20.5351

Table 2.6-18. Similar to Table 2.6-17 but for the Lorentzian peak

H	*p*				
	2	3	4	5	6
0.0025	0.7606	0.9734	1.1971	1.4289	1.6793
	8.4377	13.1050	17.6283	22.1047	26.5195
0.005	0.7807	1.0046	1.2413	1.4994	1.7420
	8.0934	12.6423	17.0340	21.3340	25.6825
0.01	0.8027	1.0395	1.2903	1.5470	1.8220
	7.7334	12.1529	16.4081	20.6208	24.7697
0.02	0.8278	1.0779	1.3441	1.6243	1.9057
	7.3547	11.6378	15.7430	19.7716	23.7978

Table 2.6-19. Parameters $M_c^{(2)}$, $M_s^{(2)}$ and $M_i^{(2)}$(Eq. (2.6-26)) for the Gaussian peak

p	ξ_0				
	10	20	50	100	200
	x10^3	x10^4	x10^5	x10^5	x10^6
2	2.6041	1.0109	0.6064	2.3884	0.9434
	0.1274	0.02448	0.005849	0.01147	0.002258
	0.009560	0.001002	0.0001060	0.00011035	0.00001147
3	3.1556	1.2026	0.7251	2.7766	1.0862
	0.1544	0.02916	0.006964	0.01338	0.002617
	0.01384	0.0014364	0.0001505	0.0001555	0.00001604
4	3.6729	1.3765	0.8048	3.1132	1.2079
	0.1806	0.03365	0.007819	0.01500	0.002908
	0.01798	0.001857	0.0001933	0.0001988	0.00002043
5	4.2314	1.5706	0.8765	3.4403	1.3191
	0.2088	0.03845	0.008643	0.01668	0.003191
	0.02211	0.002277	0.0002359	0.0002422	0.00002480
6	4.7710	1.7837	1.0038	3.7319	1.4420
	0.2364	0.04373	0.009811	0.01826	0.003498
	0.02624	0.002700	0.0002791	0.0002856	0.00002921

Table 2.6-20. Similar to Table 2.6-19 but for the Lorentzian peak

p	ξ_0				
	10	**20**	**50**	**100**	**200**
	x10^3	**x10^4**	**x10^5**	**x10^5**	**x10^6**
2	3.1382	1.1803	0.6898	2.6532	1.02468
	0.1535	0.02878	0.006695	0.01282	0.002469
	0.01109	0.001160	0.0001225	0.0001272	0.00001318
3	4.0921	1.5262	0.8792	3.3251	1.2701
	0.2014	0.03739	0.008572	0.01617	0.003079
	0.01651	0.001718	0.0001800	0.0001858	0.00001914
4	5.1340	1.8857	1.07325	4.0308	1.5298
	0.2534	0.04642	0.01051	0.01969	0.003720
	0.02185	0.002265	0.0002365	0.0002435	0.00002503
5	6.1715	2.2649	1.1988	4.7163	1.8368
	0.3058	0.05597	0.01206	0.02313	0.004453
	0.02715	0.0028130	0.0002924	0.0003011	0.00003095
6	7.2000	2.6098	1.4396	5.5431	2.1173
	0.3586	0.06462	0.01428	0.02718	0.005144
	0.03246	0.003355	0.0003492	0.0003593	0.00003687

APPENDIX J

LOCK-IN AMPLIFIER

We will provide a simplified explanation of the operational principles of the lock-in amplifier (LIA) (Fig. J-1). The interested reader is directed to the monograph [1].

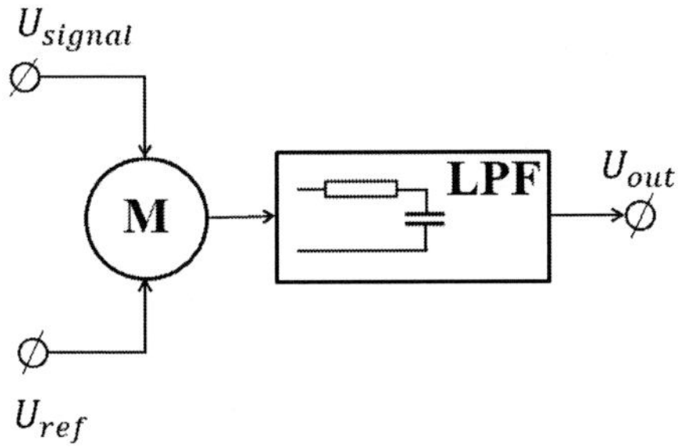

Figure J-1. The lock-in amplifier.
M is a multiplier; LPF is the low-pass filter.

Suppose that the input signal is a sum of the sinusoids (harmonics), $U_{signal} = \sum_{k=1}^{N} V_k sin(\omega_k t + \varphi_k)$, and the reference signal is $U_{ref} = V_{ref}(\omega_{ref} t + \varphi_{ref})$. Then the output signal of the multiplier (M): $U_M = U_{signal} U_{ref} = \sum_{k=1}^{N} V_k sin(\omega_k t + \varphi_k) V_{ref} sin(\omega_{ref} t + \varphi_{ref})$= $0.5V_{ref}\{\sum_{k=1}^{N} V_k \langle cos([\omega_k - \omega_{ref}]t + \varphi_k - \varphi_{ref}) - cos([\omega_k + \omega_{ref}]t + \varphi_k + \varphi_{ref})\rangle\}$.

The output signal of the multiplier U_M is passed through a low pass filter (LPF), which removed all harmonics with the frequencies $\omega_k - \omega_{ref}$ and $\omega_k + \omega_{ref}$. However, if $\omega_{ref} = \omega_k$, then a direct current signal $U_M = 0.5V_{ref} V_k cos(\varphi_k - \varphi_{ref})$ is proportional to the amplitude V_k. By adjusting the phase of the reference signal to the phase of the kth-harmonic, U_M has its maximum value $0.5V_{ref}V_k$.

References

1. Kloos, G. (2018). Applications of Lock-in Amplifiers in Optics. https://spie.org/Publications/Book/2307758?SSO=1

APPENDIX K

LASER DIODES

We believe that our readers have already received basic knowledge of semiconductors from high school courses. However, in order to understand the principles of modulation laser spectroscopy, a deep penetration into laser theory is not required. Interested readers are directed to the references.

Light-emitting diodes emit spontaneous optical radiation that originates because of the electron-hole recombination (annihilation of positively charged holes and negatively charged electrons). However, exciting photons of some resonant frequency may cause forced recombination by creating new photons. In this case, the diode produces the stimulated emission, where the photons have a phase, frequency, polarization, and direction identical to the exciting photons.

The central part of the laser diode is a very thin plate (optical waveguide), which is the p-n junction (Fig. K-1). Parallel mirror sides of the plate are the optical resonator (Fabry-Perot interferometer). Repeatedly reflected spontaneous emission photons cause forced recombination. A laser begins to emit if the intensity of the stimulated emission is stronger than the resonator's light absorption.

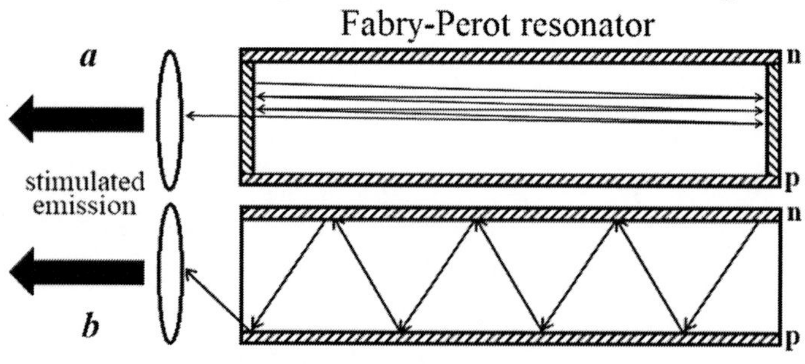

Figure K-1. Simplified diagram of the laser diode: longitudinal (a) and transverse (b) modes.

The directions of the spontaneous and stimulated emissions may be the same (Fig. K-1a) and perpendicular (Fig. K-1b). In the first case, the stimulated emission, which is called longitudinal mode, is parallel to the plate. In the second case, the laser diode generates light in a transverse mode. In both cases, multi-modes are possible: that is, there are many standing waves in the resonator with different wavelengths and different wave vector directions that satisfy the resonance condition. The number of modes depends on the plate's thickness. If the thickness is comparable to the emission wavelength, only the single-mode exists.

The laser's oscillation bands exist in many longitudinal modes if the longitudinal mode spacing is relatively narrow. To improve the longitudinal mode selection, the distributed Bragg reflection and distributed feedback lasers are standard diode lasers used in the near-infrared spectral region [2].

Many researchers have tried to improve the performance of mid-infrared lasers or develop new type lasers for higher sensitivity (e.g., a diode laser with the external cavity, interband, and quantum cascade lasers) [2]. The optical properties of the simple laser diodes (Fig. K-1) were significantly improved by introducing semiconductor double heterostructures:

"including Quantum Wells, Wires, and Dots [...] It can be said, that if the possibility to control the type of conductivity of a semiconductor material by doping with various impurities and the idea to inject non-equilibrium charge carriers were the seeds from which semiconductor electronics developed, heterostructures could make the solution of a considerably more general problem possible, that of controlling the fundamental parameters inside the semiconductor crystals and device: namely band gaps, effective masses of the charge carriers and the mobilities, refractive indices, electrons energy spectrum, and so forth"[3].

The double heterostructure (DHS) lasers are "sandwiches", which include gallium arsenide (GaAs) and gallium aluminum arsenide (GaAlAs) (Fig. K-2). The inner layer (GaAs) is between two outer layers (claddings) (GaAlAs). Two junctions GaAlAs-GaAs form DHS. In DHS laser, the thin regions contain the majority of the electron-hole pairs, which can contribute to amplification. Also, light reflection occurs in these regions.

Blue-emitting gallium nitride semiconductor lasers emit short wavelengths - near 400 nm, which is half that of GaAs lasers (Fig. K-3) [4].

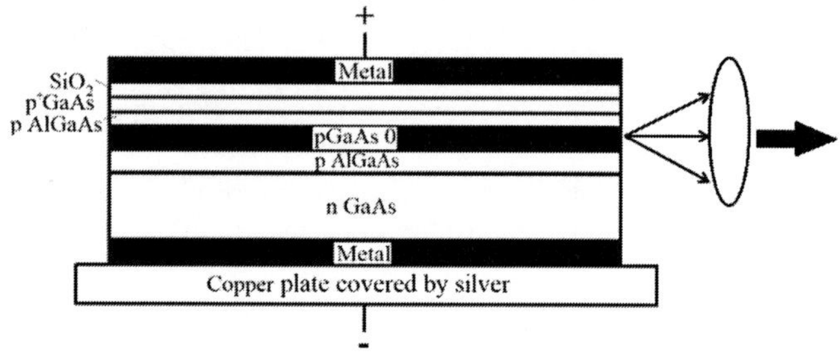

Figure K-2 (adapted from [3]). Schematic structure of the DHS laser.

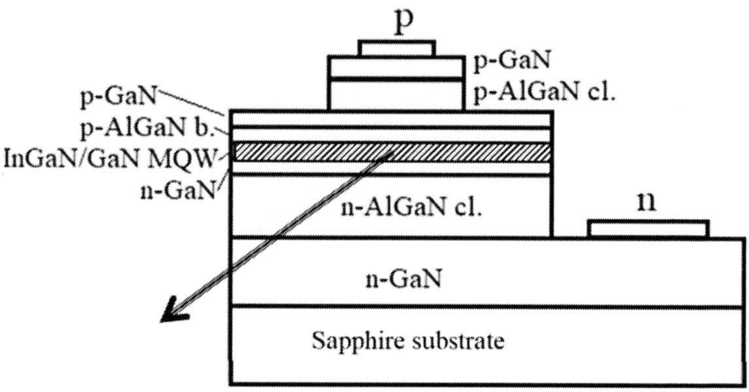

Figure K-3 (adapted from [4]). Schematic structure of the edge-emitting blue multiple quantum well diode laser. “cl” and “b” stand for “cladding” and “barrier”, respectively.

The semiconductor quantum cascade laser (QSL) usually emits mid-infrared light. The QSL involves the inter-sub-band transitions of the semiconductor structure (Fig. K-4). A cascade includes many quantum wells in a series. In each period of the structure, the electron undergoes the laser transition in a quantum well, and stimulated emission occurs. Then it performs non-radiative transition to the next quantum well. Due to this process, QSL has a high optical output. QSL is capable of operating at room temperature.

A distributed feedback lasers sharpen the output of conventional Fabry-Perot lasers, which emit a "comp" of wavelengths. Technically, a corrugated (crankle) structure is added above the active layer of the laser. Particular structure refractive index and the spacing of the crankles reflect only a specific wavelength beam.

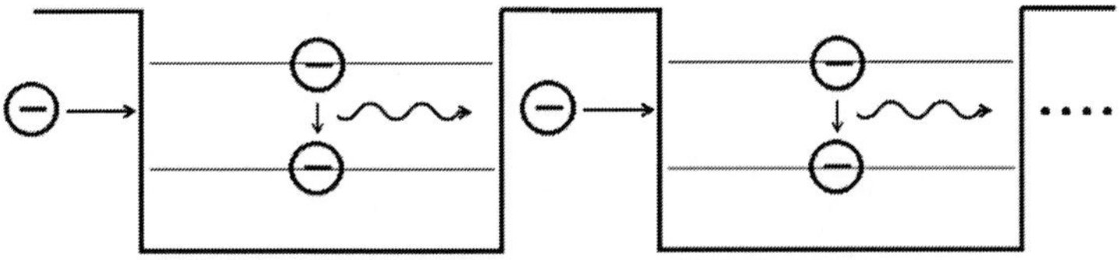

Figure K-4 (adapted from [5]). Schematic structure of the semiconductor quantum cascade laser.

An optical parametric oscillator (OPO) is a light source, similar to a laser. In OPO, the optical gain is due to the parametric amplification in a nonlinear crystal. OPO allows emitting wavelengths (e.g., in the mid-infrared, far-infrared, or terahertz spectral region) that are difficult (maybe impossible) to obtain from any laser. A singly resonant OPO has a resonator that is only resonant at singe frequency.

Reference

1. Sands, D. *Diode Lasers*. CRC Press. 2004.
2. Wang, F., Jia, S., Wang, Y., Tang, Z. (2019). Recent developments in modulation spectroscopy for methane detection based on tunable diode laser. Applied Science, 9, 2816. https://www.mdpi.com/2076-3417/9/14/2816/pdf
3. Alferov, Z. (2001). The Double Heterostructure: The Concept and its Applications in Physics, Electronics, and Technology (Nobel Lecture). CHEMPHYSCHEM, 2, 500 – 513. https://onlinelibrary.wiley.com/doi/epdf/10.1002/1439-7641%2820010917%292%3A8/9%3C500%3A%3AAID-CPHC500%3E3.0.CO%3B2-X
4. Johnson, N. M., Nurmikko, A. V., DenBaars, S. P. (2000). Blue diode lasers. Physics Today, 31-36. https://www.researchgate.net/publication/243390759_Blue_Diode_Lasers
5. Paschotta, R. Quantum Cascade Lasers. RP Photonics Encyclopedia. https://www.rp-photonics.com/quantum_cascade_lasers.html
6. Paschotta, R. Optical Parametric Oscillators. RP Photonics Encyclopedia. https://www.rp-photonics.com/optical_parametric_oscillators.html

APPENDIX L

SIGNAL SHAPES AND THEIR DERIVATIVES IN ELECTROCHEMICAL ANALYSIS

A part of this appendix is the author's translation of a modified section of article [1] from Russian into English. All technical details can be found in electrochemical textbooks.

Reversible electrode transfer (Nernstian) processes in constant current and normal impulse polarography based on the dropping mercury electrode

The current

$$i = i_l/[1 + exp(x)], \qquad (L-1)$$

where $i_l = i$ when $x = C_1(E - E_0) \to -\infty$; C_1 is a constant; E is the potential; E_0 is the half-wave potential, that is, $i = i_l/2$ if $E = E_0$.

The derivatives:

$$di/dE \sim exp(x)/[(1 + exp(x))]^2 \sim 1/[ch(x/2)]^2, \qquad (L-2)$$

where $ch(y) = 0.5[exp(y) + exp(-y)]$ is the hyperbolic cosine.

$$d^2i/dE^2 \sim exp(x)[exp(x) - 1]/[(1 + exp(x))]^3. \qquad (L-3)$$

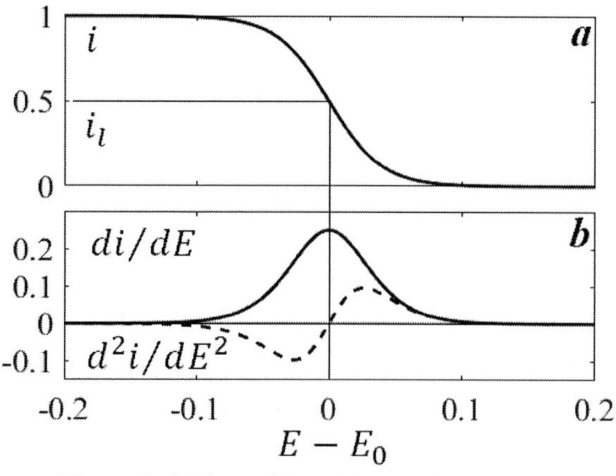

Figure L-1. Plots of Eqs. (L-1 - L-3). $C_1 = 50$.

Study [2] published equations of the signal shape and its 1st- and 2nd-order derivatives for the non-Nernstian system with a slower electron transfer. These very bulky formulas are only interesting from a theoretical point of view.

An equation, similar to Eq. (L-2), described the fractional 0.5-order derivative $d^{0.5}i/dt^{0.5}$ of the reversible processes [3].

Sinusoidal alternating current polarography for reversible processes

This method involves sinusoidal current modulation (Eq. (L-1)). All current harmonics were estimated using a cumbersome integral equation. If the modulation amplitude is small, then Eq. (L-2) defines the 1st-order derivative.

Linear scan voltammetry with a spherical electrode[4]

$$i = C_2(x) + C_3\Phi(x), \qquad (L-4)$$

where C_2 and C_3 depend on the radius and electrode area, and concentration of the oxidizer, diffusion coefficient; $\chi(x)$ and $\Phi(x)$ are cumbersome expressions.

The first term is different for reversible and irreversible processes. For the plane stationary electrode, the second term is zero [5].

The derivatives of Eq. (L-4), given in the tables in [4], demonstrated no significant narrowing.

Chronopotentiometry

For the reversible processes:

$$E - E_0 \sim ln(\sqrt{\tau/t} - 1), \qquad (L-5)$$

where τ is the transition time.

If $\tau/t = 4$, $E = E_0$.

Differentiation of the theoretical chronopotentiograms (Eq. (L-5)) has no advantage due to the shape transformations (Fig. L-2).

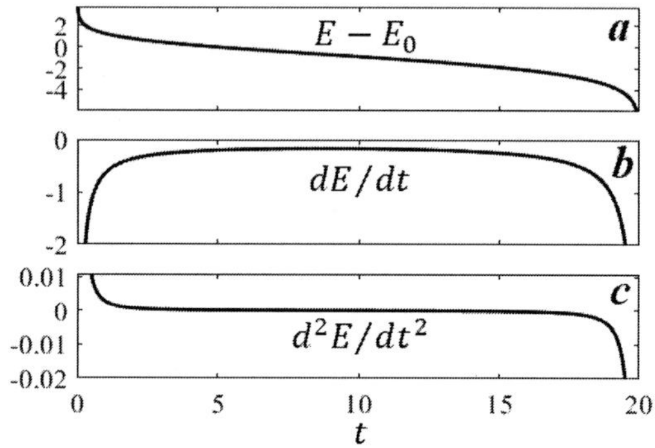

Figure L-2. (a) Plots of Eq. (L-5) and (b, c) their SG-derivatives. $\tau = 20$.

References

1. Dubrovkin, J. (1988). Chemical analysis with the use of the mathematical differentiation. Zhurnal Analiticheskoi Khimiya, 43, 965-979. [Russian].
2. Kim, M.-H., Smith, V. P., Hong, T.-K. (1993). First- and second-order derivative polarography/voltammetry for reversible, quasi-reversible, and irreversible electrode processes. Journal of Electrochemical Society, 140, 712-721.
3. Palys, M., Korba, T., Bos, M., Van Der Linden, W.E. (1991). The separation of overlapping peaks in cyclic voltammetry by means of semi-differential transformation. Talanta, 38, 723-733.
4. Perone, S. P., Mueller T. R., (1965). Application of derivative techniques to stationary electrode polarography. Analytical Chemistry, 37, 2-9.
5. Perone, S. P., Evins, C. V. (1965). Theory of derivative voltammetry with irreversible systems. Analytical Chemistry, 37, 1061-1063.

APPENDIX M

SPECTROSCOPY

Optical spectroscopy

This section will familiarize readers with the primary optical spectroscopy concepts in a very brief, concise form. Very often, analytical chemists with extensive practical experience in the spectrometric analysis have forgotten the theory of the methods that they studied at university. We hope that this discussion will revive their memory and facilitate understanding of the main text. Interested readers are also referred to the fundamental book [1], popular review [2], and references therein.

A bit of theory

Optical spectroscopy is a branch of the natural sciences studying the interaction of light and matter. The light that is absorbed, emitted, or reflected by sample contents information about its physical-chemical properties, including qualitative and quantitative composition, and interactions between components.

Here, light is defined, in a broad sense, as the electromagnetic waves whose properties depend on the frequency range (Fig. M-1).

The frequency:

$$\nu = v/\lambda, \qquad (M-1)$$

where v is the light velocity in the sample, and λ is the wavelength. Its units are micrometers (μm=10^{-6}m) or nanometers (nm=10^{-9}m).

In the IR-region, a variable "wavenumber" $\tilde{\nu} = 1 / \lambda$, measured in reciprocal centimeters $[cm^{-1}]$ is in common use.

Optical spectroscopy received its name due to the use of classical optics elements (lenses and mirrors) in the spectral instrument. It covers the near UV, VIS, and IR regions. The last portion of the electromagnetic spectrum includes the near- (NIR), mid-, and far-infrared (FIR), named for their relation to the VIS (Table M-1). The last column of this table requires an explanation for a non-professional reader.

According to the quantum theory, any anomic or molecular systems are only stable in some stationary states related to a discrete or continuous system energy set. The ground state is the lowest energy level with minimum energy; the upper levels correspond to the excited states (Fig. M-2).

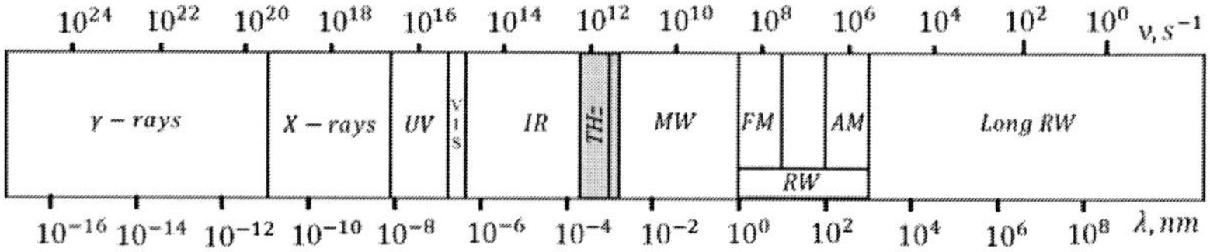

Figure M-1. The electromagnetic spectrum (adapted from [1]): UV, ultraviolet; VIS, visible; IR, infrared; THz, terahertz; MW, microwaves; FM, frequency Modulation; AM, amplitude modulation; and RW-radio waves.

Table M-1. Characteristics of optical spectroscopy

Range	Interval	Object of study
UV, VIS	780-2500 nm	Electronic transitions
NIR	14000–4000 cm^{-1}	Molecular overtone and combination vibrations
Mid-IR	4000–400 cm^{-1}	Fundamental vibrations and associated rotational-vibrational structure
FIR	400–10 cm^{-1}	Rotational spectroscopy

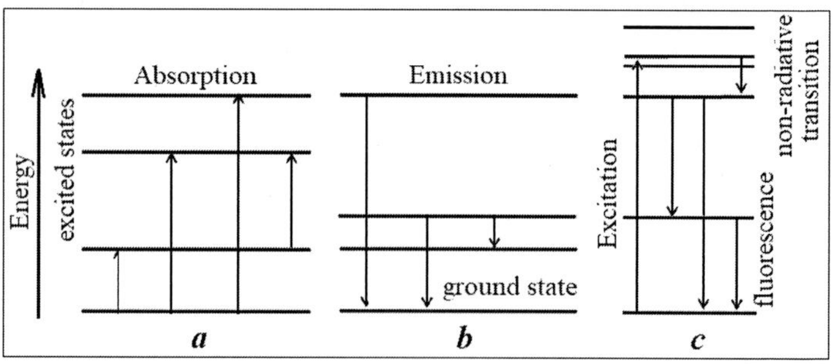

Figure M-2. Transitions between energy states which cause absorption, emission, and fluorescence spectra.

Light radiation absorbed by a sample may cause transitions from a lower energy state (including the ground state) to a high energy state of atoms or molecules (Fig. M-2a). The absorption spectrum of a chemical element or a chemical compound is the spectrum of the frequencies of this absorbance radiation. The emission spectrum is the spectrum of frequencies of the radiation emitted because of the transition from a high energy state to a lower energy state (Fig. M-2b).

The frequency of the radiation, absorbed or emitted because of the transition between the energy states E_{high} and E_{lower}, is

$$\nu = \Delta E/h, \qquad (M-2)$$

where $\Delta E = E_{high} - E_{lower}$; h is the Planck constant.

However, in addition to the "pure" frequencies (Eq. (M-2)), multiples of ν (overtones) and combination frequencies may be observed due, for example, to intramolecular interactions.

Unfortunately, a spectrum is not "a comb with infinitely narrow teeth", which is located at their transition frequency (Eq. (M-2)) and has a specific intensity. Practically, infinitely narrow lines broaden because of physical and instrumental factors. These factors are discussed in Chapter 1.1 and the next section.

The single line has a bell-shaped profile described by the Gaussian, Lorentzian, or Voigt mathematical functions.

Figure M-3 shows the overlapped finite widths peaks of the benzene UV-VIS electronic absorption spectrum in the gas phase.

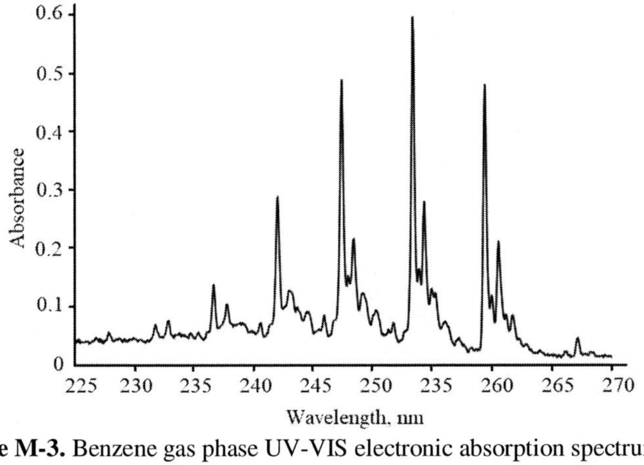

Figure M-3. Benzene gas phase UV-VIS electronic absorption spectrum [3].

Some substances that have absorbed electromagnetic radiation may emit light that has a longer wavelength than the absorbed radiation due to the non-radiative transitions (Fig. M-2c). This phenomenon is named fluorescence. In non-radiative transitions, excess energy is dissipated, e.g., by heating the sample.

The origin of atomic spectra is electronic transitions. Molecular spectra are substantially different since the movement of the molecular particles is very complicated. The whole pattern includes the electrons' movements, nuclear vibrations, and molecule rotation, which produce electronic, vibrational, and rotational spectra, respectively. However, there are complex interactions between different types of movements.

The optical spectroscopy technique

Optical spectra are measured by spectrometers. In a broad sense, an optical spectrometer separates the mixture of the light's spectral components according to the length of their waves and measures some parameters of the components. As the simplest example, imagine the spatial decomposition of white light into the rainbow colors in the sky and estimating their intensity by eye.

Let us consider block-diagrams of absorption spectrometers (Fig. M-4, panels a and b). In the single-beam mode (panel a), the radiation emitted by the radiation source is interrupted by a low-frequency chopper (I_0). Then the beam passes through the sample, which absorbs the incident light. Therefore, the light intensity after the sample $I < I_0$. We also suppose that the source radiation has a continuous frequency spectrum according to the spectrometer type (Table M-1).

The absorbance process may cause energy transitions that produce a spectrum of frequencies for this absorbance radiation. The frequency component must be detected to record the absorption spectrum. For this goal, the monochromator decomposes the light, which is passed through the sample into a set of consecutive frequencies (or wavelength).

Figure M-5 shows the block diagram of a grating monochromator with mirrors. The light image of the input rectangle narrow slit is directed by the collimator mirror on the diffraction grating. An excellent property of the grating is that its blazed (ruled) surface reflects off the light at different angles corresponding to different frequencies (wavelengths). Therefore, the camera mirror projects a set of the input slit images with different wavelengths at the plane of the narrow output slit. A small angular rotation of the grating allows illuminating the output slit with sequential images with the central wavelengths $\lambda_{1c}, \lambda_{2c}, \dots, \lambda_{Nc}$ (Fig. M-6a). This process is named "scanning of spectra". The light intensity distribution over the wavelength in each image has a bell-shaped profile (Fig. M-6b) since the monochromator mixes beams with close wavelengths. In other words, it has a finite resolution. Also, the images are overlapped (Fig. M-6a). So, the monochromator disturbs the true spectrum of the radiation source. Mathematically, this distortion is expressed by the instrumental function of the monochromator.

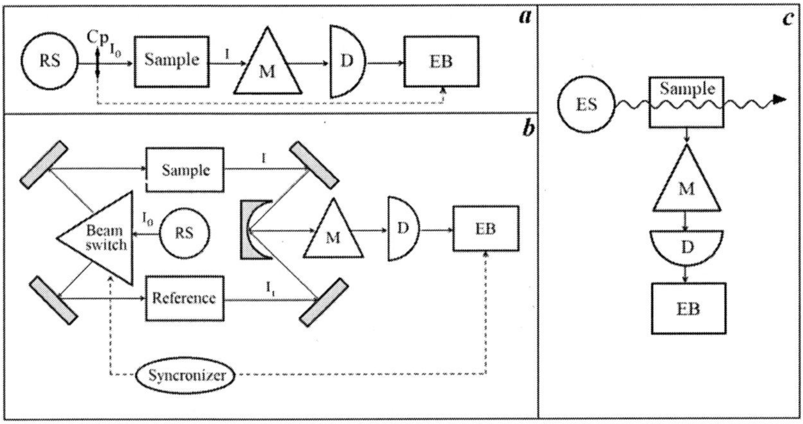

Figure M-4. Figure M-4. Block-diagrams of the single (a), double-beam (c) absorption, and fluorescence (c) spectrometers: RS, radiation source; Cp, chopper; M, monochromator; D, detector; EB, electronic block; and ES, excitation source.

To decrease instrumental distortion, one can narrow the slits. However, in this case, the light intensity decreases, and the impact of the noise on spectral data increases.

Laser radiation sources emit monochromatic beams. Therefore, monochromators are not needed (see Chapter 2.10).

The electronic block (EB) includes a lock-in amplifier (Appendix J).

The main disadvantage of the single-beam spectrometer is the problem of the estimation of the sample absorbance:

$$A = -logT = log(I/I_0), \qquad (M-3)$$

where $T = I/I_0$ is the transmittance, and I_0 and I are the light intensities incident on the sample and passed through it, respectively.

The estimation of A (Eq. (M-3)) requires two separate measurements: of I and I_0 (the empty channel) or, generally, of the reference signal I_r produced by the reference (blank) cell using for comparison. The double-beam spectrometer (Fig. M-4b) eliminates this drawback by introducing the two-channel scheme.

The radiation source alternately illuminates the sample and the reference cells using the beam switch. Correspondingly, each channel's light intensity is measured, and the absorbance is calculated in the electronic block. A synchronizer controls this process.

The fluorescence spectrum is usually measured at the right angle relative to the direction of the excitation radiation (Fig. M-4c). This geometry avoids the impact of the transmitted excitation light on the detector.

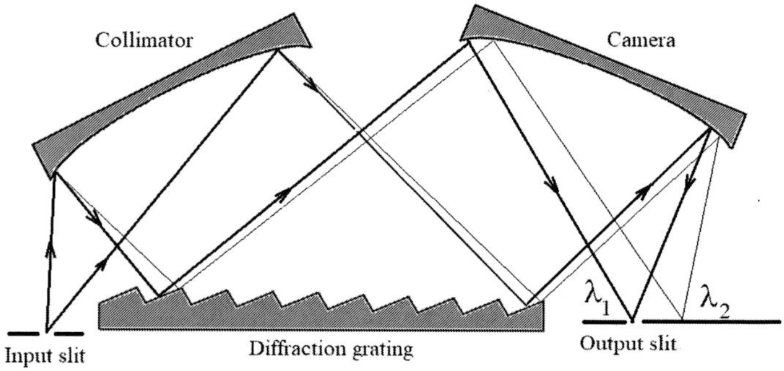

Figure M-5. Diagram of a grating monochromator with mirrors. The scale of the elements does not correspond to a real device. The grating does not reflect beams but, instead, diffracts them.

The electronic block of spectrometers includes a high gain low noise amplifier, a low-pass filter, and a computer unit. From a spectroscopic point of view, the most important parameter of this block is the time constant. It defines the inertial distortions of a spectrum similar to those of the integration RC circuit (Appendix E).

The radiation sources and detectors of absorption optical spectrometers depend on their wavelength range (Table M-2).

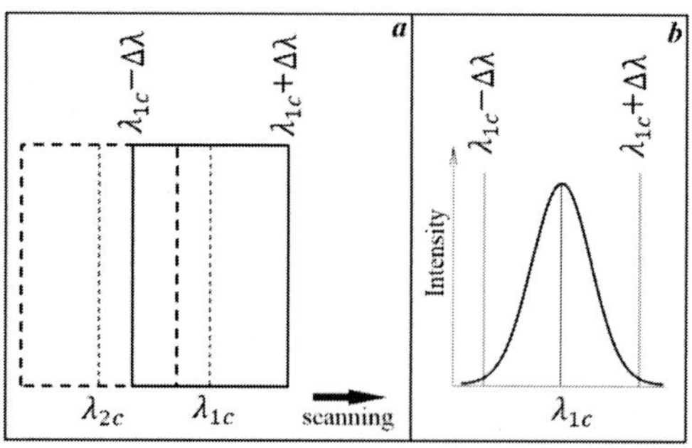

Figure M-6. (a) Overlapped images of the input slit. (b) The light intensity distribution over the wavelength of the input slit image.

Table M-2. Elements of absorption optical spectrometers

Range	Radiation source	Detector
UV, VIS	Tungsten filament (300-2500 nm) Deuterium arc lamp (190-400 nm) Xenon arc lamp (160-2000 nm) Light emitting diodes (VIS)	Photomultiplier tube Photodiode arrays* Charge-coupled device (CCD)*
NIR	Quartz halogen light bulbs Light emitting diodes Lasers	Silicon-based CCD* InGaAs detectors
Mid-IR	Hot inert solids (1500-2000K) (Globar)	Bolometers** Photovoltaic***
FIR	Globar High-pressure mercury lamp	Golay cell (optoacoustic detector)**** Pyroelectric bolometer Liquid helium cooled bolometers

* A group of these types of detectors, combined into one or two-dimensional arrays, is used in non-scanning monochromators.
** The thermal bolometer measures radiation via heating a material with a temperature-dependent electrical resistance. The pyroelectric bolometer contains a crystal material, which generates a voltage when its temperature is changed.
*** The photovoltaic detector is a semiconductor on which p-n junction photoelectric current appears upon illumination.
**** The Golay cell includes a gas-filled cell with an IR absorbing material and a flexible membrane. Light reflected off the membrane is detected by a photodiode. The absorbed light heats the gas, which expands and deforms the membrane. The membrane deformation changes the signal on the photodiode.

Fourier Transform IR spectrometers

Fourier transforms IR (FTIR) spectroscopy is based on the Fourier transform (FT) of the time-domain interferogram into the IR spectrum, which is similar to that recorded by the single-beam spectrometer. The essential part of a Fourier spectrometer is a Michelson interferometer (Fig. M-7). The beam splitter splits the radiation passed through a sample into two halves (I_1 and I_2). A moving mirror changes the path length difference (d) between the beams. Therefore, two light waves (with frequency f and wavelength λ) incident on the detector:

$$E_1 = E_{10}\cos(2\pi ft), E_2 = E_{20}\cos(2\pi ft + \varphi), \qquad (M-4)$$

have a phase difference $\varphi = 2\pi d/\lambda$.

The intensities:

$$I_1 = \overline{E_1^2} = E_{10}^2\overline{[\cos(2\pi ft)]^2} = E_{10}^2/2\,;\; I_2 = \overline{E_2^2} = E_{20}^2/2, \qquad (M-5)$$

where bar designates the mean value.

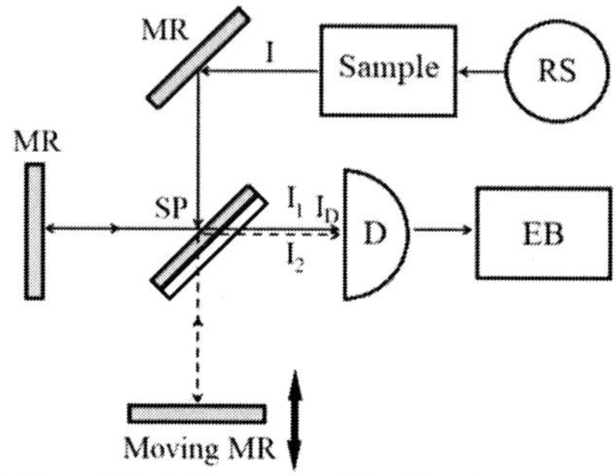

Figure M-7. Simplified diagram of the Fourier spectrometer based on the Michelson interferometer: RS, radiation source; MR, mirror; SP, beam splitter; D, detector; EB, electronic block.

The light waves interfere with each other producing a combination, which depends on the angle φ (Fig. M-8). The result is

$$I_D = \overline{(E_1 + E_2)^2} = I_1 + I_2 + 2E_{10}E_{20}\,\overline{\cos(2\pi ft)\cos(2\pi ft + \varphi)} =$$
$$I_1 + I_2 + E_{10}E_{20}\,[\cos(4\pi ft) + \cos(\varphi)] =$$
$$I_1 + I_2 + 2\sqrt{I_1 I_2} cos(2\pi d/\lambda). \qquad (M-6)$$

If $I_1 = I_2 = I/2$, then

$$I_D = I[1 + cos(2\pi d/\lambda)]. \qquad (M-7)$$

Since the light beam is a superposition of different wavelengths, the detector's output signal is obtained by integration over the wavelength interval $[\lambda_1, \lambda_2]$:

$$S_{output}(\Delta) = \int_{\lambda_1}^{\lambda_2} I_\lambda [1 + cos(2\pi d/\lambda)] \mathcal{F}_\lambda \, d\lambda, \qquad (M-8)$$

where $\mathcal{F}_\lambda = S_\lambda / I_\lambda$ is the detector's sensitivity.

The curve $S_{output}(\Delta)$ is called "interferogram". It is the Fourier cosine transform the function $I_\lambda \mathcal{F}_\lambda$ (Appendix A). The inverse FT of the interferogram allows estimating the spectrum I_λ.

We will now provide some technical details without a rigorous mathematical explanation.

Suppose that the monochromatic beam with the wavenumber $\tilde{\nu}_0$ passes through the FTIR spectrometer. In this case, I_λ (Eq. (M-8)) is the delta impulse. Then, the inverse FT of the output signal, called "the instrumental function", of the spectrometer is

$$IF(x) = sin(x d_{max})/x, \qquad (M-9)$$

where $x = \pi(\tilde{\nu} - \tilde{\nu}_0)$; $[-d_{max}, d_{max}]$ is the path length difference interval.

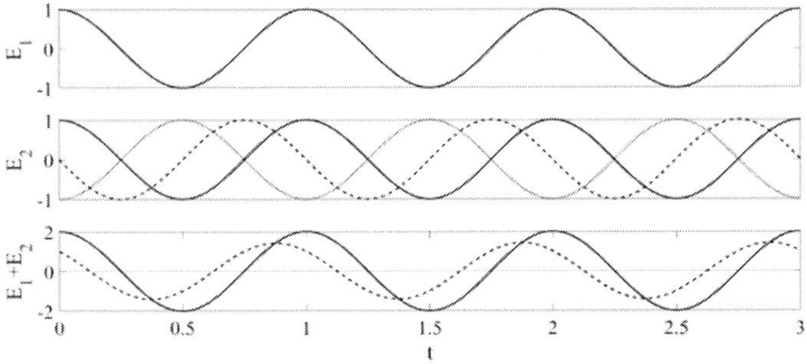

Figure M-8. The interference. $E_1 = \cos(2\pi ft)$; $E_2 = \cos(2\pi ft) + \varphi)$; $\varphi = 0, \pi/2$, and π (solid, dashed, and dotted curves, respectively).

Two spectral lines are resolved if the maximum of the second line coincides with the zero of the first line (Fig. M-8). Therefore, from Eq. (M-9) we have

$d_{max}\Delta x = \pi$ or $\Delta\tilde{\nu} = 1/d_{max}$. Since $\Delta\tilde{\nu}/\tilde{\nu} = \Delta\lambda/\lambda$, $\Delta\tilde{\nu} = \Delta\lambda/\lambda^2$. Finally, we obtain

$$\Delta\lambda = \lambda^2/d_{max}. \qquad (M-10)$$

The side petals of the instrumental function (Fig. M-9) are due to the Gibbs phenomenon (Appendix A). Apodization ("removing the foot") is a technique that allows decreasing these unwanted pulses.

FTIR spectrometry has many advantages over the direct method:

1. It avoids power losses at a monochromator's input slit.
2. The Fellgett advantage: significant improvement of the signal-to-noise ratio if the thermal detector noise is dominant over the shot noise. This improvement increases spectral resolution.
3. FTIR is simpler to implement because the required spatial resolution is lower than in the visible and ultraviolet region.

An in-depth study of the FTIR technique requires professional knowledge in physics and mathematics. Therefore, interested readers are referred to monograph [5].

Atomic absorption spectrometers

A simplified diagram of the atomic absorption spectrometer (AAS) is, in principle, a single-beam device (Fig. M-4a) [6]. The radiation source (e.g., hollow cathode and electrodeless discharge lamps) has a continuous or discrete spectrum in the VIS-IR region. The sample is the atomization system, which generates atomic vapor from the solution. In flame AAS

(FAAS), the atomization is performed by a flame. Graphite furnace atomic absorption spectrometry (GFAAS) uses electrothermal atomization. GFAAS is a high sensitivity method, but an effective background correction is needed as opposed to FAAS.

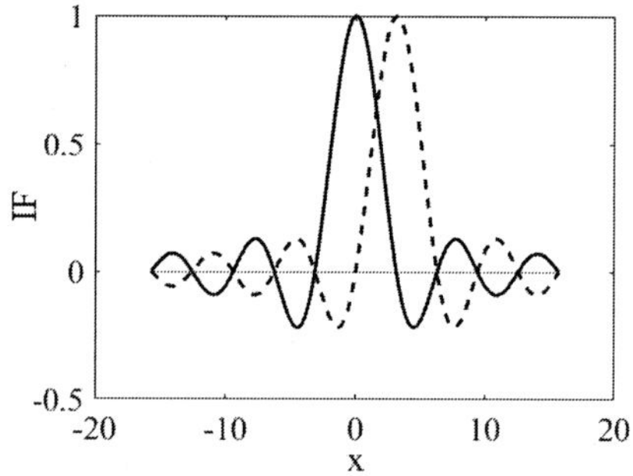

Figure M-9. Resolution limit of FTIR. Solid and dashed lines represent $IF(x)$ and $IF(x-\pi)$. $d_{max}=1$.

The background correction methods are a two-line deuterium lamp, Zeeman correction, a pulsed lamp, and wavelength modulation (Chapters 2.9 and 2.10).

For the technical details and applications of AAS, see review [6].

References

1. Sharma, B. K. *Spectroscopy.* 19th. ed. India, Meerut-Delhy: Goel Publishing House. 2007.

2. https://chem.libretexts.org/Courses/Northeastern_University/10%3A_Spectroscopic_Methods/10.1%3A_Overview_of_Spectroscopy

3. http://www.colby.edu/chemistry/PChem/notes/benzeneUVVis.pdf

4. https://chem.libretexts.org/Bookshelves/Physical_and_Theoretical_Chemistry_Textbook_Maps/Supplemental_Modules_(Physical_and_Theoretical_Chemistry)/Spectroscopy/Fundamentals_of_Spectroscopy/The_Power_of_the_Fourier_Transform_for_Spectroscopists

5. Griffiths, P.R., De Haseth, J. A. *Fourier transform infrared spectroscopy*. Wiley. 2007.

6. Filho, H. J. I., Salazar, R. F. S., Capri1, M.R., et al. (2012). Chapter: State-of-the-Art and Trends in Atomic Absorption Spectrometry. In book: Atomic Absorption Spectroscopy. INTECH, Ed. M. A. Farrukh, 13-36.

APPENDIX SW

MATLAB PROGRAMS

This section includes some files which can be readily modified by beginners in MATLAB programming. The project "Derivative Spectroscopy" (https://www.researchgate.net/profile/Joseph_Dubrovkin) provides additional open-source software.

SW1. For Exercise 1.1-1

The readers are invited to study: The dependence of the Voigt peak width on the ratio w_L/w_G, using the % program VoigtTest.m. For the rough approximation of Voigt peak: $F_L * F_G$ the program VoigtTest.m. must be slightly modified.

```
%VoigtTest.m
% © Copyright by Joseph Dubrovkin

% Generation of Voigt profile in the range 1:h:limit
limit=401;

%Amplitude
A=1;
%In the centre
imax=(limit-1)/2;
%For good wings the peak width wV <=100
wV=100;
%The widths of the Lorentzian and Gaussian components
wL=30;
wG=sqrt((wV-wL)*(wV-0.0692*wL));

coef=[1,imax,wL,wG];

%For interpolation
%stepNew can be less or more than one
stepNew=1;
y=1:stepNew:limit;
F=VoigtNew(coef, y, limit);
```

```
x=(y-imax)/wV;
plot(x,F,'-k','LineWidth',2)
xlabel('x')
ylabel('Intensity')
axis([-2,2,0, 1])

wGp=100;
Fg = gauss(A, y, imax, wGp);
hold on
plot(x,Fg,':k','LineWidth',2)
hold on
Fl = lorentz(A, y, imax, wGp);
plot(x, Fl,'--k','LineWidth',2)

function Fg = gauss(ampl, x, imax, width)
 %x-range
 arg=(x-imax).*(x-imax)/width/width;
 Fg=ampl*exp(-4*log(2)*arg);
end

function Fl = lorentz(ampl, x, imax, width)
 %x-range
 arg=(x-imax).*(x-imax)/width/width;
 Fl=ampl*1./(1+4*arg);
end

function F = VoigtNew(coef, Ynew, limit)
% © Copyright by Joseph Dubrovkin
% Generation of Voigt profile in the range 1:limit
%coef include the peak parameters
% we suppose that the peak width <= 100 for good wings
% Ynew - new range for interpolation

h=2;% For integration

%Peak parameters
A=coef(1);
imax=coef(2);
wL=coef(3);
wG=coef(4);
```

```
mnL=2/(wL*h);
mnG=2*sqrt(log(2))/(wG*h);

rangeI=1:h:h*limit;
rangeJ=-2*limit:2*h:2*limit;

F=zeros(1,length(rangeI));

c=1;
for i=rangeI %1:2:401*2
 s=0;
for j=rangeJ %-2*401:2*2:2*401
 y=mnG*(j-imax);
 x=mnL*(i-j-imax);
 s=s +exp(-y*y)/(1+x*x);
end

 F(c)=s;
 c=c+1;
end

F=A*F/max(F);
%Interpolation
rngY=1:length(F);
F=interp1(rngY, F, Ynew);
end
```

SW2. FFT smoothing and differentiation

```
%% function fractDeriv
% © Copyright by Joseph Dubrovkin
%% x=1:range;
%% y is data to be processed
%% im=range/2 is the middle point
%% vectorOrder is an array of the derivative orders
%% flagNorm=1 Normalization for plots

function [matrix_derivatives, FFT] =…
fractDeriv(y, vectorOrder, x, flagNorm ,alpha, power)
```

```
if nargin==4
 alpha=0;
 power=0;
end

range=length(x);
im=(range)/2;

Kdif=2*pi/range;

matrix_derivatives=ones(length(vectorOrder),range);

FFT=fft(y);

freq=[0:im-1 0 -im+1:-1];
len=length(freq);

mn=(1i*freq)*Kdif;

fourier_filterSmooth=ones(1,len);

c=1;
for t=vectorOrder
if t==0
 al=alpha;
 else
 al=alpha*t^power;
end
if nargin==6
 fourier_filterSmooth=1./(ones(1,len)+al*freq.^power);
end

fourier_filter=fourier_filterSmooth.*(mn.^t);
Fourie_D=real(ifft( FFT.*fourier_filter));

%%Normalization
if flagNorm==1
 Fourie_D=Fourie_D/(max(Fourie_D)-min(Fourie_D));
end
if flagNorm==2
 Fourie_D=Fourie_D/(max(abs(Fourie_D)));
```

```
end
matrix_derivatives(c,:)=Fourie_D;
c=c+1;
end
end
```

Chapter 2.4

```
% © Copyright by Joseph Dubrovkin
% This function performs FFT smoothing and differentiation.
% FourieSmooth is the smoothed (or derivative) vector.
% vector is input data
% alpha is the regularization parameter
% power is the smoothing filter order
% n is the differentiation order
% the smoothing is performed for the number of the arguments is three.

function FourieSmooth = FourSmoothFun(vector, alpha, power, n)
%Copyright by Joseph Dubrovkin
len=length(vector);
if mod(len,2)==1
 len=len+1;
end
len1=len/2;
freq=[1:len1,-len1+1:1:0];
FFT=fft(vector);
M=alpha*freq.^power;
den=(ones(1,len)+M);

if nargin==3
 fourier_filter=1./den.^2;
end
if nargin==4
 difOp=(1i*freq).^n;
 fourier_filter=difOp./den.^2;
end

fourier_filter=fourier_filter(1:len-1);
FFTsolut=FFT.*fourier_filter;
FourieSmooth=real(ifft(FFTsolut));
end
```

SW3.

Exercise 2.4-1

The readers are invited to study the FTT smoothing of the Gaussian peak corrupted by the normal noise with the mean zero and the standard deviation σ. Remark. In some case, the minimum of the functional $\psi(\alpha, p)$ is not observed.

```
% studySmoothFunct.m
% © Copyright by Joseph Dubrovkin

range=1000;
step=1;
ampl=1;
width=200;
[x, y] = gaussPeak(range, step, ampl, width);

sigma=0.02;
NOISE=sigma*randn(1,length(x));
yn=y +NOISE;

power=4;

rngAlpha=1e-6:1e-6:1e-3;
vectorFunct=ones(1,length(rngAlpha));

c=1;
for alpha=rngAlpha
 vectorFunct(c)=FourFunctional(yn, alpha, power);
 c=c+1;
end

[val, ind]=min(vectorFunct);

subplot(2,1,1)
plot(rngAlpha,vectorFunct,'-k','LineWidth',1)

alphaOpt=rngAlpha(ind);

ynSm = FourSmoothFun(yn, alphaOpt, power);

subplot(2,1,2)
```

```
plot(x,yn,'-k','LineWidth',1)
hold on
plot(x,ynSm,'-k','LineWidth',2)

function [x, y]= gaussPeak( range, step, ampl ,width )
 width2=width*width;
 ic=range/2;
 x=-ic :step: ic;
 y=ampl*exp(-4*log(2)*x.*x/width2);
end
```

Exercise 2.4-2

The readers are invited to study the FTT 1st and 2nd differentiation of the Gaussian peak corrupted by the normal noise with the mean zero and the standard deviation σ. Remark. In some case, the minimum of the functional $\psi\ (n, \alpha, p)$ is not observed.

```
% studyDifFunct.m
% © Copyright by Joseph Dubrovkin

range=1000;
step=1;
ampl=1;
width=200;
[x, y] = gaussPeak(range, step, ampl, width);

sigma=0.05;
NOISE=sigma*randn(1,length(x));
yn=y+ NOISE;
power=8;
rngAlpha=1e-6:1e-6:1e-3;
vectorFunct=ones(1,length(rngAlpha));

n=2;
c=1;
for alpha=rngAlpha
 vectorFunct(c)=FourFunctional(yn, alpha, power, n);
 c=c+1;
end

[val, ind]=min(vectorFunct);
```

```
subplot(2,1,1)
plot(rngAlpha,vectorFunct,'-k','LineWidth',1)
xlabel('\alpha')
ylabel('Functional')
alphaOpt=rngAlpha(ind);

ynSm = FourSmoothFun(yn, alphaOpt, power, n);

subplot(2,1,2)
plot(x,yn,'-k','LineWidth',1)
hold on
plot(x,ynSm,'-k','LineWidth',2)
xlabel('x')
ylabel('Intensity')
hold on
plot(x, zeros(1,length(x)),'-k','LineWidth',1)


% © Copyright by Joseph Dubrovkin
%% For three arguments (Eq. (2.4-4))
%% For four arguments (Eq. (2.4-8))
function functional = FourFunctional (vector, alpha, power, n)

len=length(vector);

if mod(len,2)==1
 len=len+1;
end
len1=len/2;

freq=[1:len1,-len1+1:1:0];

FFT=fft(vector);

M=alpha*freq.^power;

den=(ones(1,len)+M);
if nargin==3
 fourier_filter=M./den.^2;
end
if nargin==4
```

```
 difOp=(1i*freq).^n;
 fourier_filter=difOp.*M./den.^2;
end

fourier_filter=fourier_filter(1:len-1);
integr=FFT.*fourier_filter;
functional=sum(real(integr));
end
```

SW4. For Chapter 2.7

```
%% studySmoothSplines.m
% © Copyright by Joseph Dubrovkin

% The source of splinefit.m
%Lundgren, J. SPLINEFIT version 1.0.0 (238 KB).
% https://ww2.mathworks.cn/matlabcentral/ fileexchange/
% 71225-splinefit?focused=888438cd-bc77-4e1d-8c1a-
% bd6e763f4113&tab=function

h=0.02;
st=4;
fin=4;
x=-st: h: fin;
L=length(x);
flagModel='b';
flagNoise='pink'; %%'white';

%{
For the pink noise:
https://www.mathworks.com/matlabcentral/mlc-downloads/downloads/
submissions/42919/versions/11/previews/pinknoise.m/index.html
%}
switch flagModel

case 'a'
A1=1;
A2=1;
w1=1;
w2=1;
deltaX=0.52;
```

```
peakAmpl=1.0599;
peakAmp2=1.0599;
x0T1=-0.44;
x0T2=0.44;
%% for searching the center
limC=15;
case 'b'
A1=1;
A2=0.5;
w1=1;
w2=2;
deltaX=0.80;
peakAmpl=1.0881;
peakAmp2=0.5008;
x0T1=-0.76;
x0T2=0.78;
%% for searching the center
limC=15;

end

Fg=gaussDoublet(x,A1,w1, A2,w2, deltaX);
sigma=0.02;

switch flagNoise
case 'white'
noise=sigma*randn(1,L);
case 'pink'
noise=sigma*pinknoise(L);
end

shift=deltaX/h;
if mod(L,2)==0
 centr=L/2;
else
 centr=(L-1)/2;
end

rngCentre=centr-shift/2:centr+shift/2;
rngLeft=centr:-1:1;
rngRight=centr: L;
```

```
numbExperim=1e4;

%measured values
x01pos=ones(1,numbExperim);
x02pos=x01pos;
Ampl1=x01pos;
Ampl2=x01pos;

rngNumbPieces=20:10:90;
rngSplineOrder=4:9;
lenRngNumbP=length(rngNumbPieces);
lenRngSplOrd=length(rngSplineOrder);

%% Output Data
errGen=ones(lenRngSplOrd, lenRngNumbP);
meanX01=errGen;
stdX01=errGen;
meanX02=errGen;
stdX02=errGen;

Fg1=Fg+ noise;
cc=1;
for splineOrder=rngSplineOrder
kk=1;
for numbPieces=rngNumbPieces

splineOrder_=splineOrder
numbPieces_=numbPieces

tt=1;
%Counter "no min"
hh=1;

while tt<=numbExperim

randBreaks1= randBreaksGen(x, splineOrder, numbPieces);
pp = splinefit(x, Fg1, randBreaks1, splineOrder);
ySm = ppval(pp, x);

%Centre
[valC, indC]=min(ySm(rngCentre));
```

```
posC=x(centr-shift/2+indC-1);

%There is MIN in the center

if abs(posC)<limC*h
 %Left peak
 [valL, indL]=max(ySm(rngLeft));
 %Right peak
 [valR, indR]=max(ySm(rngRight));

 posL=x(centr-indL+1);
 posR=x(centr+indR-1);

 x01pos(tt)=posL;
 x02pos(tt)=posR;

 Ampl1(tt)=valL;
 Ampl2(tt)=valR;

 tt=tt+1;

else
 disp('NO MIN IN THE CENTRE')
end %%if abs(posC)<limC*h

end %% tt<=numbExperim

systErr1=1-mean(Ampl1)/peakAmpl;
systErr2=1-mean(Ampl2)/peakAmp2;

errGen(cc, kk)=100*sqrt(systErr1^2+systErr2^2+...
(std(Ampl1)/peakAmpl)^2+std((Ampl2)/peakAmpl)^2);

meanX01(cc, kk)=(mean(x01pos)-x0T1)/h;
stdX01(cc, kk)=std(x01pos)/h;
meanX02(cc, kk)=(mean(x02pos)-x0T2)/h;
stdX02(cc, kk)=std(x02pos)/h;
 kk=kk+1;
end
cc=cc+1;
end
```

```
errGen_=errGen
meanX01_=meanX01
stdX01=stdX01
meanX02_=meanX02
stdX02=stdX02

% © Copyright by Joseph Dubrovkin
function D= gaussDoublet(x,A1,w1, A2,w2, x0)
D=A1*exp(-4*log(2)*(x+x0).*(x+x0)/w1^2)+...
A2*exp(-4*log(2)*(x-x0).*(x-x0)/w2^2);
end

% Generation of Random Knots
function randBreaks = randBreaksGen(x, splineOrder, numbPieces)
L=length(x);
h=x(2)-x(1);
if x(1) < x(L)
 % Interpolate breaks linearly from x-data
 dx = diff(x);
 ibreaks = linspace(1, L, numbPieces+1);
 [~,ibin] = histc(ibreaks,[0,2:L-1,L+1]);
 breaks = x(ibin) + dx(ibin).*(ibreaks-ibin);
else
 breaks = x(1) + linspace(0,1,p+1);
end

 randBreaks=breaks;
 H=breaks(2)-breaks(1);
 delta=splineOrder*h;
 %Remove warnings
 randBreaks(2) = breaks(1) + H*rand;
 randBreaks(1)= (x(1)+0.02*randBreaks(2))/1.01;

 for ii=2:numbPieces-1
 randBreaks(ii+1) = breaks(ii) + H*rand;

c=1;
while randBreaks(ii+1) - breaks(ii)<=delta &&...
randBreaks(ii+1) - randBreaks(ii)<=delta

randBreaks(ii+1) = breaks(ii) + H*rand;
```

```
c=c+1;
if c>1000
 break
end
end
end
randBreaks(numbPieces)= (x(L)+0.02*randBreaks(numbPieces-1))/1.01;
end

% © Copyright by Joseph Dubrovkin
 plotSplineSmoothedDoublet.m

h=0.02;
st=4;
fin=4;
x=-st: h: fin;
L=length(x);

deltaX=0.8;
w1=1;
w2=2;
A1=1;
A2=0.5;
Fg=gaussDoublet(x,A1,w1, A2,w2, deltaX);
sigma=0.05;
noise=sigma*randn(1,L);
Fg1=Fg+noise;

%Set breaks
breakOpt=x([ 1, 90, 120:20:320, 400, L]);
splineOrder=4;
pp = splinefit(x,Fg1,breakOpt,splineOrder);
ySm = ppval(pp, x);

% The fitting error
fitErrSline=norm(Fg-ySm)

%Fourier smoothing
alpha=6e-5;
Fgf = FourSmoothFun(Fg1,alpha,splineOrder);
```

```
%The fitting error
fitErrFour=norm(Fg-Fgf)

plot(x,Fg1,'-k','LineWidth',1)
hold on
plot(x,ySm,'-k','LineWidth',1)
hold on
plot(x,Fgf,':k','LineWidth',2)
```

SW5. For Chapter 2.8
ForFig2_8_2.m

```
% © Copyright by Joseph Dubrovkin

%%CWT derivative of the Gaussian peak
numbPoint=1024*2;
%%D1PR - accurate first-order derivative
len=32;
st=len/numbPoint;
x=1:st:len;
dx=x(2)-x(1);
mn=2*sqrt(log(2));
a1=1.0; t1=len/2; w1=len/16;
arg1=mn*(x-t1)/w1;
y=a1*exp(-arg1.^2);
D1PR=-(2*mn/w1)*arg1.*y;
D1PR=D1PR/sum(abs(D1PR));

rngX=700:1300;
rngS=0.2:0.2:1;

for s=rngS
D1wt = CWTgaussD1(x, dx, y, s);
D1wt=D1wt/sum(abs(D1wt));

subplot(2,1,1)
plot(x(rngX),D1wt(rngX),'-k','LineWidth',2)
ylabel('F^{(1)}')
hold on
plot(x(rngX),zeros(1,length(rngX)),'-k','LineWidth',1)
hold on
plot(x(rngX),D1PR(rngX),'--k','LineWidth',2)
```

```
hold on

subplot(2,1,2)
plot(x(rngX),D1PR(rngX)-D1wt(rngX),'-k','LineWidth',2)
xlabel('x')
ylabel('F^{(1)}_{PR}-F^{(1)}_{CWT}')
hold on
plot(x(rngX),zeros(1,length(rngX)),'-k','LineWidth',1)
hold on
end

function WavL = CWTgaussD1(x, dx, y, s)
% © Copyright by Joseph Dubrovkin
%x is abscissa
% dx is the sampling interval
% s is the scaling factor

len=length(x);
WavL=zeros(1,len);

c=1;
for a=x
 WL=WLGD1(x,a,s);
 WavL(c)=sum(y.*WL);
 c=c+1;
end
WavL=dx*WavL;
end

% © Copyright by Joseph Dubrovkin
function WL=WLGD1(rngI, a, s)
%Gauss D1 wavelet
% rngI is abscissa
% a is dilation
% s is the scaling factor

z=(rngI-a)/s;
z2=z.*z;

%Normalization
MN=sqrt(1/s);
```

```
WL=MN*(1/s)*2*z.*exp(-z2);
end

%%For Exercise 2.8-1
%% plotD1WTSGFT.m
% © Copyright by Joseph Dubrovkin
%% Plots of the first-order derivatives obtained
%% by CWT, DWT, SG and FT methods
%% Models:
%% 'G1'-single Gaussian peak (width=w)
%%'L1'-single Lorentzian peak (width=w)
%'T1'-triangle (width=w)
%'C'-three function

model='C';
switch model
case 'C'
numbPoint=1024*5;
case {'G1', 'L1','T1'}
numbPoint=1024*2;
end

%y-noise-free signal, D1PR-the precise first-order derivative
%x-abscissa, dx-sampling interval

[x,dx,y,D1PR] = inputDataWaveletModels(numbPoint, model);

switch model
case 'C'
b1=find(x==1);
e1=find(x==30);
case {'G1', 'T1'}
b1=find(x==12);
e1=find(x==20);
case 'L1'
b1=find(x==11);
e1=find(x==21);
end

newRng=b1:e1;
 %%Normalization of precise derivative
```

```
D1PR=D1PR(newRng)/sum(abs(D1PR(newRng)));
%standard noise deviation
stdN=0.02;
%For SG differentiation filter
m=12;
%Scaling factor for CWT based on the first-order derivative of Gauss
s=0.18;
FourSmooth=0;

flagNoise='w'; %% w- white, 'p'-pink, 'b'=blue, 'r'-red
flagWT=1; %% 1-DWT, 2-CWT

len=length(y);
switch flagNoise
case 'w'
noise=stdN*randn(1,len);
case 'p'
noise=stdN*pinknoise(len);
case 'b'
noise=stdN*bluenoise(len);
case 'r'
noise=stdN*rednoise(len);
end

y1=y+ noise;


power=2;
n=1;
D1SG=derivativeSG( m, power, n, y1 ,1);

%Normalization
D1SG=D1SG(newRng)/sum(abs(D1SG(newRng)));

switch flagWT
case 2
D1wt = CWTgaussD1(x,dx,y1,s);
case 1
level=6;
D1wt=derivative_dwt(y1,'haar',level,dx,1);
end
```

```
%Normalization
D1wt=D1wt(newRng)/sum(abs(D1wt(newRng)));

figure
subplot(2,1,1)
plot(x(newRng),y1(newRng),'-k');

switch model
case 'C'
axis([1,30,-0.2, 1.2])
case {'G1', 'T1'}
axis([12,20,-0.2, 1.2])
case 'L1'
axis([11,21,-0.2, 1.2])
end

strTitle='Model-';
strTitle=strcat(strTitle, model);
strTitle=strcat(strTitle,' Noisy-');
strTitle=strcat(strTitle, flagNoise);
strTitle=strcat(strTitle,' data');

title(strTitle);

if FourSmooth
 alpha=1e-6;
 power=4;
 D1FS = FourSmoothFun(y1,alpha,power,1);
 D1FS=D1FS(newRng)/sum(abs(D1FS(newRng)));
end

subplot(2,1,2)
plot(x(newRng),D1wt,'-k','LineWidth',1)
xlabel('x')
hold on
if FourSmooth
 plot(x(newRng),D1FS,':k','LineWidth',2)
else
 plot(x(newRng),D1SG,':k','LineWidth',1)
end
hold on
```

```
plot(x(newRng),zeros(1,length(newRng)),'-k','LineWidth',1)

switch model
case 'C'
axis([1,30,-1.5e-3, 1.5e-3])

case {'G1', 'T1'}
axis([12,20,-0.006, 0.006])

case 'L1'
axis([11,21,-0.006, 0.006])

end
hold on
plot(x(newRng),D1PR,'--k','LineWidth',1);

if FourSmooth
if flagWT==1
 title('differentiation DWT (-) FS (...) D1PR (--)');
end
if flagWT==2
 title('differentiation CWT (-) FS (...) D1PR (--)');
end
else
if flagWT==1
 title('differentiation DWT (-) SG (...) D1PR (--)');
end
if flagWT==2
 title('differentiation CWT (-) SG (...) D1PR (--)');
end
end


% © Copyright by Joseph Dubrovkin
function FourieSmooth = FourSmoothFun(vector, alpha, power, n)
% n is the differentiation order
len=length(vector);

if mod(len,2)==1
 len=len+1;
end
```

```
len1=len/2;
freq=[1:len1,-len1+1:1:0];

FFT=fft(vector);
M=alpha*freq.^power;
den=(ones(1,len)+M);

if nargin==3
 fourier_filter=1./den.^2;
end
if nargin==4
 difOp=(1i*freq).^n;
 fourier_filter=difOp./den.^2;
end

fourier_filter=fourier_filter(1:len-1);
FFTsolut=FFT.*fourier_filter;
FourieSmooth=real(ifft(FFTsolut));
end

% © Copyright by Joseph Dubrovkin
function [x,dx,y,D1PR] = inputDataWaveletModels(numbPoint, model)
%%D1PR - precise first-order derivative
len=32;
st=len/numbPoint;
x=1:st:len;
dx=x(2)-x(1);
switch model
case 'G1'
 mn=2*sqrt(log(2));
 a1=1.0; t1=len/2; w1=len/16;
 arg1=mn*(x-t1)/w1;
 y=a1*exp(-arg1.^2);
 D1PR=-(2*mn/w1)*arg1.*y;
case 'L1'
 mn=2;
 a1=1.0; t1=len/2; w1=len/16;
 arg1=mn*(x-t1)/w1;
 den=1+arg1.*arg1;
 y=a1./den;
 D1PR=-(2*mn/w1)*arg1./(den.^2);
```

```
case 'T1'
 a1=1.0; t1=len/2; w1=len/16;
 x1=t1-w1;
 x2=t1+w1;
 y=zeros(1,length(x));
 D1PR=y;
 rngX1=0:st:w1;
 beg1=find(x1==x);
 fin1=find(t1==x);
 y(beg1:fin1)=(1/w1)*rngX1;
 D1PR(beg1:fin1)=w1*ones(1,length(beg1:fin1));
 fin2=find(x2==x);
 y(fin1:fin2)=(-1/w1)*rngX1+1;
 D1PR(fin1:fin2)=-w1*ones(1,length(fin1:fin2));
 y=a1*y;
case 'C'
 mn=2*sqrt(log(2));
 a1=1.0;
 t1=5;
 w1=1.3;
 arg1=mn*(x-t1)/w1;
 y1=a1*exp(-arg1.^2);
 Dy1=-(2*mn/w1)*arg1.*y1;
 a2=1.0; t2=15.0; w2=1.3;
 arg2=2*(x-t2)/w2;
 den=1+arg2.^2;
 y2=a2./den;
 Dy2=-(4*a2/w2)*arg2./den.^2;
 a3=1.0; t3=25;
 arg3=3*(x-t3);
 den1=1+exp(-arg3);
 y3=a3./den1;
 Dy3=3*a3*exp(-arg3)./den1.^2;
 y=y1+y2+y3;
 D1PR=Dy1+Dy2+Dy3;
end
end
```

SW6. For Chapter 3.4

Exercise 3.4-1 Readers are invited to inspect some triplets and quartets plotted by the program plotQuartets.m and try to visually guess the number of the peaks and their parameters. Pay attention to the possibility of the false structure in the 2nd-order derivative.

```
%%plotQuartets.m
% © Copyright by Joseph Dubrovkin

range=1000;
x=1:range;
zeroLine=zeros(1,range);
ic=range/2;
flagPeak=1; %Gauss
stdNoise=0;
n=2;
m=2;
power=2;
W = savitzkyGolayModif(m, power,n);

w=100;
I10=1;
I20=1;
I30=1;
I40=1;

%doubletParam=[relShift1, relShift2,relShift2, R1, R2, R3, R4,
%% r1, r2, r3, r4];

for panel=['a', 'b', 'c']
switch panel
case 'a'
 relParam=[ 0.85, 0.65, 1,... %%RelShift
 1, 0.8, 1, 0,... %%Intensities
 1, 0.9, 0.8, 1];%%Widths
 rngX=250:700;
case 'b'
 relParam=[1.7, 0.5, 1,...%%RelShift
 2, 1, 0.5, 0,...%%Intensities
 1, 0.5, 0.5, 1];%%Widths
 rngX=150:650;
case 'c'
```

```
 relParam=[ 1.7, 0.5, -0.5,...
 2, 1, 0.5, 0.1,...
 1, 0.5, 0.5, 1];
 rngX=150:650;
end

relShift1=relParam(1);
relShift2=relParam(2);
relShift3=relParam(3);
R1=relParam(4);
R2=relParam(5);
R3=relParam(6);
R4=relParam(7);
r1=relParam(8);
r2=relParam(9);
r3=relParam(10);
r4=relParam(11);

delta1=relShift1*w;
delta2=relShift2*w;
delta3=relShift3*w;
I1=R1*I10;
I2=R2*I20;
I3=R3*I30;
I4=R4*I40;
w1=r1*w;
w2=r2*w;
w3=r3*w;
w4=r4*w;

[d2, nrm, qvrt,y01N, y02N,y03N,y04N] =...
quartetD2(flagPeak, range, ic, I1, I2,I3,I4, delta1,delta2,delta3, w1,
w2,w3, w4, m, W,stdNoise);

subplot(2,1,1)
plot(x(rngX),qvrt(rngX),'-k','LineWidth',2)
ylabel('F^{(0)}')
hold on
plot(x(rngX),y01N(rngX),':k','LineWidth',2)
hold on
plot(x(rngX),y02N(rngX),':k','LineWidth',2)
```

```
hold on
plot(x(rngX),y03N(rngX),':k','LineWidth',2)
hold on
plot(x(rngX),y04N(rngX),':k','LineWidth',2)
hold off
subplot(2,1,2)
plot(x(rngX),-d2(rngX),'-k','LineWidth',2)
xlabel('x')
ylabel('-F^{(2)}')
hold on
plot(x(rngX),zeroLine(rngX),'-k','LineWidth',1)
hold off
pause
end

function W = savitzkyGolayModif(m, power, n)
x=-m: m;
N=2*m+1;
x=x(:);
A=ones(N,1);
for k=1:power
 A=[A x.^k];
end
matrixW = (A'*A)\A';
W=matrixW(n+1,:)*factorial(n);
end
```

SW7. For Chapter 4.3
For Exercise 4.3-2

The readers are invited to estimates the figures of merit using various sets of the quartet parameters. We recommend studying the strong overlapped peaks to compare the zero-order and derivative analysis

```
%% orthogProg.m
% © Copyright by Joseph Dubrovkin
%%Doublet parameters
range=100;
ic=range/2;
delta=7;
step=1;
ampl1=1;
```

```
ampl2=1.5;
ampl3=0.8;
ampl4=1.2;

imax1=ic-3*delta;
imax2=ic-delta;
imax3=ic+ delta;
imax4=ic+2*delta;

Imax=[imax1 ,imax2,imax3 ,imax4];

width1=10;
width2=15;
width3=12;
width4=5;

[x,y1] = gaussCurve( range,step,ampl1,imax1,width1);
[x,y2] = gaussCurve( range,step,ampl2,imax2,width2);
[x,y3] = gaussCurve( range,step,ampl3,imax3,width3);
[x,y4] = gaussCurve( range,step,ampl4,imax4,width4);

%The second-order derivative
m=2;
d1=derivativeSG( m, 2, 2, y1 ,1);
d2=derivativeSG( m, 2, 2, y2 ,1);
d3=derivativeSG( m, 2, 2, y3 ,1);
d4=derivativeSG( m, 2, 2, y4 ,1);


X=[y1; y2; y3; y4]';
XD=[d1; d2; d3;d4]';

y=y1+y2+y3+y4;
yD=d1+d2+d3+d4;
yF=y2+y3+y4;
nrm=norm(yF);
nrm2=nrm*nrm;
%%comp1_ort=comp1-%%comp2*(comp1')*comp2/norm(comp2)/
%%norm(compt2);
y1ort=y1-yF*(y1')*yF./nrm2;
```

```
[Sen,SEL1,SEL2,SEL3,SEL4,SenNAS,SEL1NAS,SEL2NAS,SEL3NAS,
SEL4NAS,y1ortM1,y1ortM2,y1ortM3,y1ortM4]...
= senSelEstim(y1,y2,y3,y4,Imax);

%% Matrix orthogonal projections
XortM=[y1ortM1,y1ortM2,y1ortM3,y1ortM4];

%% The second-order derivative

%Analytical points
Ianal=[21, imax2, 49, imax4];
[SenD,SEL1D,SEL2D,SEL3D,SEL4D] =...
senSelEstim(d1,d2,d3,d4,Ianal);

newRng=15:75;
subplot(3,1,1)
plot(x(newRng), X(newRng,:), ':k','LineWidth',2)
axis([15,75,0,1.6])
hold on
plot(x(newRng), y(newRng), '-k','LineWidth',2)

subplot(3,1,2)
plot(x(newRng), y(newRng),'-k','LineWidth',2)
axis([15,75,-0.5,1.6])
hold on
plot(x(newRng), XortM(newRng,:),':k','LineWidth',2)
hold on
plot(x(newRng), y1ort(newRng),'-k','LineWidth',1)
hold on
plot(x(newRng), yF(newRng),'--k','LineWidth',2)
hold on
plot(x(newRng), zeros(1,length(x(newRng))),'-k','LineWidth',1)

subplot(3,1,3)
XD(:,1:3)=XD(:,1:3).*10;
plot(x(newRng),-2*XD(newRng,:),':k','LineWidth',2)
axis([15,75,-0.6,2])
hold on
plot(x(newRng),-0.5*yD(newRng)/max(yD(newRng))+1,...
'-k','LineWidth',2)
hold on
```

```
plot(x(newRng),ones(1,length(x(newRng))),'-k','LineWidth',1)

%%%% Sensitivity %%%%%%%%%%
Sen_=Sen
%%-1.999--> >1000
%%SELECTIVITY
SEL1_=SEL1
SEL2_=SEL2
SEL3_=SEL3
SEL4_=SEL4
%%%% Sensitivity NAS
SenNAS_=SenNAS

%%SELECTIVITY NAS

SELNAS=[SEL1NAS,SEL2NAS,SEL3NAS,SEL4NAS]'

%%% The second-order derivative
%%%% Sensitivity
SenD_=SenD

%%SELECTIVITY
%%-1.999--> >1000
SEL1D_=SEL1D
SEL2D_=SEL2D
SEL3D_=SEL3D
SEL4D_=SEL4D
%%%% Sensitivity NAS %%%%%%%%%%
%%% a,b,c,d,e - auxiliary variables
[a,b,c,d,e,SenNASD,SEL1NASD,SEL2NASD,SEL3NASD,SEL4NASD]
= senSelEstim(d1,d2,d3,d4,Imax);
SenNASD_=SenNASD

%%SELECTIVITY NAS-differentiation
SELNASD=[SEL1NASD,SEL2NASD,SEL3NASD,SEL4NASD]'
```

SW8. For Chapter 4.4

For Exercise 4.4-1

The readers are invited to repeat the above numerical experiments by changing the model parameters the differentiation order, the width, and the polynomial power of the SG filter

```
%%informQuantAnal1.m
% © Copyright by Joseph Dubrovkin
%% SG-filter parameters
m=17;
order=2;
power=2;

[X, XD, Imax, W, range, x, errComp] = QuartetA(m, order, power);
Imax=Imax+[-8 3 0 0];

%%Analytical data in D0
Xanal=X(:,Imax);
Ianal=Imax;
%%Analytical data in D2
XderAnal=XD(:,Ianal);

%Sensitivity & Selectivity
%Spectrum
[Sen, SEL, SenNAS, SELNAS]=...
senSelEstim1(X, Imax, errComp);

%Derivative
errCompD=derivativeSG( m, power, order, errComp ,1);
[SenD, SELD, SenNASD, SELNASD]=...
senSelEstim1(XD, Imax, errCompD);

Sen
SenD
SEL
SELD
SenNAS
SenNASD
SELNAS
SELNASD

concentr=[0.7, 0.1, 0.8, 0.1];
%% Mixture spectrum with the interfering component which produces
%% systematic errors
Ypr=concentr*X+ errComp;
stdNoise=0.01;
numbRep =1e3;
```

```
errC=ones(numbRep,4);
errCD=errC;

function [X, XD, Imax, W, range, x, errComp] = QuartetA(m, order,
power)
range=100;
ic=range/2;
delta=6;
step=1;
ampl1=1;
ampl2=1.5;
ampl3=0.8;
ampl4=1.2;
imax1=ic-3*delta;
imax2=ic-delta;
imax3=ic+ delta;
imax4=ic+2*delta;
Imax=[imax1 ,imax2,imax3 ,imax4];
width1=14;
width2=11;
width3=10;
width4=16;
[x,y1] = gaussCurve( range,step,ampl1,imax1,width1);
[x,y2] = gaussCurve( range,step,ampl2,imax2,width2);
[x,y3] = gaussCurve( range,step,ampl3,imax3,width3);
[x,y4] = gaussCurve( range,step,ampl4,imax4,width4);

%The error component
ampl5=0.02;
imax5=ic+2*delta;
width5=50;
[x, errComp]=gaussCurve(range,step,ampl5,imax5,width5);

% n-order derivative
d1=derivativeSG( m, power, order, y1 ,1);
d2=derivativeSG( m, power, order, y2 ,1);
d3=derivativeSG( m, power, order, y3 ,1);
[d4, W]=derivativeSG( m, power, order, y4 ,1);

X=[y1; y2; y3; y4];
XD=[d1; d2; d3;d4];
```

```
end

function [x, y ] = gaussCurve( range, step, ampl, imax, width )
c=1;
y=1:step:range;
%x=y;
for i=1:step:range
 y(c)=gauss(ampl, i, imax, width);
 c=c+1;
end
x=1:c-1;
end

function [der2,W] = derivativeSG( m, power, n, vector, flagSGsym)

%%flagSGsym=1 symSG, [-m, m], =0, asymSG [1,N]
%power=2t
%n=dif order
%m - half-length of the filter
lenVector=length(vector);
der2=zeros(lenVector,1);
if flagSGsym==1
 W = savitzkyGolayModif(m, power, n);
end
if flagSGsym==0
 W = savitzkyGolayModifAsym(m, power, n);
end

for i=1+m:lenVector-m
 der2(i)=(vector(i-m: i+ m)*W');
end
der2=der2';
end

function W = savitzkyGolayModifAsym(m, power, n)
%x=1:N instead of [-m, m]
N=2*m+1;
x=1:N;
x=x(:);
A=ones(N,1);
for k=1:power
```

```
 A=[A x.^k];
end
matrixW = (A'*A)\A';
W=matrixW(n+1,:)*factorial(n);
end

function Cest = cramer(Yanal, Xanal)
det0=det(Xanal);
det1=det([Yanal; Xanal(2:4,:)]);
c1=det1/det0;
det2=det([Xanal(1,:);Yanal; Xanal(3:4,:)]);
c2=det2/det0;
det3=det([Xanal(1:2,:);Yanal; Xanal(4,:)]);
c3=det3/det0;
det4=det([Xanal(1:3,:);Yanal]);
c4=det4/det0;
Cest=[c1,c2,c3,c4];
end
```

SW9 For Appendix B

```
% © Copyright by Joseph Dubrovkin
function [pmg, pmg1, pmg2, pmg3, pmg4] = pmgDerivative(x, tau, r)
%For arguments (y-deltaY/2)/r
if nargin==2
 r=1;
end

[pmg, B, z] = PMG(x, tau);
%Derivatives of Z
dZ = difZ(tau, B);
%Derivatives of ARG
ARG1=-2*z.*dZ(1,:)/r;
ARG2=-2*(dZ(1,:).^2+z.*dZ(2,:))/r^2;
ARG3=-2*(3*dZ(1,:).*dZ(2,:)+z.*dZ(3,:))/r^3;
ARG4=-2*(3*dZ(2,:).^2+4*dZ(1,:).*dZ(3,:)+z.*dZ(4,:))/r^4;

function [pmg, B, z] = PMG(x, tau)
if tau==0
 B=1;
 z=x;
else
```

```
 B=1./(1+tau*x);
 z=(1-B)/tau;
end
z2=z.*z;
pmg=exp(-z2);
end

function arrZ = difZ(tau, B)
len=length(B);
%Derivatives of z
arrZ=zeros(4,len);
fac=1;
for ii=1:4
 fac=fac*ii;
 arrZ(ii,:)=-((-1)^ii)*fac*tau^(ii-1)*B.^(ii+1);
end
end

% © Copyright by Joseph Dubrovkin
function [Fd, Fdd] = difDobosz(y, tau, r)
%For arguments (y-deltaY/2)/r
if nargin==2
 r=1;
end
len=length(y);
[Fl, Fld] = difLorentz(y, r);

if tau==0
 [Fd, Fdd] = difLorentz(y, r);
else
 C=exp(-tau)/tau;
 Ky=tau*atan(y);
 %derivatives
 Kyd=zeros(5,len);
 Kyd(1,:)=tau*Fl;
for ii=2:5
 Kyd(ii,:)=tau*Fld(ii-1,:);
end

Ryd=zeros(4,len);
Ry=C*exp(Ky);
```

```
Ryd(1,:)=(1/r)*Ry.*Kyd(1,:);
Ryd(2,:)=(1/r)*(Ryd(1,:).*Kyd(1,:)+Ry.*Kyd(2,:));
Ryd(3,:)=(1/r)*(Ryd(2,:).*Kyd(1,:)+2*Ryd(1,:).*Kyd(2,:)+...
Ry.*Kyd(3,:));
Ryd(4,:)=(1/r)*(Ryd(3,:).*Kyd(1,:)+3*Ryd(2,:).*Kyd(2,:)+...
3*Ryd(1,:).*Kyd(3,:)+Ry.*Kyd(4,:));

Fd=C*tau*exp(Ky).*Fl;

Fdd=zeros(4,len);
Fdd(1,:)=Ryd(1,:).*Kyd(1,:)+Ry.*Kyd(2,:);
Fdd(2,:)=(Ryd(2,:).*Kyd(1,:)+2*Ryd(1,:).*Kyd(2,:)+...
Ry.*Kyd(3,:));
Fdd(3,:)=Ryd(3,:).*Kyd(1,:)+3*Ryd(2,:).*Kyd(2,:)+...
3*Ryd(1,:).*Kyd(3,:)+Ry.*Kyd(4,:);
Fdd(4,:)=Ryd(4,:).*Kyd(1,:)+4*Ryd(3,:).*Kyd(2,:)+...
6*Ryd(2,:).*Kyd(3,:)+4*Ryd(1,:).*Kyd(4,:)+Ry.*Kyd(5,:);
end
end

function [Fl, Fld] = difLorentz(y, r)
%% 1-6th order Lorentz derivatives Fld
len=length(y);
%For arguments (y-deltaY/2)/r
if nargin==1
 r=1;
end

y2=y.*y;

den=1+y2;
Fl=1./den;

y3=y2.*y;
y4=y2.*y2;
y5=y4.*y;
y6=y3.*y3;

nom=zeros(6,len);
Fld=nom;
```

```
nom(1,:)=-2*y;
nom(2,:)=6*y2-2;
nom(3,:)=-24*y3+24*y;
nom(4,:)=24*(5*y4-10*y2+1);
nom(5,:)=120*(-6*y5+20*y3-6*y);
nom(6,:)=720*(7*y6-35*y4+21*y2-1);

for ii=1:6
 mn=r^ii;
 Fld(ii,:)=nom(ii,:)./((den.^(ii+1))*mn);
end
```

BIBLIOGRAPHY

Part I
Chapter One

1. Danzer, K. *Analytical chemistry. Theoretical and metrological fundamentals*. Springer. 2005.

2. Wentzell, P.D., Brown, C. D. (2000). *Signal Processing in Analytical Chemistry.* In Encyclopedia of Analytical Chemistry R.A. Meyers (Ed.) pp. 9764–9800. John Wiley & Sons Ltd, Chichester.

3. Dubrovkin, J. (1983). Effectiveness of spectral coordinate transformation method in evaluation the unknown spectral parameters. Journal of Applied Spectroscopy, 38, 191-194.

4. Dubrovkin, J. (2014). Critical analysis of spectral deconvolution methods. Journal of Emerging Technologies in Computational and Applied Sciences, 1-10, 40-49.

5. Scott, R. P. W. *Principles and practice of chromatography*. Library-Forscience, LLC. 2003.

6. Felinger, A. (1998). *Data analysis and signal processing in chromatography*. Elsevier.

7. Di Marco, V. B., Bombi, G. G. (2001). Mathematical functions for the representation of chromatographic peaks. Journal of Chromatography A, 931, 1–30.

8. Sharma, B. K. (2007). *Spectroscopy.* 19th. ed. India, Meerut-Delhy: Goel Publishing House.

9. Salzmann, D. (1998). *Atomic Physics in Hot Plasmas*. Oxford University Press.

10. Macomber, R. S. (1998). *A Complete Introduction toModern NMR Spectroscopy*. Wiley.

11. Lazarev, V. V., Krauze, A. S. (2013). The shape of contour bands in vibrational spectra of liquids. Bulletine of Ufa Scientific Centre of the Russian Academy of Science, 1, 5-10. [Russian].

12. Gradshteyn, I. S., Ryznik, I. M. *Table of Integrals, Series, and Products*. 6th ed., Academic Press. 2000.

13. Hirschman, I. I., Widder, D. V. *The Convolution Transform.* Princeton, NJ: Princeton University Press. 1955.

14. Rautian, S. G. (1958). Real spectral instruments. Uspekhi Phisicheskich Nauk, 46, 475-517. [Russian].

15. Olivero, J. J., Longbothum, R. L. (1977). Empirical fits to the Voigt line width: A brief review. Journal of Quantitative Spectroscopy and Radiative Transfer, 17, 233-236.
16. Thompson, P., Cox, D. E., Hastings, J. B. (1987). Rietveld refinement of Debye-Scherrer synchrotron X-ray data from Al2O3. Journal of Applied Crystallography, 20, 79-83.
17. Ida, T., Ando, M., Toraya, H. (2000). Extended pseudo-Voigt function for approximating the Voigt profile. Journal of Applied Crystallography, 33, 1311-1316.
18. Limandri, S. P., Bonetto, R. D., Di Rocco H. O., et al. (2008). Fast and accurate expression for the Voigt function. Application to the determination of uranium M linewidths. Spectrochimica Acta Part B, 63, 962-967.
19. CasaXPS: Processing Software for XPS, AES, SIMS and More. http: //www.casaxps.com/help_manual/line_shapes.htm
20. Harutyunian, H. A. (1996). An expansion of the Voigt function. Astrophysics, 39, 619-625.

Chapter Two

1. Abramowitz, M., Stegun, I. A. (Eds.) (1972). Orthogonal Polynomials. Ch. 22 in Handbook of Mathematical Functions with Formulas, Graphs, and Mathematical Tables. New York: Dover, pp. 771-802.
2. Salzer, H. E., Zucker, R., Capuano, R., (1952). Table of the zeros and weight factors of the first twenty Hermite polynomials. Journal of Research of the National Bureau of Standards, 48, No.2, Research Paper 2294. https://nvlpubs.nist.gov/nistpubs/jres/048/jresv48n2p111_A1b.pdf
3. Chrystal, G. *Algebra: An Elementary Text-Book.* http://djm.cc/library/ Algebra_Elementary_Text-Book_Part_I_Chrystal_edited.pdf
4. Dubrovkin, J., Belikov, V. G. Derivative Spectroscopy. *Theory, Technics, Application. Rostov University*. 1988. [Russian].
5.Di Marco, V. B., Bombi, G. G. (2001). Mathematical functions for the representation of chromatographic peaks. Journal of Chromatography A, 931, 1–30.
6. De Oliveira, E. C., Machado, J. A. T. (2014). A Review of definitions for fractional derivatives and integral. Mathematical Problems in Engineering, Article ID 238459, 6 p.
7. Toman, J. J., Brown, S. D. (1981). Peak resolution by semiderivative voltammetry. Analytical Chemistry, 53, 1497-1504.

8. Grdeń, M. (2017). Semi-differential analysis of irreversible voltammetric peaks. Journal of Solid State Electrochemistry, 21, 1045-1058.
9. Schmitt, J. M. (1998). Fractional derivative analysis of diffuse reflectance spectra. Applied Spectroscopy, 52, 840-846.
10. Kharintsev, S.S., Salakhov, M. Kh. (2003). Peak parameters determination using fractional derivative spectrometry. The Publications of the Astronomical Observatory of Belgrade, No. 76, 211-214.
11. Rzodkiewicz, W., Kulik, M., Papis, E., Szerling, A. (2009). Optical analyses of Si and GaAs semiconductors by fractional-derivative-spectrum methods. Acta Physica Polonica A, 116, S95-S98.
12. Li, Yl., Lin, S. Y. U., Gang, Z. (2007). Resolution method for overlapping peaks based on the fractional-order differential. Science in China. Series B-Chemistry, 5, 797-805.
13. Li, Yl. (2009). Using the Tsallis distribution and the fractional differentiation to resolve the overlapping bands. Analytical and Bioanalytical Chemistry, 394, 637-645.
14. Li, Yl. (2009). Fractional-order differentiation of the Gaussian function for processing overlapped peaks. Analytical Sciences, 25, 1339-1344.
15. Li, Yl., Tang, Hq, Chen Hx. (2011). Fractional-order derivative spectroscopy for resolving simulated overlapped Lorentzian peaks. Chemometrics and Intelligent Laboratory Systems, 107, 83-89.
16. Zheng, K.-Y., Zhang, X., Tong, P.-J., et al. (2015). Pretreating near infrared spectra with fractional order Savitzky-Golay differentiation (FOSGD). Chinese Chemical Letters, 26, 293-296.
17. Hong, Y., Chen, Y., Yu, L., et al. (2018). Combining fractional order derivative and spectral variable selection for organic matter estimation of homogeneous soil samples by VIS-NIR spectroscopy. Remote Sensing, 10, 2-20.
18. Tsallis, C. (1999). Nonextensive statistics: Theoretical, experimental and computational evidences and connections. Brazilian Journal of Physics, 29, 1-35.
19. Tseng, C.-C., Pei, S.-C., Hsia, S.-C. (2000).Computation of fractional derivatives using Fourier transform and digital FIR differentiator. Signal Processing, 80, 151-159.
20. Gladkina, A. Realizing fractional derivatives of elementary and composite functions through the generalized Euler's integral transform and integer derivative series: building the mathematical framework to model the fractional Schrodinger equation in fractional space-time. Thesis MS. (1998). https://inside.mines.edu/~lcarr/theses/gladkina_thesis_2018.

pdf
21. Jalab, H. A., Ibrahim, R. W. (2013). Texture enhancement based on the Savitzky-Golay fractional differential operator. Mathematical Problems in Engineering, ID 149289, 1-8.
22. Li, Yl., Pan, C., Meng, X., et al. (2015). Haar wavelet based implementation method of the non-integer order differentiation and its application to signal enhancement. Measurement Science Review, 15, 101-106.
23. Buslov D.K. (2004). Modification of derivatives for resolution enhancement of bands in overlapped spectra. Applied Spectroscopy, 58, 1302-1307.
24. Antonov, L. (2018). Comment on “Learning to read spectra: Teaching decomposition with Excel in a scientific writing course”. Journal of Chemical Education, 95, 9, 1679-1681.
25. Antonov, L., Stoyanov, S. (1993). Approach for increased information from the second-derivative spectra in UV-Vis absorption spectroscopy. Applied Spectroscopy, 47, 1712-1715.
26. Antonov, L., Stoyanov, S. (1996). Step by step filter - an approach for noise reduction in the derivative UV-visible spectra. Analytica Chimica Acta, 324, 77-83.
27. Antonov, L., Stoyanov, S. (1996). Noise reduction in second derivative UV-VIS spectroscopy. Spectroscopy Letters, 29, 231-239.
28. Antonov, L. (1997). Drawbacks of the present standards for processing absorption spectra recorded linearly as a function of wavelength. TrAC Trends in Analytical Chemistry, 16, 536-543.
29. Antonov, L. (1997). Fourth derivative spectroscopy - a critical view. Analytica Chimica Acta, 349, 295-301.
30. Petrov, V., Antonov, L., Ehara, H., Harada, M. (2000). Step by step filter based program for calculations of highly informative derivative curves. Computers and Chemistry, 24, 561-569.
31. Antonov, L, Nedeltcheva, D. (2000). Resolution of overlapping UV-Vis absorption bands and quantitative analysis. Chemical Society Reviews, 29, 217-227.
32. Antonov, L. (2017). An alternative for the calculation of derivative spectra in the near-infrared spectroscopy. Journal of Near Infrared Spectroscopy, 25, 145-148.

Chapter Three

1. Mark, H., Workman, Jr. *Chemometrics in Spectroscopy*, Elsevier.

2007.
2. Galbán, J., de Marcos, S., Sanz, I., et al. (2007). Uncertainty in modern spectrophotometers. Analytical. Chemistry, 79, 4763-4767.
3. León-Carrión, J., León-Domínguez, U. (2012). Functional Near-Infrared Spectroscopy (fNIRS): Principles and neuroscientific applications. *Neuroimaging- Methods*. Bright, P. (Ed). ISBN: 978-953-51-0097-3, InTech.
4. Brigadoi, S., Ceccherini, L., Cutini, S., et al. (2013). Motion artefacts in functional near-infrared spectroscopy: a comparison of motion correction techniques applied to real cognitive data. NeuroImage, 85 (01).
5. Wentzell, P. D., Brown, C. D. (2000). In *Encyclopaedia of Analytical Chemistry*. John Wiley & Sons, 9764-9800.
6. Fernandez–Ahumada, E., Roger, J. M., Palagos, B. (2012). A new formulation to estimate the variance of model prediction. Application to near infrared spectroscopy calibration. Analytica Chimica Acta, 721, 28-34.
7. Dubrovkin, J., Anan'ev, V. (2013). Error estimation in the quantitative spectrochemical analysis of multicomponent mixtures with randomly varying spectral baseline. Bulletin of Kemerovo State University, 55, 232-236. [Russian].
8. Seber, G. A. F., Lee, A. J. *Linear Regression Analysis*, 2nd ed., Wiley. 2003.

Chapter Four

1. Den Dekker, A. J., Van Den Bos, A. (1997). Resolution: a survey. Journal of the Optical Society of America A, 14, 547-557.
2. Jones, A. W., Bland-Hawthorn, J., Shopbell, P. L. (1995). Towards a general definition for spectroscopic resolution. Astronomical Data Analysis Software and Systems IVASP Conference Series, 77, 503-506.
3. Unser, M., Trus, B. L., Steven, C. (1987). A new resolution criterion based on spectral signal-to-noise ratios. Ultramicroscopy, 23, 39-52.
4. Meilan, P. F., Garavaglia, M. (1997). Rayleigh resolution criterion for light sources of different spectral composition. Brazilian Journal of Physics, 27, 638-643.
5. Sparrow, C. M. (1916). On spectroscopic resolving power. Astrophysical Journal, 44, 76-86.
6. Grechushnikov, B. N., Koragin, V. F. (1970). Noise impact on the resolution of the spectrum lines using different registration means. Journal of Applied Spectroscopy, 13, 569-570. [Russian].

7. Petrash, G. G. (1964). Study of the instrumental distortions and their correction in IR-spectroscopy. Studies on molecular spectroscopy. Proceedings of FIAN, Moscow, 27, 3-62. [Russian].
8. Kozlov, V. P. (1964). Resolution properties of the spectral instruments. 1. Formulation of the problem and a resolution criterion. Optica i Spectroscopiya, 16, 501-506. [Russian].
9. Dubrovkin, J. (1986). Limits of mathematical methods of improving spectral resolution. Journal of Applied Spectroscopy, 43, 991-995.
10. Dubrovkin, J. (2013). Resolution limit of asymmetrical doublets. International Journal of Emerging Technologies in Computational and Applied Sciences, 1-5, 25-31.
11. Dubrovkin, J. *Mathematical Processing of Spectral Data in Analytical Chemistry: A Guide to Error Analysis*. Cambridge Scholars Publishing. 2018.
12. Czarnecki, M. A. (2015). Resolution enhancement in second-derivative spectra. Applied Spectroscopy, 69, 67-74.
13. Dubrovkin, J., Belikov, V. G. *Derivative Spectroscopy. Theory, Technics, Application*. Rostov University. 1988. [Russian].
14. Dubrovkin, J. (2014). Mathematical analysis of asymmetrical spectral lines, International Journal of Emerging Technologies in Computational and Applied Sciences, 1-8 , 27-36.
15. Michelson, A. A. *Light waves and their uses*. Chicago. The University of Chicago Press. 1903.

Chapter Five

1. Eckschlager, K., Danzer, K. (1994). *Information theory in analytical chemistry*. Wiley.
2. Apyari, V.V. (2017). An entropy based approach to estimation of analytical information. A hypothesis. Chemometrics and Intelligent Laboratory Systems, 168, 38-44.
3. Kullback, S. *Information Theory and Statistics*. N.Y.: Dover Publication, Inc. 1997.
4. Dubrovkin, J. (1981). Change in the amount of information contained in a spectrum on differentiation. Journal of Applied Spectroscopy, 35, 1149-1153.
5. Tarasov, K. I. *Spectral instruments*. Mashinostroenie, Leningrad. 1968. [Russian].
6. Goormaghtigh, E., Ruysschaert, J.-M., Raussens, V. (2006). Evaluation of the information content in infrared spectra for protein secondary structure determination. Biophysical Journal, 90, 2946–2957.

Part II
Chapter One

1. Dubrovkin, J., Belikov, V. G. *Derivative Spectrometry. Theory, Technics, Application.* Rostov University. 1988. [Russian].
2. Courant, R. *Differential and Integral Calculus*. Vol. 1. Ishi Press. 2013.
3. Dubrovkin J. (2015). Mathematical analysis of asymmetrical spectral lines. Case study: Exponentially modified functions. International Journal of Emerging Technologies in Computational and Applied Sciences, 3-11, 194-201.
4. Santiago, J. *Circuit Analysis for Dummies*. John Wiley & Sons, Inc. 2013.

Chapter Two

1. Courant, R. *Differential and Integral Calculus*. Vol. 1. Ishi Press. 2013.
2. Korn, G. A., Korn, T. M. *Mathematical handbook for scientists and engineer: definitions, theorem, and formulas for reference and review*. N. Y.: McGraw-Hill, 1968.
3. Dubrovkin, J., Belikov, V. G. *Derivative Spectrometry. Theory, Technics, Application*. Rostov University. 1988. [Russian].
4. Gradshteyn, I. S., Ryznik, I. M. *Table of Integrals, Series, and Products*. 7th ed. Elsevier. 2007.
5. Smith, S. W. *The Scientist and Engineer's Guide to Digital Signal Processing*. 2nd ed. San Diego: California Technical Publishing. 1997-1999.
6. Hamming, R. *Digital Filters*. Dover Publication. 1997.
7. Rinnan, Å., Van den Berg , F., Engelsen, S., B. (2009). Review of the most common pre-processing techniques for near-infrared spectra. TrAC Trends in Analytical Chemistry, 28, 1201-1222.
8. Norris, K. H., Williams, P. C. (1984). Optimization of mathematical treatments of raw near-infrared signal in the measurement of protein in hard red spring wheat. I. Influence of particle size. Cereal Chemistry, 61,158-165.
9. Dubrovkin, J. (2017). Use of a hypothetically optimal digital smoothing filter for comparative study of signal denoising methods in analytical chemistry. International Journal of Emerging Technologies in Computational and Applied Sciences, 1-21, 5-10.

Chapter Three

1. Savitzky, A., Golay, M. J. E. (1964). Smoothing and differentiation of data by simplified least squares procedures. Analytical Chemistry, 36, 1627-1639.

2. Steinier, J., Termonia, Y., Deltour, J. (1972). Comments on smoothing and differentiation of data by simplified least square procedures. Analytical Chemistry, 44, 1906-1909.

3. Madden, H. H. (1978). Comments on Savitzky-Golay convolution method for least-squares fit smoothing and differentiation of digital data. Analytical Chemistry, 50, 1383-1386.

4. Leach, R. A., Carter, C. A., Harris, J. M. (1984). Least-squares polynomial filters for initial point and slope estimation. Analytical Chemistry, 56, 2304-2307.

5. Baedecker, P. A. (1985). Comments on least-squares polynomial filters for initial point and slope estimation. Analytical Chemistry, 57, 1477-1479.

6. Gorry, P. A. (1990). General least-squares smoothing and differentiation by the convolution (Savitzky-Golay) method. Analytical Chemistry, 62, 570-573.

7. Lanczos, C. *Applied Analysis*. Dover Books on Mathematics. 2010.

8. Dubrovkin, J. (1984). Choice of an optimal scheme of numerical differentiation of spectra. Journal of Applied Spectroscopy, 41, 789-794.

9. Dubrovkin, J. (1985). Theory of smoothing and differentiation Savitzky-Golay filters. Journal of Applied Spectroscopy, Deposited at VINITI 26-09-1985. [Russian].

10. Dubrovkin, J. (1986). Processing of spectrophotometric data by digital filters. Accuracy of smoothing and differentiation using Savitzky-Golay filters. The Bulletin of the North Kavkaz Scientific Centre of Rostov University. Russia. Deposited at 28-03-1986. [Russian].

11. Dubrovkin, J., Belikov, V. G. *Derivative spectrometry. Theory, Technics, Application*. Rostov University Publishing. Russia. 1988. [Russian].

12. Menon, S. V., Seelamantula. C. S. (2014). Robust Savitzky-Golay filters. Proceedings of the 19th International Conference on Digital Signal Processing.

13. Persson, P.-O., Strang. G. (2003). Smoothing by Savitzky-Golay and Legendre filters. Mathematical Systems Theory in Biology, Communications, Computation, and Finance. The IMA Volumes in Mathematics and its Applications, vol. 134. Springer, New York, NY, 301-315.

14. Tikhonov, A. N., Arsenin. V. Y. *Solutions of Ill-Posed Problems*. Winston and Sons, Washington DC, 1977.
15. Sadeghi, M., Behnia, F. Optimum window length of Savitzky-Golay filters with arbitrary order. https://arxiv.org/abs/1808.10489.
16. Krishnan, S. R., Seelamantula, C. S. (2013). On the selection of optimum Savitzky-Golay filters. IEEE Transactions on Signal Processing, 61. 380-391.
17. Schafer, R. W. What Is a Savitzky-Golay filter? IEEE signal processing magazine [111] July 20111053.
18. Dubrovkin. J. *Mathematical Processing of Spectral Data in Analytical Chemistry: A Guide to Error Analysis*. Cambridge Scholars Publishing. 2018.
19. Dubrovkin, J. (2017). Use of a hypothetically optimal digital smoothing filter for comparative study of signal denoising methods in analytical chemistry, International Journal of Emerging Technologies in Computational and Applied Sciences, 1-21, 5-10.
20. Hershey, H. C., Zakin, J. L., Simha, R. (1967). Numerical differentiation of equally spaced and not equally spaced experimental data. Industrial & Engineering Chemistry Fundamentals, 6, 413-421.
21. Antonov, L., Stoyanov, S. (1996). Step by step filter - an approach for noise reduction in the derivative UV-visible spectra. Analytica Chimica Acta, 324, 77-83.
22. Petrov, V., Antonov, L., Ehara, H., Harada, M. (2000). Step by step filter based program for calculations of highly informative derivative curves. Computers and Chemistry, 24, 561-569.

Chapter Four

1. Heideman, M., Johnson, D., Burrus. C. (1984). Gauss and the history of the fast Fourier transform. IEEE ASSP Magazine, 1, 14-21.
2. Lanczos. C. *Applied Analysis*. Dover Books on Mathematics. 2010.
3. Tikhonov, A. N., Arsenin, V.Y. *Solutions of Ill-Posed Problems*. Winston and Sons, Washington DC, 1977.
4. Angelini, A., De Canditiis, D. (2000). Fourier frequency adaptive regularization for smoothing data. Journal of Computational and Applied Mathematics, 115, 35-50.
5. Betty, K. R., Horlick. G. (1976). A Simple and versatile Fourier domain digital filter. Applied Spectroscopy, 30, 23-27.
6. Dubrovkin, J. Zubarev, S. (1978). Fourier transform derivative IR-spectra. Zavodskay Laboratory, 4, 434-435. [Russian].

7. Savitzky, A., Golay, M. J. E. (1974). Smoothing and differentiation of data by simplified least squares procedures. Analytical Chemistry, 36, 1627-1639.
8. Dubrovkin, J. *Mathematical Processing of Spectral Data in Analytical Chemistry: A Guide to Error Analysis*. Cambridge Scholars Publishing. 2018.
9. Whittaker. E. T. (1922). On a new method of graduation. Proceedings of the Edinburgh Mathematical Society, 41, 63-75.
10. Hansen, P. C. Regularization Tools. A MATLAB package for analysis and solution of discrete ill-posed problems. http://www.imm.dtu.dk/~pcha/Regutools/
11. Zhivomirov, H. https://www.mathworks.com/matlabcentral/answers/429107-how-can-i-generate-a-pinknoise-with-a-certain-duration
12. Taniguchi, M., Du, H., Lindsey. J. S. (2018). PhotochemCAD 3: Diverse modules for photophysical calculations with access to multiple spectral databases. Journal of Photochemistry and Photobiology A: Chemistry, 94, 277-289.
13. Taniguchi, M., Lindsey. J. S. (2018). Database of absorption and fluorescence spectra of >300 common compounds for use in PhotochemCAD. Journal of Photochemistry and Photobiology A: Chemistry, 94, 290-327.
14. Dubrovkin, J. (2017). Use of a hypothetically optimal digital smoothing filter for comparative study of signal denoising methods in analytical chemistry, International Journal of Emerging Technologies in Computational and Applied Sciences, 1-21, 5-10.

Chapter Five

1. Dubrovkin, J. (2017). Use of a hypothetically optimal digital smoothing filter for comparative study of signal denoising methods in analytical chemistry, International Journal of Emerging Technologies in Computational and Applied Sciences, 1-21, 5-10.
2. Savitzky, A., Golay, M. J. E. (1964). Smoothing and differentiation of data by simplified least squares procedures. Analytical Chemistry, 36, 1627-1639.
3. Whittaker, E. T., Robinson. G. *The Calculus of Observations. A Treatise on Numerical Mathematics,* Old Bailey, London, 1924.
4. Lanczos, C. *Applied Analysis*, Prentice Hall, 1956.
5. Bromba, M. U. A., Ziegler, H. (1979). Efficient computation of polynomial smoothing digital filters. Analytical Chemistry, 51, 1760-1762.

6. Dubrovkin, J., Belikov, V.G. *Derivative Spectrometry: Theory, Technology, and Application.* Rostov: Rostov University, 1988. [Russian].
7. Gorry, P. A. (1990). General least-squares smoothing and differentiation by the convolution (Savitzky-Golay) method. Analytical Chemistry, 62, 570-573.
8. Chen, K., Zhang, H., Wei, H., Li, Y. (2014). Improved Savitzky-Golay-method-based fluorescence subtraction algorithm for rapid recovery of Raman spectra. Applied Optics, 53, 5559-5569.
9. Bromba, M.U.A., Ziegler, H. (1981). Application hints for Savitzky-Golay digital smoothing filters. Analytical Chemistry, 53, 1583-1586.
10. Ziegler, H. (1981). Properties of digital smoothing polynomial (DISPO) filters. Applied Spectroscopy, 35, 88-92.
11. Bromba, M .U. A., Ziegler, H. (1983). Digital smoothing of noisy spectra. Analytical Chemistry, 55, 648-653.
12. Tikhonov, A. N., Arsenin, V. Y. *Solutions of Ill-posed Problems*. VH Winston and sons, 1977.
13. Eilers, P. H. C. (2003). A perfect smoother. Analytical Chemistry, 75, 3631-3636.
14. Stickel, J. J. (2010). Data smoothing and numerical differentiation by a regularization method. Computers & Chemical Engineering, 34, 467-475.
15. Palacio, C., Pascual, C., Suarez, F., Lloret, I. (2002). Smoothing of digital spectroscopic data by using a Chebyshev filter. Vacuum, 64, 481-485.
16. López-Camacho, E., García-Cortés, A., Palacio, C. (2006). A family of smoothing algorithms for electron and other spectroscopies based on Chebyshev filter. Thin Solid Films, 513, 72-77.
17. Bromba, M. U. A., Ziegler, H. (1983). On Hilbert space design of least-weighted-squares digital filters. International Journal of Circuit Theory and Applications, 11, 7-32.
18. Jusman, Y., Isa, N. A. M., Ng, S. C., et al. (2015). Quadratic of half ellipse smoothing technique for cervical cells FTIR spectra in a screening system. Procedia Computer Science, 59, 133-141.
19. Jordan, R., Dyer, S.A. (1989). Adaptive signal processing techniques: an application to Raman spectroscopy. Applied Spectroscopy, 43, 750-758.
20. Kalambet, Y., Maltsev, S., Kosmin, Y. The best method of noise filtering. Winter School on Chemometrics, 02/2012 (Russia). http://wsc.chemometrics.ru/wsc8/

21. Smith, S. W. *The Scientist and Engineer's Guide to Digital Signal Processing*. 2nd Ed., California Technical Publishing, 1999.
22. Urbas, A. A., Choquette, S. J. Automated spectral smoothing with spatially adaptive penalized least-squares. http://ws680.nist.gov/publication/get_pdf.cfm?pub_id=904908.
23. Kauppinen, J. K., Moffatt, D. J., Mantsch, H. H., Cameron. D.G. (1982). Smoothing of spectral data in the Fourier domain. Applied Optics, 21, 1866-1872.
24. Abdel-Hafiez, A. (2007). Fourier transformation methods in the field of gamma spectrometry. International Journal of Pure and Applied Physics, 3, 132-141.
25. Frasso, G., Eilers, P.H. C. (2015). L- and V-curves for optimal smoothing. Statistical Modelling, 15, 91-111.
26. Walczak, B., Massart, D. L. (1997). Tutorial. Noise suppression and signal compression using the wavelet packet transform. Chemometrics and Intelligence Laboratory Systems, 36, 81-94.
27. Pasti, L., Walczak, B., Massart, D. L., Reschiglian, P. (1999). Optimization of signal denoising in discrete wavelet transform. Chemometrics and Intelligence Laboratory Systems, 48, 21-34.
28. Iwata, T., Koshoubu, J. (1994). New method to eliminate the background noise from a line spectrum. Applied Spectroscopy, 48, 1453-1456.
29. Iwata, T., Koshoubu, J. (1996). Minimization of noise in spectral data. Applied Spectroscopy, 50, 747-752.
30. Zaharov, V. V., Farahi, R. H., Snyder, P. J., et al. (2014). Kahrunen-Loève treatment to remove noise and facilitate data analysis in sensing, spectroscopy and other applications. Analyst, 139, 5927-5935.
31. Antonacopoulos, A., Economou, A. (1998). A structural approach for smoothing noisy peak-shaped analytical signals. Chemometrics and Intelligence Laboratory Systems, 41, 31-42.
32. Engel, J., Gerretzen, J., Szymańska, E., et al. (2013). Breaking with trends in pre-processing? TrAC Trends in Analytical Chemistry, 50, 96-106.
33. Brown, C. D., Wentzell, P .D. (1999). Hazards of digital smoothing filters as a preprocessing tool in multivariate calibration. Journal of Chemometrics, 13, 133-152.
34. Simonoff, J. S. *Smoothing Methods in Statistics, Springer Series in Statistics*, 1996.
35. Kabanikhin, S. I. (2008). Definitions and examples of inverse and ill-posed problems. Journal of Inversed and Ill-Posed Problems, 16, 317-357.

36. Seber, G. A. F., Lee, A. J. *Linear Regression Analysis*, 2th edition, Wiley, 2003.
37. https://www.researchgate.net/post/How_can_one_analytically_inverse_the_sum_of_the_singular_and_diagonal_matrices/1 [accessed July 1, 2020].
38. Zhivomirov, H. Pink, red, blue and violet noise generation with Matlab implementation. https://www.mathworks.com/matlabcentral/fileexchange/42919-pink--red--blue-and-violet-noise-generation-with-matlab-implementation?focused=6636118&tab=function

Chapter Six

1. Dubrovkin. J. *Mathematical Processing of Spectral Data in Analytical Chemistry: A Guide to Error Analysis.* Cambridge Scholars Publishing. 2018.
2. Dubrovkin, J., Belikov, V. G. *Derivative Spectrometry: Theory, Technology, and Application.* Rostov University Publishing. Russia. 1988. [Russian].

Chapter Seven

1. Gans, P., Gill, J. B. (1984). Smoothing and differentiation of spectroscopic curves using spline functions. Applied Spectroscopy, 38, 370-376.
2. Rzhevskii, A. M., Mardilovich, P. P. (1994). Generalized Gans-Gill method for smoothing and differentiation of composite profiles in practice. Applied Spectroscopy, 48, 13-20.
3. Voscoboinikov, U. E. (1986). Frequency approach to the estimation of the accuracy of the experimental data smoothing and differentiation using smoothing splines. Autometriya, 1, 38-43. [Russian].
4. Lundgren, J. SPLINEFIT version 1.0.0 (238 KB). https://ww2.mathworks.cn/matlabcentral/fileexchange/71225-splinefit?focused=888438cd-bc77-4e1d-8c1a-bd6e763f4113&tab=function

Chapter Eight

1. Daubechies, I. (1996). Where do wavelets come from? Proceedings of the IEEE, 84, 510-513.
2. Chau, F. T., Leung, A. K. M., Gao, J. B. (1998). A review on application of wavelet transform techniques in chemical analysis: 1989-1997. Chemometrics and Intelligent Laboratory Systems, 43, 165-184.

3. Chau, F. T., Leung, A. K. M., Gao, J. B., Shao, X. G. *Chemometrics from basics to wavelet transform. Chemical analysis*. A series of monograph on analytical chemistry and its applications. Ed. Winefordner. Vol. 164. Wiley. 2004.
4. Leung, A. K. M., Chau, F. T., Gao, J. B. (1998). Wavelet transform: a method for derivative calculation in analytical chemistry. Analytical Chemistry, 70, 5222-5229.
5. Shao, X., Ma, C. (2003). A general approach to derivative calculation using wavelet transform. Chemometrics and Intelligent Laboratory Systems, 69, 157-165.
6. Zhao, F., Wang, J., Wang, A. (2016). An improved spectral background subtraction method based on wavelet energy. Applied Spectroscopy, 70, 1994-2004.
7. Sundling, C. M., Sukumar, N., Zhang, H., et al. (2006). Wavelets in chemistry and cheminformatics. Reviews in Computational Chemistry, 22, Ch. 5, 295-329.
8. Dinç, E., Yazan, Z. (2018). Wavelet transform-based UV spectroscopy for pharmaceutical analysis. Frontiers in Chemistry, 6.
9. Kharintsev, S. S., Kamalova, D.I., Salakhov, M. Kh., Sevastianov, A.A. (2005). Resolution enhancement of composite spectra using wavelet-based derivative spectrometry. Spectrochimica Acta A, 61,149-156.
10. Süslü İ., Dinç E., Altinöz, S. (2007). An application of continuous wavelet transform to electrochemical signals for the quantitative analysis. In: Taş, K., Machado, T., J.A. Baleanu, D. (Ed.) Mathematical Methods in Engineering. Springer, Dordrecht. 303-313.
11. Torrence, C., Compo, G. P. (1998). A Practical guide to wavelet analysis. Bulletin of the American Meteorological Society, 79, 61-78.
12. Walnut, D. F. *An Introduction to wavelet analysis*. Birkhäuser. 2004
13. Mallat, S. G. (1989). A theory for multiresolution signal decomposition: the wavelet representation. IEEE Transactions on Pattern Analysis and Machine Intelligence, 2, 674-693.
14. Antoniadis, A., Fan, J. (2001) Regularization of wavelet approximations. Journal of the American Statistical Association, 96, 939-967.
15. Fu, C.-L., Feng, X. L., Qian, Z. (2010). Wavelets and high order numerical differentiation. Applied Mathematical Modelling, 34, 3008-3021.
16. Zhivomirov, H. (2018). A method for colored noise generation. Romanian Journal of Acoustics and Vibration, 15, 14-19.

Chapter Nine

1. Zhao, Z., Bai, L., Han, J., Yue, J. High-SNR snapshot multiplex spectrometer with sub-Hadamard-S matrix coding. https://arxiv.org/ftp/arxiv/papers/1905/1905.07543.pdf
2. Dubrovkin, J., Belikov, V. G. *Derivative Spectrometry. Theory, Technics, Application*. Rostov University. 1988. [Russian].
3. El-Malla, S., Kamal, A., Hammad, S. (2016). A review on UV spectrophotometric methods for simultaneous multicomponent analysis. European Journal of Pharmaceutical and Medical Research, 3, 348-360.
4. Galitskas, A. A. (1977). Identification of the derivatives of the complicated characteristic using a deep modulation of its argument. Autometriya, 6, 41-44. [Russian]. https://www.iae.nsk.su/images/stories/5_Autometria/5_Archives/1977/6/41-44.pdf
5. Supplee, M. J., Whittaker, E. A., Lenth, W. (1994).Theoretical description of frequency modulation and wavelength modulation spectroscopy. Applied Optics, 33, 6294-6302.
6. Hager, R. N., Anderson, R. C. (1970). Theory of the derivative spectrometer. Journal of the Optical Society of America, 60, 1444-1449.
7. Korn, G. A., Korn, T. M. *Mathematical handbook for scientists and engineer: definitions, theorem, and formulas for reference and review*. N. Y.: McGraw-Hill. 1968.
8. Hangauer, A. (2012). Detection schemes, algorithms and device modeling for tunable diode laser absorption spectroscopy. Doctor diss. Technische universitat Munchen.. Lehrstuhl fur Halbleitertechnologie am Walter Schottky Institut. https://d-nb.info/1037820614/34.
9. Dyroff, C. *Tunable Diode-Laser Absorption Spectroscopy for Trace-Gas Measurements with High Sensitivity and Low Drift*. Karlsruhe Series in Photonics & Communications, Vol. 5, Eds. J. Leuthold, W. Freude. 2005. https://pdfs.semanticscholar.org/42d4/ f8b65aac06a7d7c64723a6f099e565c8f284.pdf
10. Green, G. L., O'Haver, T. C. (1974). Derivative luminescence spectrometry. Analytical Chemistry, 46, 2191-2196.
11. Wang, Z., Fu, P., Chao, X. (2019). Laser absorption sensing systems: challenges, modeling, and design optimization. Applied Science, 9, 2723. https://www.mdpi.com/2076-3417/9/13/2723/pdf-vor
12. Hammond, V. J., Price, W. C. (1953). A new system for the elimination of scattered light effects in spectrophotometers. Journal of the Optical Society of America, 43, 924-924.
13. McWilliam, I. G. (1959). An oscillating-plate differentiator for spectrophotometry. Journal of Scientific Instruments, 36, 51-52.

14. Snelleman, W., Rains, T. C., Yee, K. W., et al. (1970). Flame emission spectrometry with repetitive optical scanning in the derivative mode. Analytical Chemistry, 42, 394-398.
15. Epstein, M. S., Rains, T. C., O'Haver, T. C. (1976). Wavelength modulation for background correction in graphite furnace atomic emission spectrometry. Applied Spectroscopy, 30, 324-329.
16. Zander, A. T. (1977). Waveform effects in wavelength modulation spectrometry. Analytical Chemistry, 49, 458-461.
17. Koirtyohann, S. R., Glass, E. D., Yates, D. A., et al. (1977). Effect of modulation wave form on the utility of emission background corrections obtained with an oscillating refractor plate. Analytical Chemistry, 49, 1121-1126
18. Michel, R. G., Sneddon, J., Hunter, J. K., et al. (1981). A novel method of wavelength modulation for atomic spectrometry - some preliminary experiments. Analyst, 106, 288-298.
19. McCaffrey, J. T., Michel, R. G. (1983). Sectored wheel wavelength modulation for flame atomic fluorescence spectrometry. Analytical Chemistry, 55, 488-492.
20. Sydor, R. J., Hieftje, G. M. (1976). Inexpensive, wavelength-modulated filter flame photometer. Analytical Chemistry, 48, 535-541.
21. Aramu, F., Rucci, A. (1966). Self-modulating derivative densitometer. Review of Scientific Instruments, 37, 1696-1698.
22. Cook, T. E., Pardue, H. L., Santini, R.E. (1976). Derivative spectrometry with a vidicon detector. Analytical Chemistry, 48, 451-452.
23. Cook, T. E., Santini, R.E., Pardue, H. L. (1977). Design and evaluation of a vidicon based derivative spectrometer. Analytical Chemistry, 49, 871-877.
24. Baldini, G., Grilli, E., Guzzi, M. (1975). Simple derivative optical spectrometer. Applied Optics, 14, 2687-2690.
25. Bonfiglioli, G., Brovetto, P. (1964). Principles of self-modulating derivative optical spectroscopy. Applied Optics, 3, 1417-1424.
26. Montegu, B., Laugier, A., Fournier, J. M. (1979). Symmetrical wavelength modulation spectrometer. Journal of Physics E: Scientific Instruments, 12, 1153-1159.
27. Shaw, S.-Y., Lue, J.-T. (1981). An absolute measurement of optical reflectivity by modification of a WMS system. Journal of Physics E: Scientific Instruments, 14, 1135-1137.
28. Wynn, T.F., Clardy, P., Vaughn, L., et al. (1981). Wavelength-modulated, continuum-source excited atomic fluorescence spectrometric system for wear metals in jet engine lubricating oils using electrothermal atomization. Analytica Chimica Acta, 124, 155-161.

29. Zucca, R., Shen, Y. R. (1973). Wide-range wavelength modulation spectrometer. Applied Optics, 12, 1293-1298.

Chapter Ten

1. Zare, R. N. (1984). Laser chemical analysis. Science, 226, 298-303.
2. Wright, J. C., Wirth, M. J. (1980). Principles of Lasers. Analytical Chemistry, 52, 1087A-1095A.
3. Hieftje, G. M., Travis, J. L., Fred, E. *Lasers in Chemical Analysis.* Springer. 1981.
4. Ferraro, J. R., Nakamoto, K., Brown, C. W. *Introductory Raman Spectroscopy*. Elsevier Inc. 2003.
5. Nwaboh, J. A., Desbois, T., Romanini, D., et al. (2011). Molecular laser spectroscopy as a tool for gas analysis applications. International Journal of Spectroscopy, 2011, ID 568913.
6. Schilt, S., Thevenas, L., Robert, P. (2003). Wavelength modulation spectroscopy: combined frequency and intensity laser modulation. Applied Optics, 42, 6728-6738.
7. Reid, J., Labrie, D. (1981). Second-harmonic detection with tunable diode lasers-Comparison of experiment and theory. Applied Physics, B 26, 203-210.
8. Di Rosa, M. D., Reiten, M.T. (2015). Computational expressions for signals in frequency-modulation spectroscopy. Applied Optics, 54, 5031-5036.
9. Kramer, M. A., Boyd, R. W., Hillman, L. W., Stroud, Jr., C. R. (1985). Propagation of modulated optical fields through saturable-absorbing media: a general theory of modulation spectroscopy. Journal of the Optical Society of America B, 2, 1444-1455.
10. Dubrovkin, J. *Mathematical Processing of Spectral Data in Analytical Chemistry: A Guide to Error Analysis*. Cambridge Scholars Publishing. 2018.
11. Werle, P., Slemr, F. (1991). Signal-to-noise ratio analysis in laser absorption spectrometers using optical multipass cells. Applied Optics, 30, 430-434.
12. Kluczynski, P., Gustafsson. J., Gustafsson, J., et al. (2001). Wavelength modulation absorption spectrometry-An extensive scrutiny of the generation of signals. Spectrochimica Acta Part B, Atomic Spectroscopy, 56, 1277-1354.
13. Mei, L., Svanberg, S. (2015). Wavelength modulation spectroscopy-digital detection of gas absorption harmonics based on Fourier analysis. Applied Optics, 54, 2234-2243.

14. Gregory, E. H., North, S. W. (2000). Transient laser frequency modulation spectroscopy. The Annual Review of Physical Chemistry, 51, 243-274.
15. Paschotta, R. Electro-optic Modulators. RP Photonics Encyclopedia. https://www.rp-photonics.com/electro_optic_modulators.html
16. Namjou, K., Cai, S., Whittaker, E. A., et al. (1998). Sensitive absorption spectroscopy with a room-temperature distributed-feedback quantum-cascade laser. Optics Letters, 23, 219-221.
17. Arslanov, D. D., Spunei, M., Ngai A. K. Y., et al. (2011). Rapid and sensitive trace gas detection with continuous wave optical parametric oscillator-based wavelength modulation Spectroscopy. Applied Physics B, 103, 223–228.
18. Lindsay, I. D., Groß, P., Lee, C. J., et al. (2006). Mid-infrared wavelength- and frequency modulation spectroscopy with a pump-modulated singly-resonant optical parametric oscillator. Optics Express, 14, 12341-12346.
19. Ye, J., Long-Sheng Ma, L.-S., Hall, J. L. (1998). Ultrasensitive detections in atomic and molecular physics: demonstration in molecular overtone spectroscopy. Journal of the Optical Society of America A B, 15, 6-15.
20. Mandom, J., Guelachvili, G., Picque, N. (2007). Frequency-modulation Fourier transform spectroscopy: a broadband method for measuring weak absorption and dispersions. Optics Letters, 32, 2206-2209.
21. Khazimullin, M. V., Lebedev, Yu. A. (2010). Fourier transforms approach in modulation technique of experimental measurements. Review of Scientific Instruments, 81, 043110.

Part III
Chapter One

1. Dubrovkin, J. (2014). Evaluation of the peak location uncertainty in second-order derivative spectra. Case study: Symmetrical lines. Journal of Emerging Technologies in Computational and Applied Sciences, 3-9, 248-255.
2. Dubrovkin, J., Belikov, V. G. *Derivative Spectroscopy. Theory, Technics, Application.* Rostov University. 1988. [Russian].
3. Sharma, B. K. *Spectroscopy.* 19th. ed. India, Meerut-Delhy: Goel Publishing House. 2007.

4. Smeller, L. (1998). How precise are the positions of computer-determined peaks? Applied Spectroscopy, 52, 1623-1626.
5. Witjes, H., Pepers, M., Melssen, W. J., Buydens, L. M. C. (2001). Modelling phase shifts, peak shifts and peak width variations in spectral data sets: its value in multivariate data analysis. Analytica Chimica Acta, 432, 113-124.
6. Ryu, S. R., Noda, I. Jung, Y. M. (2010). What is the origin of the positional fluctuation of spectral features: true frequency shift or relative intensity changes of two overlapped bands? Applied Spectroscopy, 64, 1017-1021.
7. Gurevich, A. L., Rusinov, L. A., Syagaev, N. A. *Automatic Chromatographic Analysis*. Leningrad, Khimiya. 1980. [Russian].
8. Zhang, J. (2009). Review of peak detection algorithms in liquid-chromatography-mass spectrometry. Current Genomics, 10, 388-401.
9. Kline, M. *Calculus: An intuitive and physical approach*. 2nd Ed. Dover books on mathematics. 1998.
10. Di Marco, V. B., Bombi, G. G. (2001). Mathematical functions for the representation of chromatographic peaks. Journal of Chromatography A, 931, 1–30.
11. Buslov, D. K. (1984). Selection of zero approximations in decomposition of complex contours. Journal of Applied Spectroscopy, 37, 459-462.
12. Dubrovkin J. *Mathematical processing of spectral data in analytical chemistry: A guide to error analysis.* Cambridge Scholars Publishing. 2018.
13. Korobkov, M. E. (1978). Informational content of derivative spectra. Journal of Applied Spectroscopy, 28, 716–718.
14. Wahab, M. F., O'Haver, T. C., Grittic, F., et al. (2019). Increasing chromatographic resolution of analytical signals using derivative enhancement approach. Talanta, 192, 492-499.

Chapter Two

1. Dubrovkin, J. (1981). Change in the amount of information contained in a spectrum on differentiation. Journal of Applied Spectroscopy, 35, 1149-1153.
2. Dubrovkin J. (2014). Evaluation of the peak location uncertainty in second-order derivative spectra. Case study: symmetrical lines. International Journal of Emerging Technologies in Computational and Applied Sciences, 3-9, 248-255.

3. Dubrovkin, J., Belikov, V. G. *Derivative Spectroscopy. Theory,Technics, Application*. Rostov University. 1988.

Chapter Three

1. Dubrovkin, J., Belikov, V. G. *Derivative Spectrometry: Theory, Technology, and Application.* Rostov: Rostov University.1988. [Russian].
2. Li, Yl., Tang, Hq, Chen, Hx. (2011). Fractional-order derivative spectroscopy for resolving simulated overlapped Lorentzian peaks. Chemometrics and Intelligent Laboratory Systems, 107, 83-89.
3. Zhang, B, Yu, H., Sun, L., et al. (2013). A Method for resolving overlapped peaks in Laser-Induced Breakdown Spectroscopy (LIBS). Applied Spectroscopy, 67, 1087-1097.
4. Kharintsev, S.S., Salakhov, M. Kh. (2003). Peak parameters determination using fractional derivative spectrometry. Publications of the Astronomical Observatory of Belgrade, No. 76, 211-214.
5. Buslov, D. K. (1984). Selection of zero approximations in decomposition of complex contours. Journal of Applied Spectroscopy, 37, 459-462. [Russian].
6. Morrey, J. R. (1968). On determining spectral peak positions from composite spectra with a digital computer. Analytical Chemistry, 40, 905-914.
7. Lischka, H., Derflinger, G. (1968). Zur Analyse von Bandenspektren, 2. Mitt. Monatshefte für Chemie, 99, 2450-2459.
8. Koryukina, E. V., Revalde, G. (2006). Separation of the overlapping spectral lines using derivative spectra. Russian Physics Journal, 49, 435-441.

Chapter Four

1. Dubrovkin. J. *Mathematical Processing of Spectral Data in Analytical Chemistry: A Guide to Error Analysis.* Cambridge Scholars Publishing. 2018.
2. Bazaraa, M. S., Hanif, D. Sherali, H. D., Shetty, C. M. (1993). *Nonlinear Programming: Theory and Algorithms*. 2nd ed. Wiley. 1993.
3. Pitha, J., Jones, R.N. (1966). A comparison of optimization methods for fitting curves to infrared band envelopes. Canadian Journal of Chemistry, 44, 3031-3050.
4. Yang, C. M., Beck, J. L. (1998). Generalized trajectory methods for finding multiple extrema and roots of functions. Journal of Optimization Theory and Applications, 97, 211-227.

5. Gribov, L. A., Baranov, V. I., Elyashberg, M. E. *Standardless molecular spectral analysis, Theoretical fundamentals.* Moscow: Editorial URSS. 2002.
6. Dubrovkin, J., Belikov, V. G. *Derivative Spectrometry: Theory, Technology, and Application.* Rostov: Rostov University. 1988. [Russian].

Part IV
Chapter One

1. O'Haver, T. C., Green, G. L. (1976). Numerical error analysis of derivative spectrometry for the quantitative analysis of mixtures. Analytical Chemistry, 48, 312-318.
2. Dubrovkin, J., Belikov, V. G. *Derivative Spectroscopy. Theory, Technics, Application*. Rostov University. 1988. [Russian].
3. Collier, G., L., Panting, A. C. M. (1959). The use of derivative spectroscopy for determining methyl groups in polythene. Spectrochimica Acta, 14, 104-118.
4. Epstein, M. S., O'Haver, T. C. (1975). Improvements in repetitive scanning techniques for reducing spectral interferences in flame emission spectrometry. Spectrochimica Acta Part B: Atomic Spectroscopy, 30, 135-146.
5. McWilliam, I. G. (1969). Derivative spectroscopy and its application to the analysis of unresolved bands. Analytical Chemistry, 41, 674-676.
6. Defrancesco, F., Seppi, A. (1979). Rapporti fra superfici differenziali e concentrazione in spectrofotometria d'assorbimento derivative. La Rivista Della Società Italiana di Scienza dell'Alimentazione. 8, 111.
7. Kalmanovskyi, V. I. (1982). Chromatographic analysis based on the signal differentiation. Journal of Analytical Chemistry, 307, 1309-1315. [Russian].
8. Meyer, V.R. (1995). Errors in the area determination of incompletely resolved chromatographic peaks. Journal of Chromatographic Science, 33, 26-33.
9. Dubrovkin, J. (1982). Investigation of the filtering properties of the algebraic background correction method. Journal of Applied Spectroscopy, 36, 106-109.
10. Mark, H., Workman, Jr. *Chemometrics in Spectroscopy*. London: Elsevier. 2007.
11. Mukhtarov, R. I., Nikolaev, A. N. (1979). Application of frequency modulation in laser gas analyzers. Journal of Applied Spectroscopy, 30, 725-730.

12. Farooq, J. B., Jeffries, J. B., Hanson, R. K. (2009). Sensitive detection of temperature behind reflected shock waves using wavelength modulation spectroscopy of CO2 near 2.7 μm. Applied Physics, B 96, 161-173.
13. Weitkamp, C. (1984). Calibration of diode-laser second-derivative modulation spectrometry with a reference cell. Applied Optics, 23, 83-86.
14. Mucha, J. A. (1982). Standard addition technique for quantitative trace gas analysis using derivative infrared diode laser spectroscopy. Applied Spectroscopy, 36, 393-400.
15. Mucha, J. A. (1984). Correction of nonlinear derivative diode laser data in standard addition analyses. Applied Spectroscopy, 38, 68-73.
16. Zhimin, P., Yanjun, D., Lu, C., et al. (2011). Calibration-free wavelength modulated TDLAS under high absorbance conditions. Optics Express, 19, 23, 23104.
17. Dubrovkin, J. (1983). Theory of quantitative multicomponent analysis from derivative spectra. Journal of Applied Spectroscopy, 38, 947-951.
18. Seber, G. A. F., Lee, A. J. *Linear Regression Analysis*, 2nd ed., Wiley. 2003.
19. Trakhtman, A.M. *Introduction to Generalized Spectral Theory of Signals.* Soviet Radio, Moscow. 1972. [Russian].
20. Dubrovkin, J. (1985). Theory and use of the method of linear transformation of spectral coordinates in physicochemical studies. Izvestia Severo-Kavkazskogo Nauchnogo Centra Vysšey Školy, Yestestvennye Nauki, 2, 51-57. [Russian].
21. Martens, H., Næs, T. *Multivariate Calibration*, New York: Wiley. 1992.
22. Bogomolov, A., Melenteva, A. (2013). Scatter-based quantitative spectroscopic analysis of milk fat and total protein in the region 400-1100 nm in presence of fat globule size variability. Chemometrics and Intelligent Laboratory Systems, 126, 129-139.
23. Dubrovkin J. (2014). Error propagation of quantitative analysis based on ratio spectra. International Journal of Emerging Technologies in Computational and Applied Sciences, 1-9, 11-20.

Chapter Two

1. Seber, G. A. F., Lee, A. J. *Linear Regression Analysis*, 2nd ed., Wiley. 2003.
2. Dubrovkin, J., Sagdeev, R. S., Sobolev, A. S. (1979). Determination of a component in the presence of impurity. Zavodskay Laboratoriy, 9, 828. [Russian].

3. Dubrovkin, J., Sagdeev, R. S., Sobolev, A. S. (1978). RC-derivative UV-spectroscopic quantitative analysis. Zavodskay Laboratoriy, 6, 685-686. [Russian].
4. Dubrovkin, J. (1983). Application of digital filters to spectrophotometry data treatment. 2. Quantitative analysis based on the generalized Fourier transform. Izvestia Severo-Kavkazskogo Nauchnogo Centra Vysšey Školy, Yestestvennye Nauki. Full text: VINITI 4956-B83, 15 p. [Russian].
5. U.S. National Library of Medicine. National Center for Biotechnology Information. https://pubchem.ncbi.nlm.nih.gov/compound/Butyl-2_4-dichlorophenoxyacetate.
6. Dubrovkin, J., Belikov, V. G. *Derivative Spectroscopy. Theory, Technics, Application*. Rostov University. 1988. [Russian].

Chapter Three

1. Verbic, T., Dorko, Z Horvai, G. (2013). Selectivity in analytical chemistry. Revue Roumaine de Chimie, 58, 569-575.
2. Vessman, J., Stefan, R. I., Van Staden, J. F., et al. (2001). Selectivity in analytical chemistry (IUPAC Recommendations 2001). Pure and Applied Chemistry, 73, 1381-1386.
3. Danzer, K. *Analytical Chemistry. Theoretical and Metrological Fundamentals*. Springer. 2007.
4. Olivieri, A.C., Faber, N. M., Ferre, J. (2006). Uncertainty estimation and figures of merit for multivariate calibration (IUPAC Recommendations. (2006). Pure and Applied Chemistry, 78, 633-661.
5. Morgan, D. R. (1977). Spectral absorption pattern detection and estimation. I. Analytical techniques. Applied Spectroscopy, 31, 404-415.
6. Lorber, A. (1986). Error propagation and figures of merit for quantification by solving matrix equations. Analytical Chemistry, 58, 1167-1172.
7. Ferré, J., Faber, N. M. (2003). Net analyte signal calculation for multivariate calibration. Chemometrics and Intelligent Laboratory Systems, 69, 123-136.
8. Short, S. M., Cogdill, R. P., Anderson, C. A. (2008). Determination of figures of merits for NIR and Raman spectrometers by net analyte signal analysis. CAS 2008. http://www.eigenvector.com /Docs/CAC2008_7_3_2008_Short.pdf
9. Dubrovkin, J. (1989). Quantitative analytical spectrometry of multicomponent systems with known quantitative composition using the

orthogonal projection method. Journal of Applied Spectroscopy, 50, 861-864. [Russian].

10. Seber, G. A. F., Lee, A. J. *Linear Regression Analysis*, 2nd ed., Wiley. 2003.

11. Kvaratskhely, U. K., Demin, U. V. (1983). Concerning expediency of choice expediency of choice of the nth - order derivatives in spectrophotometry. Journal of Analytical Chemistry, 38, 1427-1435. [Russian].

Chapter Four

1. Eckschlager, K., Danzer, K. *Information theory in analytical chemistry*. Wiley. 1994.

2. Dubrovkin, J. (1981). Change in the amount of information contained in a spectrum on differentiation. Journal of Applied Spectroscopy, 35, 1149-1153.

3. Dubrovkin, J., Belikov, V.G. *Derivative Spectroscopy. Theory, Technics, Application*. Rostov University. 1988.

4. Shores, T. S. *Applied Linear Algebra and Matrix Analysis*. Springer Science & Business Media. 2007.

Part V
Introduction

1. Dubrovkin, J., Belikov, V. G. *Derivative Spectroscopy. Theory, Technics, Application*. Rostov University. 1988. [Russian].

2. Talsky, G. *Derivative Spectrophotometry. Low and Higher Order*. Germany, Weinheim: VCH Verlagsgesellschaft. 1994.

3. Saakov, V. S., Drapkin, V. Z., Krivchenko, A. I., et al. *Derivative Spectrophotometry and Electron Spin Resonance (ESR) Spectroscopy for Ecological and Biological Questions*. Springer. 2013.

4. Saakov, V. S., Krivchenko, A. I., Rozengart, E. V., Danilova, I. G. *Derivative Spectrophotometry and PAM-Fluorescence in Comparative Biochemistry*. Springer. 2015.

5. Dubrovkin, J. (1988). Chemical analysis with the use of the mathematical differentiation. Journal of Analytical Chemistry, 43, 965-979. [Russian].

6. Karpinska, J. (2012). Basic principles and analytical application of derivative spectrophotometry. In Macro to Nano Spectroscopy. Ed. J. Uddin. 253-268. InTech, Croatia.

7. Bosch-Ojeda, C., Rojas, F. S. (2004). Recent developments in derivative ultraviolet/visible absorption spectrophotometry. Analytica Chimica Acta, 518, 1-24.
8. Rojas, F. S., Bosch-Ojeda, C. (2009). Recent developments in derivative ultraviolet/visible absorption spectrophotometry: 2004-2008: A review. Analytica Chimica Acta, 635, 22-44.
9. Bosch-Ojeda, C., Rojas, F. S. (2013). Recent applications in derivative ultraviolet/visible absorption spectrophotometry: 2009-2011: A review. Microchemical Journal. 106, 1-16.
10. Lotfy, H. M., Saleh, S. S. (2016). Recent development in ultraviolet spectrophotometry through the last decade (2006-2016): A review. International Journal of Pharmacy and Pharmaceutical Sciences. 8, 40-56.
11. Redasani, V. K., Patel, P. R., Marathe, D. Y., et al. (2018). A review on derivative UV-spectrometry analysis of drugs in pharmaceutical formulations and biological samples. Journal of the Chilean Chemical Society, 63, no. 3.
12. Parmar, A., Sharma, S. (2016). Derivative UV-vis absorption spectra as an invigorated spectrophotometric method for spectral resolution and quantitative analysis: Theoretical aspects and analytical applications: A review. TrAC Trends in Analytical Chemistry, 77, Pages 44-53.

Chapter One

Atomic spectroscopy

1. Sneddon, J. Advances in Atomic Spectroscopy, Vol, 3. Elsevier.1997.
2. Welz, B., Sperling, M. *Atomic Absorption Spectrometry*, 3rd Ed. WILEY-VCH Verlag GmbH. 1999.

Atomic emission spectroscopy

1. He, M., Hu, B., Chen, B., Zucheng, J. (2017). Inductively coupled plasma optical emission spectrometry for rare earth elements analysis. Physical Sciences Reviews, 2, 1.
2. Epstein, M. S., Rains, T. C., O'Haver, T. C. (1976). Wavelength modulation for background correction in graphite furnace atomic emission spectrometry. Applied Spectroscopy, 30, 324-329.
3. Epstein, M. S., O'Haver, T.C. (1975). Improvements in repetitive scanning techniques for reducing spectral interferences in flame emission spectrometry. Spectrochimica Acta Part B: Atomic Spectroscopy, 30, 135-146.
4. Snelleman, W., Rains, T. C., Yee, K. W., et al. (1970). Flame emission spectrometry with repetitive optical scanning in the derivative mode. Analytical Chemistry, 42, 394-398.

5. Lépine, L., Provencher, M., Thammavong, K. et al (1992). Dynamic background correction by wavelength modulation in ICP atomic emission spectrometry (AES) and its application to Flow-Injection-ICP-AES. Applied Spectroscopy, 46, 864-872.
6. Nakamura Y., Takahashi, K., Hasegava, R. (1996). Improvement of sensitivity in axially-viewed horizontal ICP-AES by the use of a long torch. Bunseki kagaku, 45, 1005-1012.
7. Zolfonoun, E. (2015). Determination of ^{6}Li and ^{7}Li abundances by inductively coupled plasma-optical emission spectrometry combined with derivative spectroscopy, multivariate curve resolution and multivariate calibration methods. Journal of Analytical Atomic Spectrometry, 30, 2003-2009.

Atomic fluorescence spectroscopy

1. Wynn, T. F., Clardy, P., Vaughn, L., et al. (1981). Wavelength-modulated, continuum-source excited atomic fluorescence spectrometric system for wear metals in jet engine lubricating oils using electrothermal atomization. Analytica Chimica Acta, 124, 155-161.
2. Fowler, W. K., Knapp, D. O., Winefordner, J. D. (1974). Double modulation atomic fluorescence flame spectrometry. Analytical Chemistry, 46, 601-602.
3. Goff, D. A, Yeung, E. S. (1978). Atomic fluorescence spectrometry with a wavelength-modulated continuous wave dye laser. Analytical Chemistry, 50, 625-627.
4. Su, E. G., Irwin, R. L., Liang, Z, Michel, R. G. (1992). Background correction by wavelength modulation for pulsed-laser-excited atomic fluorescence spectrometry. Analytical Chemistry, 64, 1710-1720.

Laser-induced Breakdown Spectroscopy

1. Sneddon, J. *Advances in Atomic Spectroscopy*, Vol, 3. Elsevier. 1997.
2. Bauer A. J. R., Buckley, S. G. (2017). Novel applications of Laser-Induced Breakdown Spectroscopy. Applied Spectroscopy, 71, 553-566.
3. Devey, K., Mucalo, M., Rajendram, G., Lane, J. (2014). Pasture vegetation elemental analysis by Laser-Induced Breakdown Spectroscopy. Communications in Soil Science and Plant Analysis, 46, 72-80.
4. Franco, M. A, M, Milori, D. M .B .P., Boas, P. R. V. (2018). Comparison of algorithms for baseline correction of LIBS spectra for quantifying total carbon in Brazilian soils. https://www. research gate.net/publication/325075446_Comparison_of_Algorithms_for_Baselin

e_Correction_of_LIBS_Spectra_for_Quantifying_Total_Carbon_in_Braz ilian_Soils.

Atomic absorption spectrophotometry

1. Sneddon, J. Advances in Atomic Spectroscopy, Volume 3. 1st Ed., Elsevier. 1997.
2. Welz, B., Sperling, M. Atomic Absorption Spectrometry, 3rd Ed. WILEY-VCH Verlag GmbH. 1999.
3. Snelleman, W. (1968). An a.c. scanning method with increased sensitivity in atomic absorption analysis using a continuum primary source. Spectrochimica Acta Part B: Atomic Spectroscopy, 23, 403-411.
4. Elser, R. C., Winefordner, J. D. (1972). Double modulation-optical scanning and mechanical chopping-in atomic absorption spectrometry with using a continuum source. Analytical Chemistry, 44, 698-709.
5. Zander, A. T, O'Haver, T. C., Keliher, P. N. (1976). Continuum source atomic absorption spectrometry with high resolution and wavelength modulation. Analytical Chemistry, 48, 1166-1175.
6. Lehnert, R., Quick, L., Rump, T., et al. (1993). Background correction in frequency modulated simultaneous atomic absorption spectrometry. Fresenius' Journal of Analytical Chemistry, 346, 392-395.
7. Zhang, D. Q., Yang, L. L., Sun, H. W. (1999). Determination of mercury by cold vapour atomic absorption spectrometry with derivative signal processing. Analytica Chimica Acta, 395, 173-178.
8. Sun, H., Yang, L., Zhang, D., Sun, J. (1997). Direct determination of cadmium at parts-per-billion level in waters by derivative atomic absorption spectrometry using atom trapping technique. Talanta, 44, 1979-1986.
9. Sun, H., Kang, W., L., Liang, S. (2003). Determination of chromium (III) and total chromium in water by derivative atomic absorption spectrometry using flow injection on-line preconcentration with double microcolumn. Analytical Science, 19, 589-592.
10. Sun, W., Yang, L. L., Zhang, D. G., et al. (1997). Determination of lead traces in water and liqueurs by derivative atom trapping flame atomic absorption spectrometry. Fresenius' Journal of Analytical Chemistry, 358, 646-651.
11. Sun, W., Yang, L. L., Zhang, D. G., et al. (2001). Derivative hydride generation atomic absorption spectrometry and determination of lead traces in waters. Fresenius' Journal of Analytical Chemistry, 371, 1154-1157.

12. Ha, J., Sun, H. W., Sun, J. M., et al. (2001). Determination of tellurium in urine by hydride generation atomic absorption spectrometry with derivative signal processing. Analytica Chimica Acta, 448, 145-149.
13. Sun, H. W., Li, L. Q. (2005). Derivative flame atomic absorption spectrometry and its application in trace analysis. Journal of the Iranian Chemical Society, 2, 268-276.
14. Sun, H. W., Li, L. Q. (2006). Methodology and application of the derivative atomic absorption spectrometry. Current Analytical Chemistry, 2, 331-339.
15. Sun, H. W., Ai, L., Li, L. Q. (2006). Derivative vapor generation atomic absorption spectrometry and its application in trace analysis. Chemical Journal on Internet, 8, 28. https://www.mdpi.org/cji/cji/2006/
16. Ataman, O. Y. (2008).Vapor generation and atom traps: Atomic absorption spectrometry at the ng/L level. Spectrochimica Acta, Part B, Atomic Spectroscopy 63, 825-834.
17. Dedina, J., Tsalev, D, L. *Hydride Generation Atomic Absorption Spectrometry*. Wiley. 1995.
18. Varzhapetyan, T. S., Hakhumyan, G. T., Babushkin, V. V., et al. (2007). Study of atomic spectral lines in a magnetic field with use of a nanocell with the thickness L = λ. Journal of Contemporary Physics (Armenian Academy of Sciences), 42, 223-229.
19. Sargsyan, A., Amiryan, A., Pashayan-Leroy, Y., et al. (2019). Approach to quantitative spectroscopy of atomic vapor in optical nanocells. Optics Letters, 44, 5533-5536.

Molecular spectroscopy

Infrared absorption spectrophotometry

1. Dubrovkin, J., Belikov, V. G. *Derivative Spectroscopy. Theory, Technics, Application*. Rostov University. 1988. [Russian].
2. Rinnan, Å., Van den Berg, F. Engelsen, F. S. (2009). Review of the most common pre-processing techniques for near-infrared spectra. TrAC Trends in Analytical Chemistry, 28, 1201-1222.
3. Rinnan, Å. (2014). Pre-processing in vibrational spectroscopy - when, why and how. Analytical methods, 6 (18).
4. De Aragão, B. J. G., Messaddeq, Y. (2008). Peak separation by derivative spectroscopy applied to FTIR analysis of hydrolized silica. Journal of the Brazilian Chemical Society, 19, 1582-1594.
5. Cooper, C., Elwell, C., Meek, J., et al. (1996). The noninvasive measurement of absolute cerebral deoxyhemoglobin concentration and mean optical path length in the neonatal brain by second derivative near infrared spectroscopy. Pediatric Research, 39, 32-38.

6. Koutsiaris, A. G. (2017). Deep tissue near infrared second derivative spectrophotometry for the assessment of claudication in peripheral arterial disease. Clinical Hemorheology and Microcirculation, 65, 275-284.
7. Nasr, J. J. M., Al-Shaalan, N. H., Shalan, S. M. (2020). Sustainable environment-friendly quantitative determination of three anti-hyperlipidemic statin drugs and ezetimibe in binary mixtures by first derivative Fourier transform infrared (FTIR) spectroscopy. Spectrochimica Acta Part A: Molecular and Biomolecular Spectroscopy, 237, 118332.
8. Gallignani, M., Garrigues, S., La Guardia, M. (1994). Derivative Fourier transform infrared spectrometric determination of ethanol in alcoholic beverages. Analytica Chimica Acta, 287, 275-283.
9. Nasr, J. J. M., l-Shaalan, N. H., Shalan, S. M. (2020). Sustainable environment-friendly quantitative determination of three anti-hyperlipidemic statin drugs and ezetimibe in binary mixtures by first derivative Fourier transform infrared (FTIR) spectroscopy. Spectrochimica Acta Part A: Molecular and Biomolecular Spectroscopy, 237, 118332.
10. Rieppo, L., Saarakkala, S., Narhi, T. (2012). Application of second derivative spectroscopy for increasing molecular specificity of Fourier transform infrared spectroscopic imaging of articular cartilage.
11. Park, S. M., Yu, H.-Y., Chun, H. S., et al. (2019). A second derivative Fourier-transform infrared spectroscopy method to discriminate perilla oil. Journal of Oleo Science, 68, 389-398.

UF-VIS absorption spectrophotometry

1. Aspnes, D. E. (1977). Modulation spectroscopy with synchrotron radiation. Festkörperprobleme, XVII, 235-259.
2. Müllerová, J., Šutta, P., Prušáková, L. (2018). Derivative optical spectroscopy of thin films of alkaline-earth titanates: Critical points. AIP Conference Proceedings 1996, 020030.
3. Millie, D., Schofield, O., Kirkpatrick, G., et al. (2002) Using absorbance and fluorescence spectra to discriminate microalgae. European Journal of Phycology, 37, 313-322.
4. Hu, Y., Zhao, D., Qin, Y., Wang, X. (2020). An order determination method in direct derivative absorption spectroscopy for correction of turbidity effects on COD measurements without baseline required. Spectrochimica Acta Part A: Molecular and Biomolecular Spectroscopy, 226, 117646.

5. Wood, R., Foster, L., Damant, A., Key, P. *Analytical methods for food additives*. CRC Press, 2004.
6. Lotfy, H. M., Saleh, S. S. (2016). Recent development in ultraviolet spectrophotometry through the last decade (2006-2016): A review. International Journal of Pharmacy and Pharmaceutical Sciences. 8, 40-56.
7. Redasani, V. K., Patel, P. R., Marathe, D. Y., et al. (2018). A review on derivative UV-spectrometry analysis of drugs in pharmaceutical formulations and biological samples. Journal of the Chilean Chemical Society, 63, no. 3.
8. Sowjanya, G., Ganapati, S., Almas, S. (2019). Derivative UV spectroscopic approaches in multicomponent analysis-A review. International Journal of Pharmacy and Pharmaceutical Sciences, 11, 1-11.
9. Cielecka-Piontek, J., Paczkowska, M., Zalewski, P., et al. (2015). Tebipenem pivoxyl. Derivative spectroscopy study of stability of the first oral carbapenem. Spectrochimica Acta Part A: Molecular and Biomolecular Spectroscopy, 135, 14-19.
10. Eldin, A. S., Azab, S. M., Shalaby, A., El-Maamly, M. (2017). The Development of a new validated HPLC and spectrophotometric methods for the simultaneous determination of Daclatasvir and Sofosbuvir: Antiviral Drugs. Journal of Pharmacy and Pharmacology Research, 1, 28-42.
11. Lotfy, H. M., Mohamed, D., Elshahed, M. (2017). Different mathematical processing of absorption, ratio and derivative spectra for quantification of mixtures containing minor component: An application to the analysis of the recently co-formulated antidiabetic drugs; canagliflozin and metformin. Spectrochimica Acta Part A: Molecular and Biomolecular Spectroscopy, 189, 100-109.
12. Das Chagas da Costa Lopes, F., Fonseca, L., Moita, G. C., De Moura Ribeiro M., V. (2018). Development and validation of methods using derivative spectrophotometry for determination of dipyrone in pharmaceutical formulations. International Journal of Pharmaceutical Sciences and Research, 9, 2201-2210.
13. Darweesh, S.A., Khalaf, H. S., Al-Khalisy, R.S., et al. (2018). Advancement and validation of new derivatives spectrophotometric method for individual and simultaneous estimation of Diclofenac Sodium and Nicotinamide. Oriental Journal of Chemistry, 34, 1625-1632.
14. Lotfy, H., M., Ahmed, D. A., Khaled, M., Weshahy, S.A.F. (2019). Study of efficiency and spectral resolution for mathematical filtration technique using novel unlimited derivative ratio and classical univariate spectrophotometric methods for the multicomponent determination-stability analysis. Heliyon, 5, e01669.

15. Pandey, R., Patel, P. O., Patil, M. U., et al. (2012). Quantitative estimation of diacerein in bulk and in capsule formulation using hydrotropic solubilizing agents by UV-spectrophotometry and the first order derivative using the area under curve method. Pharmaceutical Methods, 3, 4-8.
16. Vargas, A. M. M., Martins, A. C., Almeida, V. C. (2012). Ternary adsorption of acid dyes onto activated carbon from flamboyant pods (Delonix regia): Analysis by derivative spectrophotometry and response surface methodology. Chemical Engineering Journal, 195-196, 173-179.
17. Chalikwar, S. S., Shirkhedkar, A. A., Bagul, M. A., et al. (2012). Development and validation of zero and first-order derivative area under curve spectrophotometric methods for the determination of entacapone in bulk material and in tablets. Pharmaceutical Methods, 3, 14-17.
18. Varghese, A., George, L. (2012). Simultaneous first order derivative spectrophotometric determination of vanadium and zirconium in alloy steels and minerals. Spectrochimica Acta Part A: Molecular and Biomolecular Spectroscopy, 95, 46-52.
19. Ghaedi, M., Hajati, S., Barazesh, B., et al. (2013). Saccharomyces cerevisiae for the biosorption of basic dyes from binary component systems and the high order derivative spectrophotometric method for simultaneous analysis of Brilliant green and Methylene blue. Journal of Industrial and Engineering Chemistry, 19, 227-233.
20. Darwish, H. W., Hassan, S. A., Salem, M. Y. El-Zeany, B. A. (2013). Comparative study between derivative spectrophotometry and multivariate calibration as analytical tools applied for the simultaneous quantitation of Amlodipine, Valsartan and Hydrochlorothiazide. Spectrochimica Acta Part A: Molecular and Biomolecular Spectroscopy, 113, 215-223.
21. Xiao, N., Deng, J., Huang, K. et al (2014). Application of derivative and derivative ratio spectrophotometry to simultaneous trace determination of rhodamine B and rhodamine 6G after dispersive liquid–liquid microextraction. Spectrochimica Acta Part A: Molecular and Biomolecular Spectroscopy, 128, 312-318.
22. Jitian, S., White, S. R., Yang, H.-H. W, Weisz, A. (2014). Conventional high-performance liquid chromatography versus derivative spectrophotometry for the determination of 1, 3, 6-pyrenetrisulfonic acid trisodium salt and 1, 3, 6, 8-pyrenetetrasulfonic acid tetrasodium salt in the color additive D&C Green No. 8 (Pyranine). Journal of Chromatography A, 1324, 238-241.
23. Hajati, S., Ghaedi, M., Barazesh B. et al (2014). Application of high order derivative spectrophotometry to resolve the spectra overlap between

BG and MB for the simultaneous determination of them: Ruthenium nanoparticle loaded activated carbon as adsorbent. Journal of Industrial and Engineering Chemistry, 20, 2421-2427.

24. Shetty, P. R., Patil, D. D. (2014). Applications of simultaneous equation method and derivative method for the determination of rabeprazole sodium and levosulpiride in pharmaceutical dosage form and dissolution samples. Journal of the Association of Arab Universities for Basic and Applied Sciences, 15, 53-60.

25. Sohrabi, M.., Zarkesh, M. T. (2014). Spectra resolution for simultaneous spectrophotometric determination of lamivudine and zidovudine components in pharmaceutical formulation of human immunodeficiency virus drug based on using continuous wavelet transform and derivative transform techniques. Talanta, 122, 223-228.

26. Tabbouche, O. S., Soukkariyyeh, I. (2014). First-order derivative ultraviolet spectrophotometry of imipenem–cilastatin formulations. Journal of Taibah University Medical Sciences, 9, 178-181.

27. Hoang, V. D., Loan, N. T., Tho, V. T., Nguyen, H. M. T. (2014). UV spectrophotometric simultaneous determination of cefoperazone and sulbactam in pharmaceutical formulations by derivative, Fourier and wavelet transforms. Spectrochimica Acta Part A: Molecular and Biomolecular Spectroscopy, 121, 704-714.

28. Zeinali, N., Ghaedi, M., Shafie, G. (2014). Competitive adsorption of methylene blue and brilliant green onto graphite oxide nano particle following: Derivative spectrophotometric and principal component-artificial neural network model methods for their simultaneous determination. Journal of Industrial and Engineering Chemistry, 20, 3550-3558.

29. Zolgharnein, J., Bagtash, M., Shariatmanesh, T. (2015). Simultaneous removal of binary mixture of Brilliant Green and Crystal Violet using derivative spectrophotometric determination, multivariate optimization and adsorption characterization of dyes on surfactant modified nano-γ-alumina. Spectrochimica Acta Part A: Molecular and Biomolecular Spectroscopy, 137, 1016-1028.

30. Salem, H., Lotfy, H. M., Hassan, N. Y. et al (2015). A comparative study of different aspects of manipulating ratio spectra applied for ternary mixtures: Derivative spectrophotometry versus wavelet transform. Spectrochimica Acta Part A: Molecular and Biomolecular Spectroscopy, 135, 1002-1010.

31. Hoang, V. D., Hue, N. T., Tho, N. H., Nguyen, H. M. T. (2015). Simultaneous determination of chloramphenicol, dexamethasone and naphazoline in ternary and quaternary mixtures by RP-HPLC, derivative

and wavelet transforms of UV ratio spectra. Spectrochimica Acta Part A: Molecular and Biomolecular Spectroscopy, 139, 20-27.
32. Korany, M. A., Abdine, H. H., Ragab, M. A. A., Aboras, S. I. (2015). Application of derivative spectrophotometry under orthogonal polynomial at unequal intervals: Determination of metronidazole and nystatin in their pharmaceutical mixture. Spectrochimica Acta Part A: Molecular and Biomolecular Spectroscopy, 143, 281-287.
33. Elzanfaly, E. S., Hassan, S. A., Salem, M.Y., El-Zeany, B. A. (2015). Continuous Wavelet Transform, a powerful alternative to Derivative Spectrophotometry in analysis of binary and ternary mixtures: A comparative study. Spectrochimica Acta Part A: Molecular and Biomolecular Spectroscopy, 151, 945-955.
34. Stolarczyk, M., Maślanka, A., Apola, A., et al. (2015). Derivative spectrophotometric method for simultaneous determination of zofenopril and fluvastatin in mixtures and pharmaceutical dosage forms. Spectrochimica Acta Part A: Molecular and Biomolecular Spectroscopy, 148, 66-71.
35. Marković, B., Ignjatović, J., Vujadinović, M. et al (2015). Inter-laboratory verification of European pharmacopoeia monograph on derivative spectrophotometry method and its application for chitosan hydrochloride. Spectrochimica Acta Part A: Molecular and Biomolecular Spectroscopy, 150, 792-798.
36. Takegami, S., Kitamura, K., Ohsugi, M., et al. (2015). Partitioning of organophosphorus pesticides into phosphatidylcholine small unilamellar vesicles studied by second-derivative spectrophotometry. Spectrochimica Acta Part A: Molecular and Biomolecular Spectroscopy, 145,198-202.
37. Bagheri, A. R., Ghaedi, M., Asfaram, A., et al. (2016). Modeling and optimization of simultaneous removal of ternary dyes onto copper sulfide nanoparticles loaded on activated carbon using second-derivative spectrophotometry. Journal of the Taiwan Institute of Chemical Engineers, 65, 212-224.
38. Ali, O. I. M., Ismail, N., S., Elgohary, R. M. (2016). Validated derivative and ratio derivative spectrophotometric methods for the simultaneous determination of levocetirizine dihydrochloride and ambroxol hydrochloride in pharmaceutical dosage form. Spectrochimica Acta Part A: Molecular and Biomolecular Spectroscopy, 153, 605-611.
39. Ansari, F., Ghaedi, M., Taghdiri, M., Asfaram, A. (2016). Application of ZnO nanorods loaded on activated carbon for ultrasonic assisted dyes removal: Experimental design and derivative spectrophotometry method. Ultrasonics Sonochemistry, 33, 197-209.

40. Tehrani, M. S., Zare-Dorabei, R. (2016). Competitive removal of hazardous dyes from aqueous solution by MIL-68(Al): Derivative spectrophotometric method and response surface methodology approach. Spectrochimica Acta Part A: Molecular and Biomolecular Spectroscopy, 160, 8-18.
41. Dil, E. A., Ghaedi, M., Ghaedi, A. M. et al (2016). Modeling of quaternary dyes adsorption onto ZnO-NR-AC artificial neural network: Analysis by derivative spectrophotometry. Journal of Industrial and Engineering Chemistry, 34,186-197.
42. Pooralhossini, J., Ghaedi, M., Zanjanchi, M. A., Asfaram, A. (2017). Ultrasonically assisted removal of Congo Red, Phloxine B and Fast green FCF in ternary mixture using novel nanocomposite following their simultaneous analysis by derivative spectrophotometry. Ultrasonics Sonochemistry, 37, 452-463.
43. Dil, E. A., Ghaedi, M., Asfaram, A. (2017). Optimization and modeling of preconcentration and determination of dyes based on ultrasound assisted-dispersive liquid–liquid microextraction coupled with derivative spectrophotometry. Ultrasonics Sonochemistry, 34, 27-36.
44. Sharifpour, E., Haddadi, H., Ghaedi, M. (2017). Optimization of simultaneous ultrasound assisted toxic dyes adsorption conditions from single and multi-components using central composite design: Application of derivative spectrophotometry and evaluation of the kinetics and isotherms. Ultrasonics Sonochemistry, 36, 236-245.
45. Asfaram, A., Ghaedi, M., Ghezelbash, G. R., Pepec, F. (2017). Application of experimental design and derivative spectrophotometry methods in optimization and analysis of biosorption of binary mixtures of basic dyes from aqueous solutions. Ecotoxicology and Environmental Safety, 139, 219-227.
46. Dil, E. A., Ghaedi, M., Ghezelbash, G. R., Asfaram, A. (2017). Multi-responses optimization of simultaneous biosorption of cationic dyes by live yeast Yarrowia lipolytica 70562 from binary solution: Application of first order derivative spectrophotometry. Ecotoxicology and Environmental Safety, 139, 158-164.
47. Fahimirad, B., Asghari, A., Rajabi, M. (2017). Photo-degradation of basic green 1 and basic red 46 dyes in their binary solution by La2O3-Al2O3 nanocomposite using first-order derivative spectra and experimental design methodology. Spectrochimica Acta Part A: Molecular and Biomolecular Spectroscopy, 179, 58-65.
48. Bagheri, A. R., Ghaedi, M., Asfaram, A. et al. (2017). Design and construction of nanoscale material for ultrasonic assisted adsorption of

dyes: Application of derivative spectrophotometry and experimental design methodology. Ultrasonics Sonochemistry, 35, Part A, 112-123.
49. Causse, J., Thomas, O., Jung, A.-V., Thomas, M.-F. (2017). Direct DOC and nitrate determination in water using dual pathlength and second derivative UV spectrophotometry. Water Research, 108, 312-319.
50. Do, T. T. T., Dao, U. P. N., Bui, H. T., Nguyen, T. T. (2017). Effect of electrostatic interaction between fluoxetine and lipid membranes on the partitioning of fluoxetine investigated using second derivative spectrophotometry and FTIR. Chemistry and Physics of Lipids, 207, Part A, 10-23.
51. Sversut, R. A., Alcântara, I. C., Rosa, A. M., et al. (2017). Simultaneous determination of gatifloxacin and prednisolone acetate in ophthalmic formulation using first-order UV derivative spectroscopy. Arabian Journal of Chemistry, 10, 604-610.
52. Elimam, M. M, Shantier, S. W., Gadkariem, E. A. et al. (2017). Stability studies on florfenicol using developed derivative spectrophotometric methods. Annales Pharmaceutiques Françaises, 75, 40-44.
53. Sadeghi, S., Nasehi, Z. (2018). Simultaneous determination of Brilliant Green and Crystal Violet dyes in fish and water samples with dispersive liquid-liquid micro-extraction using ionic liquid followed by zero crossing first derivative spectrophotometric analysis method. Spectrochimica Acta Part A: Molecular and Biomolecular Spectroscopy, 201, 134-142.
54. Hassan, S. A., Abdel-Gawad, S .A. (2018). Application of wavelet and Fourier transforms as powerful alternatives for derivative spectrophotometry in analysis of binary mixtures: A comparative study. Spectrochimica Acta Part A: Molecular and Biomolecular Spectroscopy, 191, 365-371.
55. Sherovski, P., Stojković, G., Ristovska, N. (2018). Development, validation and application of first derivative spectroscopy ratio method for estimation of Bradford assay. Analytical Biochemistry, 558, 35-40.
56. Fujiyoshi, T., Ikami, T., Kikukawa, K., et al. (2018). Direct quantitation of the preservatives benzoic and sorbic acid in processed foods using derivative spectrophotometry combined with micro dialysis. Food Chemistry, 240, 386-390.
57. Mirzajani, R., Karimi, S. (2019). Ultrasonic assisted synthesis of magnetic Ni-Ag bimetallic nanoparticles supported on reduced graphene oxide for sonochemical simultaneous removal of sunset yellow and tartrazine dyes by response surface optimization: Application of derivative spectrophotometry. Ultrasonics Sonochemistry, 50, 239-250.

58. Mabrouk, M. M., Hammad, S. F., Mansour, F. R., El-Khateeb, B. Z. (2019). Simultaneous determination of diclofenac and esomeprazole by reversed phase liquid chromatography, dual wavelength and derivative spectrophotometry. Journal of Analytical Chemistry, 74, 458–466.

59. Amin, K. F. M., Abdullah, A. M., Fakhre, N. A. (2019). Mean centering-triple divisor and ratio derivative-zero crossing for simultaneous determination of some diabetes drugs in their quaternary mixture with severely overlapping spectra. Spectrochimica Acta Part A: Molecular and Biomolecular Spectroscopy, 222, 117261.

60. Ji, R., Zhao, Z., Yu, X., Chen, M. (2019). Determination of rhodamine B in capsicol using the first derivative absorption spectrum. Optik, 81, 796-801.

61. Yuan, D., O'Riordan, E. D., Jacquier, J. C. (2019). Development of a first order derivative spectrophotometry method to rapidly quantify protein in the presence of chitosan and its application in protein encapsulation systems. Food Chemistry, 289, 1-6.

62. Simion, I. M., Sârbu, C. (2019). The impact of the order of derivative spectra on the performance of pattern recognition methods. Classification of medicinal plants according to the phylum. Spectrochimica Acta Part A: Molecular and Biomolecular Spectroscopy, 219, 91-95.

63. Lavrinenko, I. A., Holyavka, M. G., Chernov, V. E., Artyukhov, V. G. (2020). Second derivative analysis of synthesized spectra for resolution and identification of overlapped absorption bands of amino acid residues in proteins: Bromelain and ficin spectra in the 240–320 nm range. Spectrochimica Acta Part A: Molecular and Biomolecular Spectroscopy, 227, 117722.

64. Rastgordani, M., Zolgharnein, Z., Mahdavi, V. (2020). Derivative spectrophotometry and multivariate optimization for simultaneous removal of Titan yellow and Bromophenol blue dyes using polyaniline@SiO2 nanocomposite. Microchemical Journal, 155, 104717.

65. El-Kimary, E.I., Khamis, E. F., Belal, S. F., Moneim, M. M. A. (2020). Fourier convolution versus derivative spectrophotometry: Application to the analysis of two binary mixtures containing tamsulosin hydrochloride as a minor component. Annales Pharmaceutiques Françaises, 78, 129-141.

66. Luthra, A., Denisov, I. G., Sligar, S. G. (2011). Temperature derivative spectroscopy to monitor the autoxidation decay of cytochromes P450. Analytical Chemistry, 83, 5394-5399.

67. Bagtash, M., Zolgharnein, J. (2018). Removal of brilliant green and malachite green from aqueous solution by a viable magnetic polymeric nanocomposite: Simultaneous spectrophotometric determination of 2

dyes by PLS using original and first derivative spectra. Journal of Chemometrics, 32, e3014.

Laser Modulation Spectroscopy

1. Sharp, B. L. (1976). Lasers: Their current and potential applications in analytical chemistry. Proceedings of the Analytical Division of the Chemical Society, 13, 104-107.
2. Tittel F. K., Richter D., Fried A. (2003) Mid-Infrared Laser Applications in Spectroscopy. In: Sorokina I. T., Vodopyanov K. L. (Eds) Solid-State Mid-Infrared Laser Sources. Topics in Applied Physics, 89. 445-516. Springer, Berlin, Heidelberg
3. Chen, C., Ren, Q., Wang, Y. Z. (2018). Review on multi gas detector using infrared spectral absorption technology. Applied Spectroscopy Reviews, 54, 425-444.
4. Du, Z., Zhang, S., Li, J., et al. (2019). Mid-Infrared tunable laser-based broadband fingerprint absorption spectroscopy for trace gas sensing: A Review. Applied Sciences, 9, 338.
5. Kroll, M., McClintock, J. A., Ollinger, O. (1987). Measurement of gaseous oxygen using diode laser spectroscopy. Applied Physics Letters, 51, 1465.
6. Weber, W. H., Remillard, J. T., Chase, R. E., et al. (2002). Using a wavelength-modulated quantum cascade laser to measure NO concentrations in the parts-per-billion range for vehicle emissions certification. Applied Spectroscopy, 56, 706-714.
7. Lytkine, A., Lim, A., Jäger, W., Tulip, J. (2010). Tunable diode laser spectroscopy of benzene near 1684 nm with a low-temperature VCSEL. Applied Physics B, 99, 825-832.
8. Tanaka, K., Tonokura, K. (2011). Sensitive measurements of stable carbon isotopes of CO_2 with wavelength modulation spectroscopy near 2 μm. Applied Physics B, 105, 463.
9. Chao, X., Jeffries, J. B., Hanson, R. K. (2012). Wavelength-modulation-spectroscopy for real-time, in situ NO detection in combustion gases with a 5.2 μm quantum-cascade laser. Applied Physics B, 106,987-997.
10. Kluczynski, K., Lundqvist, S., Belahsene, S., et al. (2012). Detection of propane using tunable diode laser spectroscopy at 3.37 μm. Applied Physics B, 108,183-188.
11. Sun, K., Chao, X., Sur, R., et al. (2013). Wavelength modulation diode laser absorption spectroscopy for high-pressure gas sensing. Applied Physics B, 110, 497-508.

12. Wei, W., Chang, J., Huang, Q., et al. (2015). Wavelength modulation spectroscopy with signal-reference beam method for highly sensitive gas detection. Applied Physics B, 118, 75-83.
13. Meng, Y., Liu, T., Liu, K., et al. (2015). Simultaneous detection of mixed gases based on overlapped spectra separation with SLIDT. IEEE Photonics Technology Letters, 27, 794-797.
14. Cao, Y., Sanchez, N.P., Jiang, W., et al. (2015). Simultaneous atmospheric nitrous oxide, methane and water vapor detection with a single continuous wave quantum cascade laser. Optics Express, 23, 2121-2132.
15. Cai, T., Gao, G., Wang, M. (2016). Simultaneous detection of atmospheric CH4 and CO using a single tunable multi-mode diode laser at 2.33 μm. Optics Express, 24, 859-873.
16. Huang, Q., Xu, X. M., Li, C. J., et al. (2016). Self-calibration wavelength modulation spectroscopy for acetylene detection based on tunable diode laser absorption spectroscopy. Chinese Physics B, 25, 11, 114202.
17. Tanaka, K., Akishima, K., Sekita, M., et al. (2017). Measurement of ethylene in combustion exhaust using a 3.3-μm distributed feedback interband cascade laser with wavelength modulation spectroscopy. Applied Physics B, 123, 219.
18. Lee, D. D., Bendana, F. A., Schumaker, S. A., et al. (2018). Wavelength modulation spectroscopy near 5 μm for carbon monoxide sensing in a high-pressure kerosene-fueled liquid rocket combustor. Applied Physics B, 124, 1.
19. Chen, X., Yang, C. G., Hu, M., et al. (2018). Highly-sensitive NO, NO_2, and NH_3 measurements with an open-multipass cell based on mid-infrared wavelength modulation spectroscopy. Chinese Physics B, 27, 4, 040601.
20. Chen, C., Ren, Q., Wang, Y. Z. (2018) Review on multi gas detector using infrared spectral absorption technology. Applied Spectroscopy Reviews, 54, 425-444.
21. Reid, J., Labrie, D. (1981). Second-harmonic detection with tunable diode lasers - Comparison of experiment and theory. Applied Physics B, 26, 203-210.
22. Cui, R., Dong, L., Wu, H., et al. (2018). Highly sensitive and selective CO sensor using a 2.33 μm diode laser and wavelength modulation spectroscopy. Optics Express, 19, 24318-24328.
23. Du, Y., Peng, Z., Ding, Y. (2018). Wavelength modulation spectroscopy for recovering absolute absorbance. Optics Express, 26, 9263-9272.

24. Kwong, C. C., Chan, E. A., Aljunid. S. A., et al. (2019). Large optical depth frequency modulation spectroscopy. Optics Express, 27, 32323-32336.
25. Stuhr, M., Fabheber, N., Friedrichs, G. (2019). Single-tone mid-infrared frequency modulation spectroscopy for sensitive detection of transient species. Optics Express, 27, 26499-26511.
26. Yang, C., Mei, L., Deng, H., et al. (2019). Wavelength modulation spectroscopy by employing the first harmonic phase angle method. Optics Express, 27, 12137-12146.
27. Hayden, T. R. S., Malarich, N., Petrykowski, D., et al. (2019). OH radical measurements in combustion environments using wavelength modulation spectroscopy and dual-frequency comb spectroscopy near 1491 nm. Applied Physics B, 125, 226.
28. Schwarm, K.K., Strand, C. L., Miller, V. A., Spearrin, R. M. (2020). Calibration-free breath acetone sensor with interference correction based on wavelength modulation spectroscopy near 8.2 μm. Applied Physics B, 126, 9.
29. Xu, L., Liu, N., Zhou, S., et al. (2020). Dual-frequency modulation quartz crystal tuning fork-enhanced laser spectroscopy. Optics Express, 28, 5648-5657.
30. Zou, M., Yang, Z., Sun, L., Ming, X. (2020). Acetylene sensing system based on wavelength modulation spectroscopy using a triple-row circular multi-pass cell. Optics Express, 28, 11573-11582.
31. Pavone F. S., Inguscio, M. (1993). Frequency-and wavelength-modulation spectroscopies: Comparison of experimental methods using an AlGaAs diode laser. Applied Physics B, 56, 118-122.
32. Miklós, A., Bozóki, Z., Jiang, Y., Fehér, M. (1994). Experimental and theoretical investigation of photoacoustic-signal generation by wavelength-modulated diode lasers. Applied Physics B, 58, 483-492.
33. Dharams, A. N., Bullock, A. M. (1996). Applications of wavelength-modulation spectroscopy in resolution of pressure and modulation broadened spectra. Applied Physics B, 63, 283-292.
34. Dharamsi A. N., Lu Y. (1996). Sensitive density-fluctuation measurements using wavelength-modulation spectroscopy with high-order-harmonic detection. Applied Physics B, 62, 273-278.
35. Uehara, K. (1998). Dependence of harmonic signals on sample-gas parameters in wavelength-modulation spectroscopy for precise absorption measurements. Applied Physics B, 67, 517-523.
36. Dubinsky, I., Rybak, K., Steinfeld, J. I., Field, R.W. (1998). Frequency-modulation-enhanced remote sensing. Applied Physics B, 67, 481-492.

37. Henningsen, J., Simonsen, H. (2000). Quantitative wavelength-modulation spectroscopy without certified gas mixtures. Applied Physics B, 70, 627-633.
38. Hancock, G., Kasyutich, V. L., Ritchie, G.A.D. (2002). Wavelength-modulation spectroscopy using a frequency-doubled current-modulated diode laser. Applied Physics B, 74, 569-575.
39. Sogaard, S., Henningsen, J. (2002). Thermal tuning and modulation of a DFB fibre laser with a thin-film heater. Applied Physics B, 75, 497-501.
40. Zybin, A., Kuritsyn, Y.A., Mironenko, V.R., Niemax, K. (2004). Cavity enhanced wavelength modulation spectrometry for application in chemical analysis. Applied Physics B, 78, 103-109.
41. Liu, J.T. C., Jeffries, J. B., Hanson, R.K. (2004). Wavelength modulation absorption spectroscopy with 2 f detection using multiplexed diode lasers for rapid temperature measurements in gaseous flows. Applied Physics B, 78, 503-511.
42. Borri, S., Bartalini, S., De Natale, P. et al. (2006). Frequency modulation spectroscopy by means of quantum-cascade lasers. Applied Physics B, 85, 223.
43. Bomse, D. S., Kane, D. J. (2006). An adaptive singular value decomposition (SVD) algorithm for analysis of wavelength modulation spectra. Applied Physics B, 85, 461.
44. Waechter. H., Sigrist, M.W. (2007). Mid-infrared laser spectroscopic determination of isotope ratios of N2O at trace levels using wavelength modulation and balanced path length detection. Applied Physics B, 87, 539-546.
45. Vasudev, R. (2007). Wavelength modulated cavity enhanced absorption spectroscopy of ammonia at 1994 nm. Applied Physics B, 87, 163-167.
46. Zhao, W., Gao, X., Chen, W. (2007). Wavelength modulated off-axis integrated cavity output spectroscopy in the near infrared. Applied Physics B, 86, 353-359.
47. Chen, J., Hangauer, A., Strzoda, R., Amann, M.-C. (2008). Accurate extraction method for the FM response of tunable diode lasers based on wavelength modulation spectroscopy. Applied Physics B, 90, 243-247.
48. Hangauer, A., Chen, J., Amann, M.-C. (2008). Modeling of the nth harmonic spectra used in wavelength modulation spectroscopy and their properties. Applied Physics B, 90, 249-254.
49. Shao, J., Lathdavong, L. Kluczynski, P., et al. (2009). Methodology for temperature measurements in water vapor using wavelength-

modulation tunable diode laser absorption spectrometry in the telecom C-band. Applied Physics B, 97, 727-748.
50. Farooq, A., Jeffries, J. B., Hanson, R. K. (2009) Sensitive detection of temperature behind reflected shock waves using wavelength modulation spectroscopy of CO2 near 2.7 μm. Applied Physics B, 96,161-173.
51. Hamilton, M. L., Peverall, R., Ritchie, G. A. D., et al. (2009). Wavelength modulation and cavity enhanced absorption spectroscopy using ~1.9 μm radiation produced by difference frequency generation with a MgO doped PPLN crystal. Applied Physics B, 97, 715.
52. Lins, B., Zinn, P., Engelbrecht, R., Schmauss, B. (2010). Simulation-based comparison of noise effects in wavelength modulation spectroscopy and direct absorption TDLAS. Applied Physics B, 100, 367-376.
53. Hamilton, M. L., Ritchie, G. A. D., Arita Y., Ewart, P. (2010). Multi-mode absorption spectroscopy, MUMAS, using wavelength modulation and cavity enhancement techniques Applied Physics B, 100, 665-673.
54. Khan, M.A., Mohan K., Dharamsi, A. N. (2010). Optical pathlength saturation signatures in wavelength modulation spectroscopy signals of atmospheric molecular oxygen. Applied Physics B, 99, 363-369.
55. Vanderover, Wang, W., Oehlschlaeger, M. A. (2011). A carbon monoxide and thermometry sensor based on mid-IR quantum-cascade laser wavelength-modulation absorption spectroscopy. Applied Physics B, 103, 959-966.
56. Arslanov, D. D., Spunei, M., Ngai, A. K. Y. et al. (2011). Rapid and sensitive trace gas detection with continuous wave Optical Parametric Oscillator-based Wavelength Modulation Spectroscopy. Applied Physics B, 103, 223-228.
57. Mohan, K., Khan M. A., Dharamsi A. N. (2011). Characterization of lineshape structure by wavelength modulation spectroscopy. Applied Physics B, 102, 569-578.
58. Cao, Y., Jin, W., Ho, H. L.et al. (2012). Acetylene detection based on diode laser QEPAS: combined wavelength and residual amplitude modulation. Applied Physics B, 109, 359-366.
59. Malara, P., Witinski, M. F., Capasso, F., et al. (2012). Sensitivity enhancement of off-axis ICOS using wavelength modulation. Applied Physics B, 108, 353-359.
60. Hangauer, A., Chen, J., Strzoda, R., Amann, M.-C. (2013). Multi-harmonic detection in wavelength modulation spectroscopy systems. Applied Physics B, 110, 177-185.

61. Hodgkinson, J., Masiyano, D., Tatam, R. P. (2013). Using integrating spheres with wavelength modulation spectroscopy: effect of pathlength distribution on 2nd harmonic signals. Applied Physics B, 110, 223-231.
62. Sun, K., Sur, R., Jeffries, J. B., et al. (2014). Application of wavelength-scanned wavelength-modulation spectroscopy H2O absorption measurements in an engineering-scale high-pressure coal gasifier. Applied Physics B, 117, 411-421.
63. Wang, Q., Chang, J., Wei, W., et al. (2014). Dual-beam wavelength modulation spectroscopy for sensitive detection of water vapor. Applied Physics B, 117, 1015-1023.
64. Salati S. H., Khorsandi, A. (2014). Apodized 2f/1f wavelength modulation spectroscopy method for calibration-free trace detection of carbon monoxide in the near-infrared region: theory and experiment. Applied Physics B, 116, 521-531.
65. Lan, L. J., Ding, Y. J., Peng, Z. M., et al. (2014). Calibration-free wavelength modulation for gas sensing in tunable diode laser absorption spectroscopy. Applied Physics B, 117, 1211-1219.
66. Goldenstein, C. S., Schultz, I. A., Spearrin, R. M., et al. (2014). Scanned-wavelength-modulation spectroscopy near 2.5 μm for H2O and temperature in a hydrocarbon-fueled scramjet combustor. Applied Physics B, 116, 717-727.
67. Goldenstein, C. S., Spearrin, R. M.. Jeffries J. B., Hanson, R. K. (2014). Wavelength-modulation spectroscopy near 2.5 μm for H2O and temperature in high-pressure and -temperature gases. Applied Physics B, 116, 705-716.
68. Lima, G. R., Mota, L., Miklós, A., et al. (2014) Sensitive harmonic detection of ammonia trace using a compact photoacoustic resonator at double-pass configuration and a wavelength-modulated distributed feedback diode laser. Applied Physics B, 117, 333-341.
69. Liu, C., Xu, L Li, F., et al. (2015). Resolution-doubled one-dimensional wavelength modulation spectroscopy tomography for flame flatness validation of a flat-flame burner. Applied Physics B,120, 407-416.
70. Li, J., Du, Z., An, Y. (2015). Frequency modulation characteristics for interband cascade lasers emitting at 3 μm. Applied Physics B, 121, 7-17.
71. Qu, Z., Schmidt, F. M. (2015). In situ H2O and temperature detection close to burning biomass pellets using calibration-free wavelength modulation spectroscopy. Applied Physics B. 119, 45-53.
72. Jozdani, M. M., Khorsandi, A., Sabouri, S. G. (2015). Polymeric fiber sensor for sensitive detection of carbon dioxide based on apodized wavelength modulation spectroscopy. Applied Physics B, 118, 219-229.

73. Yan-Chao, L., Yi-Qiao, W., Chun-Yu, L. et al. (2016). Linear frequency modulation multi-beam laser heterodyne measurement for the glass thickness. Applied Physics B, 122, 24,
74. Du, Y., Lan, L., Ding, Y., Peng, Z. (2017). Measurement of the absolute absorbance based on wavelength modulation spectroscopy. Applied Physics B, 123, 205.
75. Wei, M., Feng, R. F., Chen, B., et al. (2017). Calibration-free wavelength modulation spectroscopy for gas concentration measurements using a quantum cascade laser. Applied Physics B, 123, 149.
76. Liu, B., Song, C., Li, Y., Duan, Y. (2018). A frequency-modulated-continuous-wave laser detection system based on the four-quadrant photodetector. Applied Physics B, 124, 186.
77. Henderson, B., Khodabakhsh, A., Metsälä, M., et al (2018). Laser spectroscopy for breath analysis: towards clinical implementation. Applied Physics B 124, 161.
78. Wei, W., Peng, W. Y., Wang, Y., et al. (2019). Demonstration of non-absorbing interference rejection using wavelength modulation spectroscopy in high-pressure shock tubes. Applied Physics B, 125, 9.
79. Lou, X., Xu, L., Dong, Y., et al. (2019). Detection of elemental mercury using a frequency-doubled diode laser with wavelength modulation spectroscopy. Applied Physics B, 125, 62.
80. Peng, W. Y, Strand C. L., Hanson, R. K. (2020). Analysis of laser absorption gas sensors employing scanned-wavelength modulation spectroscopy with 1f-phase detection. Applied Physics B, 126, 17.
81. Schwarm, K. K., Strand, C. L., Miller, V, A., Spearrin, R. M. (2020). Calibration-free breath acetone sensor with interference correction based on wavelength modulation spectroscopy near 8.2 μm. Applied Physics B, 126, 9.
82. Tao, B., Lei, Q., Ye, J., et al. (2020). Measurements and analysis of diode laser modulation wavelength at high accuracy and response rate. Applied Physics B, 126, 88.

Raman spectroscopy

1. Galeener, F. L. (1977). FM spectroscopy: Raman scattering and luminescence. Chemical Physics Letters, 48, 7-11.
2. Levin, K. H., Tang, C. L. (1978). Wavelength-modulation Raman spectroscopy. Applied Physics Letters, 33, 817.
3. Banishev, A. F., Sukharev, A. G., Sukhodolsky, A. T. (1984). Differential active Raman spectroscopy. Letters in Physics (Kratkie soobshchenia po fizike), 3, 3-6 [Russian].

4. Levin, K. H., Tang, C. L. (1978). Wavelength-modulation Raman spectroscopy. Applied Physics Letters, 33, 817.
5. Levine, B. F., Bethea, C. G. (1980). Frequency-modulated shot noise limited stimulated Raman gain spectroscopy. Applied Physics Letters, 36, 245.
6. Bruckner, S., Jeziorowski, H., Knözinger, H. (1984). Frequency modulation Raman spectroscopy. Chemical Physics Letters, 105, 218-222.
7. Shreve, A. P., Cherepy, N. J, Mathies, R. A. (1992). Effective rejection of fluorescence interference in Raman spectroscopy using a shifted excitation difference technique. Applied Spectroscopy, 46, 707-711.
8. Sowoidnich, K., Kronfeldt, H.-D. (2012). Fluorescence rejection by Shifted Excitation Raman Difference Spectroscopy at multiple wavelengths for the investigation of biological samples. International Scholarly Research Network ISRN Spectroscopy, 2012, 256326, 11 p.
9. Mazilu, M., De Luca, A. C., Riches, A. el al. (2010). Optimal algorithm for fluorescence suppression of modulated Raman spectroscopy. Optics Express, 18, 11383-11394.
10. De Luca, A. C., Mazilua, M., Riches, A. el al. (2010). Online fluorescence suppression in modulated Raman spectroscopy. Analytical Chemistry. 82, 738-745.
11. Mazilua, M., De Luca, A. C., Riches, A. el al. (2010). Modulated Raman spectroscopy technique for 'real-time' fluorescence rejection. Proc. of SPIE, 7568, 75680M: 1-5.
12. Kasha, M. (1950). Characterization of electronic transitions in complex molecules. Discussions of the Faraday Society, 9, 14-19.
13. Matousek, P., Towrie, M., Parker, A. W. (2005). Simple reconstruction algorithm for shifted excitation Raman difference spectroscopy. Applied Spectroscopy, 59, 848-851.
14. Martins, M. A.S., Ribeiro, D. G., Pereira dos Santos, E. A., et al. (2010). Shifted-excitation Raman difference spectroscopy for in vitro and in vivo biological samples analysis. Biomedical Optics Express, 1, 617-626.
15. Dolgy, S., Appiah, B., Melnik, E.D. (2012). A novel method for practical implementation of the shifted excitation Raman difference spectroscopy (SERDS). Special Issue, 27, no. 6. http://stage-rs.www.spectroscopyonline.com/novel-method-practical-implementation-shifted-excitation-raman-difference-spectroscopy-serds?id=&pageID=1&sk=&date=

16. Sowoidnich, K, Kronfeldt, H. D. (2012). Shifted excitation Raman difference spectroscopy at multiple wavelengths for in-situ meat species differentiation. Applied Physics B, 108, 975-982.
17. Praveen, B. B., Ashok, P. C., Mazilu, M., et al. (2012). Fluorescence suppression using wavelength modulated Raman spectroscopy in fiber-probe-based tissue analysis. Journal of Biomedical Optics 17, 077006.
18. Praveen, B. B., Mazilu, M., Marchington, R. F., et al. (2013). Optimization of wavelength modulated Raman spectroscopy: Towards high throughput cell Screening. PLoS ONE 8, 6, e67211.
19. Noack, K., Eskofier, B., Kiefer, J. el al. (2013). Combined shifted-excitation Raman difference spectroscopy and support vector regression for monitoring the algal production of complex polysaccharides. Analyst, 138, 5639-5646.
20. Kiefer, J. Noack, K., Leipertz, A. (2013). Background suppression for Raman analysis of pharmaceutically active compounds in fluorescing media. American Pharmaceutical Review, 16, 32-35.
21. Volodin, B., Dolgy, S., Ban, V, S., et al. (2014). Application of the shifted excitation Raman difference spectroscopy (SERDS) to the analysis of trace amounts of methanol in red wines. Proceedings of SPIE, 8939.
22. De Luca. A. C., Dholakia, K., Mazilu, M. (2015). Modulated Raman spectroscopy for enhanced cancer diagnosis at the cellular level. Sensors, 15, 13680-13704.
23. Gebrekidan, M. T., Knipfer, C., Stelzle, F., et al. (2016). A shifted-excitation Raman difference spectroscopy (SERDS) evaluation strategy for the efficient isolation of Raman spectra from extreme fluorescence interference. Journal of Raman Spectroscopy, 47, 198-209.
24. Sowoidnich, K., Maiwald, M., Sumpf, B., et al. (2020). Charge-shifting optical lock-in detection with shifted excitation Raman difference spectroscopy for the analysis of fluorescent heterogeneous samples, Proc. SPIE 11236, Biomedical Vibrational Spectroscopy 2020: Advances in Research and Industry.

Luminescence spectroscopy

1. Green, G. L. O'Haver, T. C. (1974). Derivative luminescence spectrometry. Analytical Chemistry, 46, 2191-2196.
2. O'Haver, T. C. Modulation and derivative techniques in luminescence spectroscopy: Approaches to increased analytical sensitivity. In "Modern Fluorescence Spectroscopy." Ed. Wehry, E.L. Plenum Press, N.Y., 1976.

3. Vo-Dinh, T., Gammage, R. B. (1979). The applicability of the second-derivative method to room-temperature phosphorescence analysis. Analytica Chimica Acta, 107, 261-271.

4. Dalterio, R. A, Hurtubise, R. J. (1984). Zeroth and second derivative fluorescence and phosphorescence analysis of mixtures of hydroxyl aromatics adsorbed on filter paper. Analytical Chemistry, 56, 819-821.

5. Lloid, J. B. F. (1971). Synchronized excitation of fluorescence emission spectra. Nature Physical Science, 231, 64 - 65.

6. Li, Y. Q., Li, X. Y., Shindi, A. A. F., et al. Synchronous Fluorescence Spectroscopy and Its Applications in Clinical Analysis and Food Safety Evaluation. In book: Reviews in Fluorescence 2010, Springer.

7. Sanchez, F. G., Blanco, C.C. (1988). Spectrofluorometric determination of pesticide residue mixtures by isodifferential derivative spectroscopy. Analytical Chemistry, 60, 323-328.

8. Eiroa, A. A., Blanco, E. V., Mahía, P. L., et al. (1998). Simultaneous determination of 11 polycyclic aromatic hydrocarbons (PAHs) by second-derivative synchronous spectrofluorimetry considering the possibility of quenching by some PAHs in the mixture. Analyst, 123, 2113-2117.

9. Borders, S. P. T. L., Hernández, C. E., Roy, L. E., et al. (1999). Comparison of analytical methods: Direct emission versus first-derivative fluorometric methods for quinine determination in tonic waters. Journal of Chemical Education, 76, 1, 85.

10. Swaminadham, V., Santhamma, C., Sastry, C. V. R. el al. (1999). A new approach in identifying very prominent polyaromatic hydrocarbons (PAHs) in petroleum fractions using second derivative synchronous excitation fluorescence spectroscopy (SDSEFS) I. Indian Journal of Pure and Applied Physics, 37, 689-698.

11. Swaminadham, V., Santhamma, C., Sastry, C. V. R. el al. (2000). Second derivative synchronous excitation fluorescence spectroscopy (SDSEFS) to identify isomers in a mixture. Indian Journal of Pure and Applied Physics, 38, 158-165.

12. Mozo-Villarías A. (2002). Second derivative fluorescence spectroscopy of tryptophan in proteins. Journal of Biochemical and Biophysical Methods, 50, 163-178.

13. Nayar, S, Brahma, A, Mukherjee, C, Bhattacharyya, D. (2002). Second derivative fluorescence spectra of indole compounds. Journal of Biochemistry, 131, 427-435.

14. Maher, H. M. (2008). Simultaneous determination of naproxen and diflunisal using synchronous luminescence spectrometry. Journal of Fluorescence, 18, 909-917.

15. El-Enany N., F. Belal, F., El-Shabrawy, Y., Rizk, M. (2009). Second Derivative Synchronous Fluorescence Spectroscopy for the simultaneous determination of chlorzoxazone and ibuprofen in pharmaceutical preparations and biological fluids. International Journal of Biomedical Science, 5, 136-145.
16. Soliman, S. M., El-Agizy, H. M. Y., El-Bayoumi, A. E-A. (2014). Derivative synchronous fluorescence spectroscopy for the simultaneous determination of dapoxetine hydrochloride and vardenafil in binary mixtures. Journal of Applied Spectroscopy, 81, 509-518.
17. Hegazy, M., Kessiba, A., Abdelkawy, M., Elgendy, A. (2016). Synchronous fluorescence spectroscopy coupled with continuous wavelet transforms and Savitzky-Golay derivatization technique for the simultaneous determination of tadalafil and dapoxetine HCl. International Journal of Pharmacy and Pharmaceutical Sciences, 8, 32-38.
18. Ragab, M. A. A., Galal, S. M., Korany, M. A., Ahmed, A. R. (2017). First derivative emission spectrofluorimetric method for the determination of LCZ696, a newly approved FDA supramolecular complex of valsartan and sacubitril in tablets. Luminescence, 32, 1417-1425.
19. Soliman, S. M., El-Agizy, H. M. Y., El Bayoumi, A. E. A. (2017). Validated stability-indicating derivative spectrophotometry and synchronous fluorescence spectroscopy methods for the determination of dapoxetine hydrochloride in the presence of its degradation product and co-formulated drugs. Pharmaceutica Analytica Acta, 8, 10.
20. Salem, A. Y., Hammouda, M. E. A., El-Enin, M. A. A., El-Ashrya, S. M. (2019). Application of derivative emission fluorescence spectroscopy for determination of ibuprofen and phenylephrine simultaneously in tablets and biological fluids. Spectrochimica Acta Part A: Molecular and Biomolecular Spectroscopy, 210, 387-397.
21. Wei, C., Ren, P., Cen, Q. et al (2019). Simultaneous determination of dissolved phenanthrene and its metabolites by derivative synchronous fluorescence spectrometry with double scans method in aqueous solution. Talanta, 195, 339-344.
22. El-Shaheny, R., Belal, F. (2020). Green conventional and first-order derivative fluorimetry methods for determination of trimebutine and its degradation product (eudesmic acid). Emphasis on the solvent and pH effects on their emission spectral properties. Spectrochimica Acta Part A: Molecular and Biomolecular Spectroscopy, 226, 117603.
23. Elbalkiny, H. T., Yehia, A. M., Riad, S. M., Elsaharty, Y. S. (2020). Derivative constant wavelength synchronous fluorescence spectrometry for the simultaneous detection of cefadrine and cefadroxil in water

samples. Spectrochimica Acta Part A: Molecular and Biomolecular Spectroscopy, 229,117903.
24. Hamad, A. E., Mohammed, B. S., Derayea, S. M., El-Malla, S. F. (2020). Micelle sensitized synchronous spectrofluorimetric approaches for the simultaneous determination of simeprevir and ledipasvir: Application to pharmaceutical formulations and human plasma. Spectrochimica Acta Part A: Molecular and Biomolecular Spectroscopy,239, 118471.
25. Attala, K., Eissa, M. S., Hasan, M. A., et al. (2020). An enhanced first derivative synchronous spectrofluorimetric method for determination of the newly co-formulated drugs, amlodipine and celecoxib in pharmaceutical preparation and human plasma. Spectrochimica Acta Part A: Molecular and Biomolecular Spectroscopy, 240, 118533.

Photoacoustic spectroscopy

1. Guo, L., Tang, Z., He, Y., Zhang, H. (2007). Characterization of a derivative photoacoustic spectrometer. Review of Scientific Instruments, 78, 023104.
2. R. D. Snook, in Encyclopedia of Analytical Science (Third Edition), 2013.
3. Castleden, S. L., Kirkbright, G. F., Spillane, D. E. M. (1981). Wavelength modulation in photoacoustic spectroscopy. Analytical Chemistry. 53, 2228-2231.
4. Schilt, S., Thevenaz, L. (2006). Wavelength modulation photoacoustic spectroscopy: Theoretical description and experimental results. Infrared Physics and Technology, 48, 154-162.
5. Shao, X., Pang, C., Su, Q. (2000). A novel method to calculate the approximate derivative photoacoustic spectrum using continuous wavelet transform. Fresenius' Journal of Analytical Chemistry, 367, 525-529.
6. Webber, E. Pushkarsky, M., Patel, C. K. N. (2003). Fiber-amplifier-enhanced photoacoustic spectroscopy with near-infrared tunable diode lasers. Applied Optics, 42, 2119-2126.
7. Ng, J., Kung, A. H., Miklós, A., Hess, P. (2004). Sensitive wavelength-modulated photoacoustic spectroscopy with a pulsed optical parametric oscillator. Optics Letters, 29, 1206-1208.
8. Zheng, J., Tang, Z., He, Y., Guo, L. (2008). Sensitive detection of weak absorption signals in photoacoustic spectroscopy by using derivative spectroscopy and wavelet transform. Journal of Applied Physics 103, 093116.

9. Wang, Z. L., Tian, C. W., Liu, Q., et al. (2018). Wavelength modulation technique-based photoacoustic spectroscopy for multipoint gas sensing. Applied Optics,. 57, 2909-2914.
10. Qiao, S., Ma, Y., He, Y., et al. (2019). A sensitive carbon monoxide sensor based on photoacoustic spectroscopy with a 2.3 μm mid-infrared high-power laser and enhanced gas absorption. Sensors, 19, 3202.

Remote Sensing Data and Applications

1. https://online-learning.harvard.edu/course/remote-sensing-data-and-applications.
2. Lucas, R., Rowlands, A., Niemann, O., Merton, R. (2004). Hyperspectral Sensors and Applications. In Advanced Image Processing Techniques for Remotely Sensed Hyperspectral Data. Springer, 11-49.
3. Demetrius's-Shah, T. H., Steven, M.D., Clark, J. A. (1990). High resolution derivative spectra in remote sensing. Remote Sensing of Environment, 33, 55-64.
4. Chen, Z., Curran, P. J., Hansom, J. D. (1992). Derivative reflectance spectroscopy to estimate suspended sediment concentration. Remote Sensing of Environment, 40, 67-77.
5. Becker, B. L., Lusch, D. P., Qi, J. (2005). Identifying optimal spectral bands from in situ measurements of Great Lakes coastal wetlands using second-derivative analysis. Remote Sensing of Environment, 97, 238-248.
6. Kochubey, S. M., Kazantsev, T. A. Derivative vegetation indices as a new approach in remote sensing of vegetation. (2012). Frontiers of Earth Science, 6, 188-195.
7. Tsai, F., Philpot, W. (1998). Derivative analysis of hyperspectral data. Remote Sensing of Environment, 66, 41-51.
8. Louchard, E. M., Reid, R. P Stephens, C., F., et al. (2002). Derivative analysis of absorption features in hyperspectral remote sensing data of carbonate sediments. Optics Express, 10, 1573-1584.
9. Smith, K.L., Steven, M.D., Colls, J. J. (2004). Use of hyperspectral derivative ratios in the red-edge region to identify plant stress responses to gas leaks. Remote Sensing of Environment, 92, Pages 207-217.
10. Ruffin, C., King, R. L., Younan, N. H. (2008). A Combined Derivative Spectroscopy and Savitzky-Golay Filtering Method for the Analysis of Hyperspectral Data. GIScience & Remote Sensing, 45, 1-15.
11. Torrecilla, E., Marine, J. P. (2009). Derivative analysis of hyperspectral oceanographic data. In Advances in Geoscience and Remote Sensing. InTech, 597-618.

12. Wolanin, A., Soppa, M. A., Bracher, A. (2016). Investigation of spectral band requirements for improving retrievals of phytoplankton functional types. Remote Sensing, 8, 871.

Chapter Two

1. Dubrovkin, J. (1988). Chemical analysis with the use of the mathematical differentiation. Zhurnal Analiticheskoi Khimiya, 43, 965-979. [Russian].

Electrochemical analytical methods

2. Wang, J. *Analytical Electrochemistry*. 2nd ed. 2000. Wiley & Sons.

3. Lingane, J. J., Richard Williams, R. (1952). Derivative Polarography. I. Characteristics of the Leveque-Roth Circuit. Journal of American Chemical Society, 74, 790-796.

4. Hammerich, O., Speiser, B. *Organic Electrochemistry. Revised and Expanded.* 5th Ed. CRC Press. 2016.

5. Perone, S. P., Mueller T. R. (1965). Application of derivative techniques to stationary electrode polarography. Analytical Chemistry, 37, 2-9.

6. Perone, S. P. Birk, J. R. (1965). Application of derivative techniques to anodic stripping voltammetry. Analytical Chemistry, 37, 9-12.

7. Evins, C. V., Perone, S. P. (1967). Application of derivative readout techniques to stationary electrode polarography with kinetic systems. Analytical Chemistry, 39, 309-315.

8. Davison, W., Whitfield, M. (1977). Modulated polarographic and voltammetric techniques in the study of natural water chemistry. Journal of Electroanalytical Chemistry,75,763-789.

9. Okamura, K. (1979). Derivative electrophoretograms and their applications. Clinica Chimica Acta, 96, 273-279.

10. Skogberg, D., Richardson, T., Blasczyk, T. (1979). Automatic sampling and monitoring of potentiometric electrodes: steady-state response by first and second derivative techniques Analytical Chemistry, 51, 2054-2057.

11. Evans, M.O., Hanck, K. W. (1981). Evaluation of derivative normal pulse polarography for use Evaluation of derivative normal pulse polarography for use in metal speciation studies. Analytica Chimica Acta, 129, 79-86.

12. Bancroft, E. E., Sidwell, J. S., Blount, H. N. (1981). Derivative linear sweep and derivative cyclic voltabsorptometry. Analytical Chemistry, 53, 1390-1394.

13. Wang, J. (1982). Derivative hydrodynamic-modulation voltammetry. Talanta, 29, 805-808.
14. Lovric, M., O'Dea, J. J., Osteryoung, J. (1983). Faradaic response in derivative and differential normal pulse voltammetry. Analytical Chemistry, 55, 704-708.
15. Tacussel, J., Leclerc, P., Fombon, J. J. (1986). Pulse voltammetry and polarography: Recent progress in microprocessor instrumentation. Journal of Electroanalytical Chemistry and Interfacial Electrochemistry, 214, 79-94.
16. Zamponi, S., Czerwinski, A., Marassi. R. (1989). Thin-layer derivative cyclic voltabsorptometry. Journal of Electroanalytical Chemistry and Interfacial Electrochemistry, 266, 37-46.
17. Palys, M., Korba, T., Bos, M., Van Der Linden, W.E. (1991). The separation of overlapping peaks in cyclic voltammetry by means of semi-differential transformation. Talanta, 38, 723-733.
18. Xie, Q., Wei, W., Nie, L Yao, S. (1993). Derivative linear sweep and derivative cyclic voltabsorptometry of the long-path-length spectroelectrochemical cell: The single reversible electrode reaction. Journal of Electroanalytical Chemistry, 348, 29-47.
19. Kim, M.-H., Smith, V. P., Hong, T.-K. (1993). First- and second-order derivative polarography/voltammetry for reversible, quasi-reversible, and irreversible electrode processes. Journal of Electrochemical Society, 140, 712-721.
20. Yu, J.-S., Zhang, Z.-X. (1996). Differentiation, semidifferentiation and semi-integration of a digital signals based on Fourier transformations. Journal of Electroanalytical Chemistry, 403, 1-9.
21. Molina, A., Morales, I. (2007). Comparison between derivative and differential pulse voltammetric curves of EC, CE and catalytic processes at spherical electrodes and microelectrodes. International Journal of Electrochemical Science, 2, 386-405.
22. Zhang, J., Hu, Y., Liu, J., Hu, Z. (2009). Overlapping-peak resolution and quantification using derivative spectrophotometry in capillary electrophoresis. Microchimica Acta,164, 487-491.
23. Grdeń, M. (2017). Semi-differential analysis of irreversible voltammetric peaks. Journal of Solid State Electrochemistry, 21, 1045-1058.
24. Hong, T.-K., Rusodimos, I., Kim, M.-H. (2017). Higher order derivative voltammetry for reversible and irreversible electrode processes under spherical diffusion. Journal of Electroanalytical Chemistry, 785, 255-264.

25. Obot, I. B., Onyeachu, I.B. (2018). Review. Electrochemical frequency modulation (EFM) technique: Theory and recent practical application in corrosion research. Journal of Molecular Liquids, 249, 83-96.
26. He, Q., Tian, Y., Wu, Y. (2019). Facile and ultrasensitive determination of 4-Nitrophenol based on acetylene black paste and graphene hybrid electrode. Nanomaterials, 2019, 9, 429.
27. Espinoza, E. M., Clark, J. A., Soliman, J., et al. (2019). Practical aspects of cyclic voltammetry: How to estimate reduction potentials when irreversibility prevails. Journal of the Electrochemical Society, 166, H3175-H3187.

Chromatography

1. Tracey. M. V. (1964). Derivative recording in column chromatography. Australian Journal of Biological Sciences, 17, 792-797.
2. Kambara, T., Saitoh, K., Ozeki, K. (1967). Simple derivative method for the gas chromatograph. Analytical Chemistry, 39, 409-410.
3. Kambara, T., Tachikawa, T. (1968). A derivative method in ion exchange chromatography. Journal of Chromatography A, 32, 728-731.
4. Milano, M. J., Lam, S., Grushka, E. (1976). Rapid scanning diode array as a multi-wavelength detector in liquid chromatography. Journal of Chromatography A, 125, 315-326.
5. Milano, M. J., Grushka, E. (1977). Diode array detector in liquid chromatography: II. Enhanced sensitivity via first derivative (dA/dλ) chromatograms. Journal of Chromatography A, 133, 352-354.
6. Wheals B. B., Russell, J. R. (1979). Electronic signal differentiation as an aid for the comparison of size-exclusion chromatograms. Journal of Chromatography A, 176, 418-420.
7. Linnett, L. M., Atkinson, D. J. (1980). Electronic signal differentiation as an aid to quantitation in gas chromatography. Journal of Chromatography A, 197, 1-10.
8. Zelt, D. T., Owen, J. A., Marks, G. S. (1980). Second derivative-high-performance liquid chromatographic-fluorometric detection of porphyrins in chick embryo liver cell culture medium. Journal of Chromatography A, 189, 209-216.
9. Traveset, J., Such, V., Gonzalo, R., Gelpi, E. (1981). First and second derivative recording of thin-layer chromatograms: Application to the assay of unresolved compounds. Journal of Chromatography A, 204, 51-58.

10. Kalmanovskyi, V. I. (1982). Chromatographic analysis based on the signal differentiation. Journal of Analytical Chemistry (Russia), 307, 1309.
11. Such, S., Traveset, J., Gonzalo, R., Gelpi, E. (1982). Application of electronically differentiated high-performance thin-layer chromatographic densitograms to the essay of some preservatives used in pharmaceutical formulations. Journal of Chromatography A, 234, 77-87.
12. Traveset, J., Such, S., Gonzalo, R., Gelpi, E. (1982). A novel approach to the qualitative and quantitative assay of HPTLC plates: Second and fourth derivative recording of spectrophotodensitograms. Journal of High Resolution Chromatography, 5, 483-488.
13. Fell A. F., Scott, H. P., Gill, R., Moffat, A. C. (1983). Novel techniques for peak recognition and deconvolution by computer-aided photodiode array detection in high-performance liquid chromatography. Journal of Chromatography A, 282, Pages 123-140.
14. Grant, A., Bhattacharyya, P. K. (1985). Application of derivative spectroscopy to the determination of chromatographic peak purity. Journal of Chromatography A, 347, 219-235.
15. Grushka, E., Atamna, I. (1987). The use of derivatives for establishing integration limits of chromatographic peaks. Chromatographia, 24, 226-232.
16. Grushka, E., Israeli, D. (1990). Characterization of overlapped chromatographic peaks by the second derivative. The limit of the method. Analytical Chemistry, 62, 717-721.
17. Vivó-Truyols, G., Torres-Lapasió, J. R., Van Nederkassel, A. M. et al. (2005). Automatic program for peak detection and deconvolution of multi-overlapped chromatographic signals: Part I: Peak detection. Journal of Chromatography A, 1096, 133-145.
18. Garcia, J. A., Plaza. J. G., Pavon, J. M. C. (1994). Use of derivative spectrophotometry in the resolution of overlapped peaks in liquid chromatography and its application in the in application in the analysis of active components in insecticide formulations. Journal of Liquid Chromatography, 17, 277-285.
19. Verdú-Andrés, J., Herráez-Hernández, R., Campíns-Falcó, P. (2001). Analysis of enantiomers giving partially overlapped peaks by using different treatments of the chromatographic ultraviolet signals: quantification of pseudoephedrine enantiomers. Journal of Chromatography A, 930, 95-107.
20. Vetter, W., Gaul, S., Melcher, J. (2007). Improved quality control in gas chromatography interfaced to stable isotope ratio mass spectrometry

by application of derivative chromatography. Analytica Chimica Acta, 590, 49-54.
21. Quintás, G., Lendl, B., Pastor, A., La Guardia, M. (2008). First-order derivative resolution of overlapped PAH peaks with common mass spectra in gas chromatography–mass spectrometry. Talanta, 74, 747-752.
22. Fredriksson, M., Petersson, P., Axelsson, B.-O., Bylund, D. (2009). An automatic peak finding method for LC-MS data using Gaussian second derivative filtering. Journal of Separation Science 32, 3906-3918.
23. Stevenson, P. G., Mnatsakanyan, M., Guiochonde, G., Shalliker, R. A. (2010). Peak picking and the assessment of separation performance in two-dimensional high performance liquid chromatography. Analyst, 135, 1541-1550.
24. Stevenson, P. G., Gritti, F., Guiochon, G. (2011). Automated methods for the location of the boundaries of chromatographic peaks. Journal of Chromatography A, 1218, 8255-8263.
25. Lytle, F. E., Julian, R. K. (2016). Automatic processing of chromatograms in a high-throughput environment. Clinical Chemistry, 62, 144–153.
26. Masse, M., Genay, S., Feutry F., et al. (2017). How to solve the problem of co-elution between two compounds in liquid chromatography through the first UV derivative spectrum. A trial on alternative plasticizers to di(2-ethylhexyl) phthalate. Talanta, 162, 187-192.
27. Wahaba, M. F., O'Haver, T.C., Grittic, F., et al. (2019). Increasing chromatographic resolution of analytical signals using derivative enhancement approach. Talanta, 192, 492-499.

X-ray spectroscopy

1. Baun, W. L., Chamberlain, M. B, Solomon, J. S. (1973). Computer differentiation of DC soft X-ray appearance potential spectra. Review of Scientific Instruments 44, 1421-1422.
2. König, C. F. J., Van Bokhoven, J. A., Schildhauer, T. J., Nachtegaal, M. (2012). Quantitative analysis of modulated excitation X-ray absorption spectra: enhanced precision of EXAFS fitting. Journal of Physical Chemistry C, 116, 19857-19866.
3. Gaur, A., Shrivastava, B. D. (2012). A Comparative study of the methods of speciation using X-ray absorption fine structure. Acta Physica Polonica, 121, 647-652.
4. Gaur, A., Shivastana, B.D., Nigam, H. L. (2013). X-Ray absorption fine structure (XAFS) spectroscopy - A review. Proceedings of the Indian National Science Academy, 79, 921-966.

5. Yu, X. D., Zhang, M. Y., Zhu, M. Q., et al. (2014). Second derivative X-Ray diffraction fingerprint of Tibetan medicine snow lotus herb based on the five-point-formula method. Applied Mechanics and Materials, 543-547, 1211-1214.

6. Chiarello, G. L., Ferric, D. (2015). Modulated excitation extended X-ray absorption fine structure spectroscopy. Physical Chemistry Chemical Physics, 17, 10579-10591.

7. Zhang, Z., Yin, H., Feng, X., Tan, W. (2015). Absorption mechanisms of Cu on a biogenic bixbyite-like Mn2O3 produced by Bacillus CUA isolated from soil. Geochemical Transactions, 16-5.

Photoelectron spectroscopy

1. Burrell, M. C., Kaller, R. S., Armstrong, N. R. (1982). Data acquisition and processing modes for quantitative Auger electron spectroscopy. Analytical Chemistry, 54, 2511-2517.

2. Hoffmann, S. *Auger- and X-Ray Photoelectron Spectroscopy in Material Science. A user-oriented guide*. Springer-Verlag, Berlin, Heidelberg, 2013.

3. Smith, N. V., Traum, M.M (1970). Higher-derivative photoelectron spectra and the energy band structure of copper. Physical Review Letters, 25, 1017-1020.

4. Pireaux, J-J. (1976). Smoothing and resolution enhancement of photoelectron spectra with the aid of a programmable desk calculator. Applied Spectroscopy, 30, 219-224.

5. Pavlath, A. E., Millard, M. M. (1979). Analysis of X-ray photoelectron spectra through their even derivatives. Applied Spectroscopy, 33, 502-509.

6. Bruninx, E., Van Eenbergen, A. (1981). Computerized detection and evaluation of peaks in survey spectra from photoelectron spectroscopy. Analytica Chimica Acta, 133, 339-347.

7. Proctor, A., Sherwood, P. M. A. (1982). Data analysis techniques in x-ray photoelectron spectroscopy. Analytical Chemistry, 54, 13-19.

8. Mizokawa, Y., Miyasato, T., Nakamura, S., et al. (1987). The C KLL first-derivative X-ray photoelectron spectroscopy spectra as a fingerprint of the carbon state and the characterization of diamondlike carbon films. Journal of Vacuum Science & Technology A, 5, 2809-2814.

9. Furukawa, Y., Nagatsuka, Y., Nagasawa, Y., et al. (2008). Practical methods for detecting peaks in Auger electron spectroscopy and X-ray photoelectron spectroscopy. Journal of Surface Analysis, 14, 225-242.

10. Mezzi, A., Kaciulis, S. (2010). Surface investigation of carbon films: from diamond to graphite. Surface and Interface Analysis, 1082-1084.

Miscellaneous

1. Saakov, V. S., Drapkin, V. Z., Krivchenko, A. I., et al. *Derivative Spectrophotometry and Electron Spin Resonance (ESR) Spectroscopy for Ecological and Biological Questions*. Springer. 2013.
2. Talsi, E., Bryliakov, K. *Application of EPR and NMR spectroscopy in homogeneous catalysis*. CRC Press. 2017.
3. Hirata, H., Kuyama, T., Ono, M., Shimoyama, Y. (2003). Detection of electron paramagnetic resonance absorption using frequency modulation. Journal of Magnetic Resonance, 164, 233-241.
4. Sasidhar, N., Mallikarjunaiah, K. J., Damle, R. (2002). Modulation schemes of wide-line NMR absorption. Journal of the Instrument Society of India, 32, 1-8.
5. Alexeev, A. *Physics of coal and mining processes*. p. 150. CRC Press, 2012.
6. Yatsenko, A. V., Palatnikov, M. N., Sidorov, N. V. (2019). Investigation of the intrinsic defects of LiTaO3 crystals by NMR spectroscopy. Crystallography Reports, 64, 36-40.
7. Bressani, T., Brovetto, P., Chiavass, E. (1967). Derivative Mössbauer spectrometer. Nuclear Instruments and Methods, 47, 164-168.
8. Obenhuber, Th., Forster, A., Potzel, W., Kalvius, G. M. (1983). A microprocessor controlled spectrometer for frequency modulation Mössbauer measurements. Nuclear Instruments and Methods in Physics Research, 214, 361-379.
9. Vagizov, F., Shakhmuratov, R., Sadykov, E. (2015). Application of the Mössbauer effect to the study of opto-acoustic phenomena. Physica Status Solidi B, 252, 469-475.
10. Reguigui, N. (2006). Gamma Ray Spectrometry. Practical Information. https://www.researchgate.net/publication/259533588.
11. Beekman, S., Zhou, T., Stoller, C., Haranger, F. (2013). Method and apparatus for gain regulation in a gamma detector. Patent US 2013/0134304 A1. https://patentimages.storage.googleapis.com/6e/1a/b3/532d2bb43a3b3e/US20130134304A1.pdf
12. Mahmoud, I., I., El-Tokhy, M. S. (2014). Development of coincidence summing and resolution enhancement algorithms for digital gamma ray spectroscopy. Journal of Analytical Atomic Spectrometry, 29, 1459-1466.
13. MODULATED DSC^{TM} COMPENDIUM. Basic Theory & Experimental Considerations. Thermal Analysis & Rheology. http://www.tainstruments.com/pdf/literature/TA210.pdf

14. Gill, P.S., Sauerbrunn, S. R., Reading, M. (1993). Modulated differential scanning calorimetry. Journal of Thermal Analysis, 40, 931-939.
15. *Modulation temperature differential scanning calometry. Theoretical and practical application in polymer characterization*. Springer, 2006.
16. Bouzidi, L., Boodhoo, M., Humphrey, K. L., Narine, S. S. (2005). Use of first and second derivatives to accurately determine key parameters of DSC thermographs in lipid crystallization studies. Thermochimica Acta, 439, 94-102.
17. Lu, J., Trnka, M. J., Roh, S.-H., et al. (2015). Improved peak detection and deconvolution of native electrospray mass spectra from large protein complexes. Journal of the American Society for Mass Spectrometry, 26, 2141-2151.
18. Chen, J., Yang, C., Zhu, H., Li, Y. (2019). Adaptive signal enhancement for overlapped peaks based on weighting factor selection, Spectroscopy Letters, 52, 1.

INDEX